D1599044

Handbook
of Expert Systems
in Manufacturing

Handbook of Expert Systems in Manufacturing

Rex Maus
Jessica Keyes

Editors

McGraw-Hill, Inc.

New York   St. Louis   San Francisco   Auckland   Bogotá
Caracas   Hamburg   Lisbon   London   Madrid
Mexico   Milan   Montreal   New Delhi   Paris
San Juan   São Paulo   Singapore
Sydney   Tokyo   Toronto

Library of Congress Catalog Card Number: 90-13450

TS
176
H336
1991

1 2 3 4 5 6 7 8 9 0   DOC/DOC   9 6 5 4 3 2 1 0

ISBN 0-07-040984-6

*Composed in Ventura Publisher by Sherry Cain, Cain & Associates.
Printed and bound by R. R. Donnelley & Sons Company.*

To our families with love

Contents

Section 2   Integration of AI Into Manufacturing

Section 3   Knowledge Engineering in a Manufacturing Environment

Section 5   Simulation, Process Modeling, and Resource Allocation

Section 6   Diagnostics

Section 7   Process Control and Planning

Section 8   Design

Section 9   Quality and Safety

Section 10   Pricing, Packaging, and Customizing

Preface

The business community has embraced expert systems as a technology that brings with it competitive and strategic advantages. Nearly 80 percent of the Fortune 500 has dabbled in and dallied over this newfound art. Expert systems have been built in industrial sectors as diverse as marketing, banking, insurance, securities, and retail. But there are probably few sectors that have experienced as rapid a push towards this technology as manufacturing.

There are two major reasons for this. First, manufacturing is labor-intensive. Companies, in their quest for both quality and timeliness, are continually struggling with a workforce that is unprepared and inconsistent in skill level. For these firms, the idea of "cloning" top-achieving experts and distributing this expertise companywide is irresistible. Second, there are few sectors of our economy that are as complicated as the multifaceted face of manufacturing. From design to assembly to material handling to inventory control to testing, the tasks that form the manufacturing puzzle must be perfectly performed and integrated for a firm to compete in an increasingly global marketplace. The use of expert systems to structure and assist in these tasks serves to provide a consistency and speed-up in time-to-market that gives a great boost to a firm's competitive advantage.

It is with this in mind that *Handbook of Expert Systems in Manufacturing* was written. And it wasn't written just for the technically proficient among us. You need not possess any prior experience in expert system technology, or even any computer experience. You need only an open mind and an interest in manufacturing technology. This multiposition readership was kept vivdly in mind when we decided on a format for this handbook. A picture is worth a thousand words. Think of each case history in this book as a picture of a working manufacturing expert system. And with this picture as blueprint, we hope our readers will soon follow the lead of our many contributors and venture forth into this strategic technology.

To facilitate the use of the handbook by the reader, the chapters are arranged in ten sections. The first section, "Introduction to Expert Systems," serves as the foundation for the rest of the book. It gives the reader a bird's-eye view of the technology and how it can fit into the manufacturing industry. Not only does this section give you the whys and wherefores of the exciting link between AI and Manufacturing, this section also gives you nuts and bolts information about how you can start using AI in your company.

Sections Two through Ten are the reason why this book was published. These sections contain the fascinating accounts of expert system and neural net manufacturing applications from the top developers in the manufacturing and AI industries. These contributors were hand-selected to showcase their innovative use of this intriguing technology.

Section Two contains some interesting insights into how you can successfully integrate AI into the typical manufacturing environment. Section Three gives some hard and fast rules for performing the knowledge acquisition process. In Section Four, you'll get a good perspective on the use of this technology in the area of scheduling and forecasting. In Section Five, you'll see how two of our contributors link the exotic worlds of simulations, process modelling, resource allocation, and AI. Perhaps the most popular use of AI is in plant or machine diagnostics. In Section Six, you'll find out how to use expert systems and neural nets in some rather unique diagnostic applications.

In Section Seven, we'll see how AI can really pay off in process control and planning. The most creative use of AI is in the area of design, and in Section Eight we'll read two accounts that shed some new light on the meaning of CAD. In Section Nine, we'll tackle the quality and safety bogie. Finally, we'll tie it all together with some systems that assist manufacturers to customize, price, and package their products.

In the competitive 1990s, manufacturing companies will need some leverage to get ahead. The tried and true is just plain tired. The AI approach may be just what the doctor ordered.

We owe much to the individuals who wrote individual chapters for this *Handbook of Expert Systems in Manufacturing*. Each of these contributors dedicated considerable time and effort to this project and offered much advice regarding the direction this handbook should take. The pooled efforts of these contributors, each of whom is a recognized authority within either the expert system or manufacturing community, has led to the creation of what we all agree is the definitive resource on manufacturing expert systems.

We would also like to thank our editors and friends at McGraw-Hill who encouraged and assisted us each step along the way. Special mention is given to Gail Nalven, Bob Hauserman, and Theron Shreve, who gave selflessly of their time and experience.

Much of the credit for turning assorted floppy diskettes and paper into a real book goes to Sherry Cain, who is a whiz with desktop publishing.

Rex Maus
Jessica Keyes

Handbook
of Expert Systems
in Manufacturing

Introduction
to Expert Systems

Introduction to Expert Systems

Rex D. Maus
President, The KnowledgeBase Group
Austin, Texas

Introduction

If you have ever considered the arguments voiced by a number of economic prognosticators which discuss the idea of the U.S. economy shifting from its manufacturing base to an economy based mostly upon services and service-related industry, please rethink the premise. Manufacturing is far more important than most non-manufacturing economic experts will give credit.

It is manufacturing that made America the power she is. It is manufacturing that keeps America great and powerful. It will be manufacturing that will keep America's future bright into the next century. Manufacturing embodies the basic human emotion of pride of craftsmanship. A time-honored American tradition that has evolved over the past 216 years. Whether the product is a bolt of cloth, a computer chip, an airplane, or even a doll, the pride in making the product remains as important today as it was ten, fifty, or two hundred years ago. I know this for a fact. I have seen the pride, that look of satisfaction on the faces of hundreds of manufacturing workers, who toil daily, making the products that represent the foundation, the economic core of the U.S.A.

As president of a firm that owes its existence to purveying advanced computer technology, particularly expert systems and intelligent simulation, in the manufacturing arena I have had the point of a daily participant, not as a casual bystander. I've spent the better part of the last decade visiting and working with manufacturers of all types. From those who make the steel that finds its way into our cars and appliances, to the manufacturers of health care products who keep us well, to those involved in making the weapons we depend upon to keep America safe, you represent the American way of life.

I have been privy to the in's and out's of how manufacturing works and doesn't work. I've seen automated manufacturing methods that would capture the attention of the most ardent science fiction aficionado. I have discussed operations methods and practices with some of the smartest "unknown" experts on the continent. I've witnessed the "latest" introduction of technology that was considered fashionable and current twenty years ago. I have shared the goose-flesh producing excitement of setting manufacturing production records. I've also shared the heartbreak and disappoint when product sales didn't meet forecasts and upper management dictated that production staff (production staff to management—real people with real families to me) had to lose their jobs. I have seen some manufacturing plants close and destroy communities yet at the same time seen others expand and flourish. Manufacturing is the heart of the American economy, make no mistake.

Technology and Manufacturing

New technologies take many different forms. Relational Database, imaging, robotics, expert systems, lasers, vision systems, intelligent simulations and sensors are all excellent examples. The use of these technologies in America presents an interesting phenomena; some manufacturers use technology effectively, some abuse it and still others ignore it. The reasons behind this diverse application threshold are as varied as the many different manufacturing applications themselves.

I've observed manufacturers who depend upon technology to keep them competitive, using the latest and greatest advances as fast as the technology evolves from the research laboratories. In areas of technological use, these companies don't want to take a chance that someone, somewhere might beat them to the punch; they live the "bleeding edge" every day—and take great pride in the fact. Conversely, I've visited with industrial concerns who are so conservative in their use of technology it was embarrassing. These companies remain intent on preserving the "Halls Of Cobol" at any cost. They shy away from technological advancement almost to the point that technology becomes so commonplace that any competitive edge gained from its use is long gone. And, when these firms finally utilize these new tools, they generally do so in a "Rube Goldberg fashion," wiring old systems with new methods creating a mess that seldom works as designed. Finally, I have witnessed manufacturers who rely on technology as a savior for every type of problem imaginable. New technologies are thrown into use with little malice of forethought again achieving less than ideal results.

The hodge-podge approach manufacturers apply to using new technology is the reason this book was written. When you are very close to a technology, like my company is with expert systems and intelligent simulation, you can easily lose perspective. Everyone you encounter is not as knowledgeable nor fully understands how and where to use these productive tools. As a result, technology isn't used to its best advantage, and more often, not used at all. Face it, people will only use a tool when they feel comfortable with it and understand how to use it. It is a simple philosophy, but it applies to technology just the same as it applies to a video recorder, microwave oven, personal computer, or even a simple power tool. Just because you have access to a pneumatic nail gun doesn't mean that you would feel comfortable using it to hang a few pictures around the house does it? It would be overkill; a hammer does the same job nicely and does not require a thorough understanding of its use nor the necessity of following required safety procedures.

With expert systems, especially, the confusion is widespread. What are they, how do they work, where do they work best, how do you gain a competitive edge using them, and how do you get started using them are just a few of the questions every manufacturer I've talked with has asked. Granted, there has been considerable material published about how expert systems work in the worlds of finance, insurance, and computers. However, little if any of this work can be directly translated into productive examples that manufacturers can use. Also, manufacturers tend to be very conservative in nature and need to be shown how an expert system can be of direct benefit for their particular problem. Just because an expert system has worked in the insurance world doesn't necessarily mean it will work in manufacturing. If more manufacturers could see practical examples of how their peers efficiently use expert systems our manufacturing operations would become more productive and competitive.

This is the reason this book was developed. Jessica and I realized that if we could put together a compilation of real-world success stories of expert system use in the manufacturing environment, others could see the benefits their peers are obtaining and the use of this exciting, powerful technology would spread. We quickly became aware that telling the story from our perspective and in our own words might not be enough to convince a manufacturing executive or factory manager to start using the technology. The "show me" reality convinced us to modify our plan. Instead of telling our point of view, we went to manufacturers themselves and to the expert system tool vendors for assistance. We asked for success stories.

Particularly, we asked for stories that would show how expert systems can be productively used in manufacturing. But even more, we wanted stories written by the developers themselves. There is no better method of "spreading the expert system gospel" than to hear the story in the words of those who make their bread and pay their mortgages from their work in manufacturing. We were surprised at the number of people who wanted to participate in this book. Those who responded to our queries were far greater than the number of case studies represented in this work. But, quite a few were not allowed to tell their stories because of their company's proprietary information regulations and still others were caught-up in the daily job of getting the product out of the door and into the customer's hands and thus were not able to devote the required time to meet our publishing deadline. We want to thank all of you who took the time and responded and especially to those of you that took the time to write. We also want to thank the vendors who helped and steered us to stories that had manufacturing impact. Finally, we want to thank the open-minded manufacturing managers and executives that let their stories be told without prejudice of proprietary impact. This book would not have existed without you. You are a class example for all manufacturers to follow. Thank you!

This book has over thirty-five contributors, all with a story to tell. It doesn't address every type of manufacturing enterprise and it won't tell you everything about expert systems in productive use today. However, it is a start, and a good one at that. We are confident that you will find the book interesting and informative. We are also confident that we have a good representation of overall manufacturing expert system use as it exists today. We hope you find studies that can be directly applied to your own manufacturing activities. Further, we hope that you will be so inspired that you undertake the development of your own systems. Finally, a self serving plug. We are always interested in the success stories we have not discovered. If your company has a success and you want to share it, contact us. We want the revision of this book (of which we are confident will occur) to include you. So, here it is, a glimpse of the world of expert systems in productive, everyday use in the American manufacturing environment. Read, learn, enjoy!

Manufacturing Activities

When this book was in the planning stages, the original front matter was to contain information detailing the current plight of American manufacturing, focusing on lost market share, lost jobs, lost hope. I decided not to dwell on these inadequacies, and rather, make the direction of the book positive. After all, nobody enjoys hearing about how bad things are, people would much prefer to look on the bright side where optimism reigns. So, before getting into the heart of the subject matter let me offer a short prelude.

American manufacturing is entering a renaissance. We refused to believe those who claimed U.S manufacturing had come and gone as a world-class leader. Today, technology combined with a revitalized "can-do" spirit is turning around what ten years ago was, at best, a bleak outlook.

We are entering an information age where executives, managers and employees must react instantaneously to their business environment, and make on-the-spot decisions that will impact their firms' competitive position in the global market.

Embracing technology and using it to its utmost advantage will mean the difference between manufacturing failure and success. Remember, these words to live by—If your company can't provide a product that features quality, a competitive cost, and on-time delivery, your customers have the world to shop for a company that can.

Author Biographical Data

Rex Maus is president of KnowledgeBase Group, a leading AI consultancy specializing in manufacturing applications. Author of Robotics—A Manager's Guide" and "Expert Systems—Tools and Applications," both published by Wiley & Sons.

Prior to this, he was Director of Advanced Technologies for CAP GEMINI AMERICA. He sits on the Board of Advisors of the American Council for Manufacturing Automation.

2

Artificial Intelligence and Expert Systems: A Technological Primer

Rex D. Maus
President, The KnowledgeBase Group

Introduction

This chapter introduces the basic concepts of artificial intelligence and expert systems. It is not the definitive authority on this technology, however. There are plenty of books available that are excellent sources of information for everything you would ever want to know about artificial intelligence and expert systems this book, however, is not one of them. Here, you will learn some of the basics of how "thinking machines" work (first: they don't think on their own, that is a only a myth; and secondly: why artificial intelligence and expert systems are important.) Most of all, this chapter will set the stage for the case studies that the book features.

Artificial Intelligence—From the Laboratory to Reality

From the original introduction of computers the developers of these information processing marvels have strived to incorporate human-like thinking abilities into their mechanical creations. Only today, after more than three decades of research effort, are we seeing the fruition of their years of dreaming and designing. The concepts of artificial intelligence are rapidly evolving from the research laboratories of universities into practical everyday use. Although artificial intelligence is gaining in acceptance and use, many people remain confused as to what it is and how it can be used.

What is Artificial Intelligence?

Simply put, artificial intelligence is an area of computer science that, using a collection of different programming techniques and programming languages, enables computers to mimic human thinking and reasoning processes. Why then is artificial intelligence important and is it really advantageous to have "smart computers?" The ability of a computer to mimic forms of human thinking is very desirable since this lets computers perform more human-like operations—acting just as a human might in decision-critical situations. From computers that can understand spoken words to robots that can see and move around obstacles to computers that can help make decisions, artificial intelligence is a very useful tool that can be used to make computers more efficient and productive which in turn helps make our lives and jobs as humans more efficient and productive.

Domains of Artificial Intelligence

Broadly defined, artificial intelligence research includes five domains:

- Natural Language: A natural language is any spoken language of humans, e.g., English, French, German. In this realm of artificial intelligence researchers work to develop hardware and software that enables computers to interact with humans via spoken commands. This ability is more commonly known as "speech recognition." We've all seen the IBM television commercial where the computer is instructed to "Write this letter to Mr. Wright, right now. This is how speech recognition works the computer understands the spoken command and can determine the proper differentiation of the three forms of the same sounding word: write, Wright, and right.

- Robotics: This area of artificial intelligence uses a combination of techniques to develop "intelligent robots" that can see, move, and manipulate objects on their own in response to changing environmental conditions. The days of the fully functional robot android are still years away, however today we are seeing robots that use vision to help in inspection tasks, machines that use sensors to control and monitor tactile gripping ability, and automated guided vehicles that can see objects in their path and move around them.

- Enhanced Human Interfaces: Here computer scientist work to improve existing, conventional computer interfaces using the techniques of behavioral psychology and learning kinetics combined with artificial intelligence programming abilities. The end result is a conventional computer application with a much improved human interface that is friendlier, easier to learn, and improves productivity through ease of use.

- Exploratory Programming: This area employs artificial intelligence techniques that were originally used to further develop artificial intelligence programs. These abilities allow computers (with human assistance) to design the most efficient computer program in a given programming language and then, using a knowledge base of actual working programs, write a new program that operates as designed in the most efficient manner possible. These disciplines are commonly known as CASA—Computer Assisted Systems Analysis, and CAP—Computer Automated Programming.

- Expert Systems: Expert systems are by far the most commercially successful of all of the artificial intelligence domains. Expert systems use artificial intelligence concepts to enable computers to function in decision-support roles as advisors, personifying human expert decision-making capabilities. The following pages introduce expert systems technology and explain how these incredible tools are used and the benefits they can provide to manufacturing.

The Human Expert

What is an expert? Most would agree that an expert is a person who possesses a considerable amount of knowledge about a particular field. A doctor, lawyer, engineer, scientist are all examples of high-level experts. Experts exist elsewhere too. A mechanic is an expert, so is a carpenter, a tailor, a machine operator, a baker, even a plumber qualifies as an expert. Experts use their knowledge daily to solve problems related to their area of expertise. That is why they are considered experts! Why are experts important? Where do you go when you need help in solving a particular type of problem? You go to an expert for help. If you need help with a legal problem you go to a lawyer. If you are retooling a manufacturing line, you go to the machine operator for help in machine tool setup. If you have a toothache, you go to a dentist for help. We use experts everyday; both in our work and in our personal lives. By going directly to an expert, you can get a quick solution or suggestion for solving your problem. An expert's knowledge makes them valuable and sometimes irreplaceable. They can consistently apply their knowledge to help solve problems and make decisions quickly.

Since we all depend upon experts, having them available whenever we need them is very important. In our personal lives this availability is generally not as important as in our jobs. If our doctor is on vacation, we can usually see another doctor who is taking the calls of our personal physician. If our mechanic is not available, another mechanic can fill in with little impact. However, when the mill scheduler is sick, it is amazing how fast business operations can turn chaotic. The expert scheduler knows all of the rules-of-thumb that make the mill operate efficiently and productively. This person can prepare a workable schedule that meets all of the constraints of production in the mill and; more importantly, when a problem with the schedule pops-up, the expert scheduler can modify the schedule quickly to get the mill running smoothly again. When the expert isn't there, it is a different picture. It might take a very long time for non-experts to create a workable schedule. Non-expert schedules often don't account for all of the constraints in the milling process and cause even more problems. As a result, the mill does not operate as efficiently and sometimes may have to be shut down while a new schedule is prepared. All the while production is lost; thus making orders late; and, with the trickle-down effect, upsetting the customer. It is easy to see why an expert is important.

Since experts are generally limited in number the demand for their services is great. Naturally, an expert's resources can be spread thin through overwork and access to them may be limited. So, if the expert isn't available, what do you do? Most of the time, in business, you can't simply put off the decision until the expert becomes available. Once time is lost, it cannot be recaptured. Time resource management is the essence of gaining and keeping a competitive edge. Sometimes it is possible to have access to multiple experts, but this is seldom the case. Like in the old apprenticeship days, tomorrow's experts are made by watching and learning from today's experts. It is not possible to make experts overnight, or is it? If it were possible to "clone" an expert, we could have better access to their decision-making abilities, thus creating the potential to be more efficient in our work. The cloning of experts is possible, in fact, it is happening more and more every day. Expert systems are these clones.

Expert Systems— Electronic Decision Makers

Simply put, an expert system is a clone of a human expert in an electronic box. An expert system packages the knowledge of one or more human experts in a software format that can operate on a variety of computers from mainframes to personal computers. Why have these systems gained such notoriety? Because they provide the capability to capture knowledge and spread the knowledge to other people who need it. Think of this, you can build an expert system that contains the knowledge of how the most efficient design engineer performs a job and then let other less efficient designers use that captured knowledge so they too can attain the same efficiencies in their work. Now, instead of "design engineer" substitute banker, doctor, programmer, or any job title you wish it's all the same, an expert system can capture knowledge about how to perform a task in human terms and then make that same knowledge available to others. You can see how this could make people more productive. An expert system is always ready to help; it doesn't forget, it doesn't go on vacation, it doesn't retire. Also, as conditions change, expert systems are easy to modify so they are always up-to-date. Expert systems help manage the information overload of today's work world by translating information into usable knowledge.

The concepts of electronic decision making sound fascinating, but how do expert systems work? For a comparison, think about how software for making spreadsheets works. The spreadsheet software has been programmed to know how to perform different arithmetic and operating functions. When you use this software, you only need to input the numbers you're working with and tell the computer, via specific commands, to add, multiply, move, copy, etc. It's really pretty simple. An expert system works in very much the same way. In an expert system, the program for the operations and procedures is called the "inference engine." The information you provide (which the inference engine manipulates) is the knowledge or rules of how a process or procedure works in a step-by-step manner. The actual operating functions are a bit more complicated how the knowledge is structured, how the computer manipulates the knowledge, etc but the basic concept is straightforward.

Conventional Systems and Expert Systems—A Comparison

Conventional programming methods have been used to create massive number-crunching, database handling systems we generally think of when contemplating computers. This type of programming allows computers to manage data in different ways. Typical conventional computer programs are used to solve math problems, manipulate data bases, and compute complex spreadsheets.

The fundamental difference between artificial intelligence techniques and conventional systems can be found in the programming languages for each and how the languages are structured. Conventional languages tend to share two basic assumptions:

1) the programmer develops an algorithm to solve a problem in a step-by-step numbered approach,

2) the data needed by the program to execute is stored in a data base and called by the algorithm as it is needed when each step is executed.

Developing a program in a conventional language (COBOL, FORTRAN, C, PASCAL) requires that an easy-to-understand algorithm first be developed. To make the program to efficiently execute, the algorithm is improved, again in numbered steps, until the program executes from the first line of code through the last.

Symbolic programming provides a totally different approach to problem solving. Historically, the artificial intelligence field has used symbolic programming languages like LISP (List Processing) and PROLOG (Programmed Logic) to develop expert systems as well as expert system development tools. Today, however, most expert system software is being written in symbolic languages and then converted to a conventional language like C. This conversion facilitates development and maintenance by individuals who do not have a symbolic programming background.

Symbolic processing uses with knowledge or rules of a particular situation or domain represented in symbolic form and stored in a computer. A symbol can be a word, letter, or a number that represents objects, actions or relationships between symbols. Objects can represent just about anything people, places, events, even ideas. Symbols are created and stored in a knowledge base. With a knowledge base developed you need to be able to manipulate the symbols within it. The "inference engine" performs this function. The inference engine uses coded instructions, written in the symbolic programming language, to establish and define relationships between symbols by manipulating the symbolic information in the knowledge base through a process of pattern-matching. Using defined inputs furnished by the expert system user the inference engine searches through the knowledge base looking for matches or relationships until a solution is found. Instead of beginning at the first instruction and ending with the last instruction (as in conventional programs), symbolic programs may search, find a match, that will in turn, lead to another search and another match. Searches don't necessarily go in step-by-step sequence either. They can begin at the end of the knowledge base and search back to the beginning, repeatedly.

All artificial intelligence based programs use this pattern-matching approach and in some instances, if insufficient knowledge is provided, the symbolic effort may not arrive at a conclusion and solve a problem. This is in direct contrast with conventional programming where an algorithm always produces a solution if given input data to process. Naturally, in symbolic programming, if sufficient data is contained in the knowledge base a solution is reached much in the same manner as if a human had solved the problem.

At its simplest, conventional software performs procedural processing of data where the algorithm details specifically how to solve the problem in a step-by-step manner. Artificial intelligence software, including expert systems, performs non-procedural processing. An artificial intelligence program contains the "what" of the problem, not the "how." The what is the knowledge base and any input furnished by the user—the how is the pattern-matching that occurs as the inference program works to find its own conclusion to the problem.

Anatomy of an Expert System

By now you're interest in expert systems has hopefully increased so that you will understand how these electronic marvels are able to mimic human expertise. The following paragraphs will explain the six components of which a typical expert system is made and how these components function together. Figure 2.1 illustrates the make-up of a typical expert system.

The Inference Engine

The inference engine is the "supervisor" that directs the operation upon the knowledge contained in the expert system. It stands between the user and the knowledge base and its internal operation is transparent to the user. An inference engine is really a program that makes decisions and judgments based upon the symbolic data (computer scientist call symbolic data what the rest of us call rules-of-thumb, e.g., "closing the door to keep the heat in" is, to us a rule-of-thumb, but to a computer scientist, a form of symbolic data) contained in the knowledge base.

The inference engine performs several tasks. Its basic functions are inferencing and control. Inference engines contain program instructions that are required for any expert system to operate. These instructions include basic procedures that tell the expert system how to begin its work. After starting the expert system, the inference engine begins to search the knowledge base examining existing facts and rules and comparing these facts to the information furnished by the user. An inference engine's operation is based upon algorithms that define specific search and pattern matching techniques. Basically the inference engine compares the input from the user with the rules contained in the knowledge base looking for matches.

For example, a user provides the expert system with the part number and description of several defective parts that need to be returned to their respective manufacturers. The inference engine takes this information and for each part entered, compares it to the rules contained in the knowledge base that describe how to return the parts, where to return them, how to package them and the specific warranty for each part. This a simple example but it illustrates how pattern matching occurs.

As described above, pattern matching continues with the inference engine searching the knowledge base for matches until a solution is found. An initial search may lead to another search and then another search and so on. Searches can be chained together to emulate human-like reasoning processes. At the same time it is busy with searches, the inference engine is also responsible for asking the user for any additional information which might assist in other searches. The interaction between the user and the expert system is called a consultation. All expert systems use pattern-matching approaches, looking for links and relationships between entities, to solve problems. Generally, this process solves the problem satisfactorily, but it only succeeds when enough information entered by the user can be matched to the rules contained in the knowledge base. If there is not enough information that can be matched to the rules the problem is not solved.

The Knowledge Base

The knowledge base is the place where the facts and rules that represent a human expert's "rules-of-thumb" reside. Most expert systems use rules as the basis for their operation, hence the reason many expert systems are called "rule-based" systems." Some expert systems use other knowledge representation schemes like semantic nets or frames. These systems are hybrids on basic expert system structures discussed here. Information about these systems can be found in most books discussing all of the "ins and outs" of expert systems. It is important to remember that the knowledge in the knowledge base is independent of the inference engine.

Knowledge bases are developed by capturing the rules and procedures human experts use in solving problems. When knowledge of a human expert is captured—through an interviews or observation—it is reduced to simple statements. Even the most complex reasoning processes can be reduced to simple statements. In most expert systems knowledge is represented in the simple form of IF and THEN rules. "IF the sensor reads 250 units/second THEN the machine is functioning normally; IF the sensor reads 10 units/second, THEN check the drive belt," are good examples of how knowledge can be represented in a knowledge base. It is this IF-THEN structure that enables an expert system to make pattern matches to solve problems.

Because knowledge bases are separate entities from inference engines, the knowledge contained in the knowledge base is easy to modify. Just as in a spreadsheet, where numbers can be changed independently of the arithmetic operand programs, expert system knowledge bases can be changed just as easily. When you make changes to a knowledge base you simply add rules, change old rules, or remove obsolete rules, in any order. Because the inference engine controls how searches occur, it doesn't matter the order in which rules are structured. However, rules are most often structured in the same hierarchical order that is representative of a procedure or process as it is performed by a person.

The Developer Interface

There has to be a method for an expert system developer to "feed" knowledge into the expert system's knowledge base. This function is performed through the developer's interface. The interface is a word processor-like editor that lets a developer enter new rules, change rules or delete rules. Because it resembles a word processing program, its functions are very similar. When a developer enters data, the interface creates a file, and much like a compiler in a conventional program, this input is converted into a format that the inference engine can use.

The Explanation Subsystem

The explanation subsystem is designed to explain to the user the line of reasoning that the expert system used to reach a conclusion. Some users of expert systems may find it difficult to "trust" the reasoning that an expert system uses during consultations, just as they may question the advice of an unfamiliar human expert. The explanation subsystem solves this problem of interaction trust. Using this feature, users of the expert system can ask "Why?" or "How?" and the expert system will provide an answer. For example, an expert system may ask the user for additional input about sensor temperature. The user may wonder why the expert system needs more information and asks the system, "Why?" The expert system responds by saying it needs the temperature information so that it can evaluate a particular rule. In many systems the rules needing further evaluation are displayed. Asking the system "How?", causes the system to display all of the rules it used to arrive at a particular conclusion. Explanation subsystems are excellent mechanisms for use in instructional situations and for debugging the system during its development.

The User Interface

This part of an expert system is used to furnish a convenient means of two-way communication between the user and the system. Simply put, the interface displays all information transactions which occur during a consultation. The interface can be in the form of menus, questions, even graphical icons that are displayed on the computer screen. The user interface also displays all responses, questions and consultation results. Communication through the user interface is accomplished by using a keyboard, mouse or track-ball.

Working Memory

Working memory is a scratch-pad area in the computer's memory set aside for keeping track of inputs, in process conclusions, and final outputs. The inference engine uses working memory like a sheet of paper to record and monitor its processing. At the end of a consultation working memory contains the final results of the consultation.

Benefits From Using Expert Systems

There are numerous benefits that can be derived from using expert systems. Not exclusively of the problems that the expert system

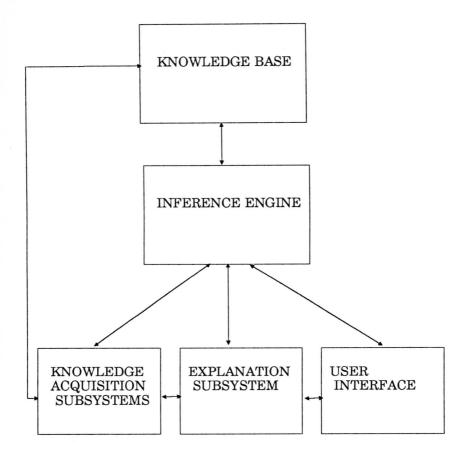

Figure 2-1   The architecture of an expert system

solves, other benefits often considered "soft benefits," add to the value of using these systems. Among these benefits are:

- Use of Most Appropriate Technology—The variety of circumstances and associated rules-of-thumb used by the humans to react to different circumstances are difficult to capture with structured algorithms used by conventional programming languages and techniques. Expert system technology is specifically designed to satisfy the requirements of heuristic situations. The rule base is separate from the program logic thus providing tremendous advantages in speed of initial development and ease with modifications. Only the rule base has to be modified when any changes occur. Rules can be added or deleted as needed without other modifications to the program. This permits quick reaction to changing requirements.

■ Clarity and Consistency—Rules and procedures can be programmed in the way a person thinks about them. Rules are programmed in English-like syntax that a user can understand without having to be a programmer. It is easy to determine if rules of any expert system are being followed and applied consistently, because you can see the rules and the logic of how each rule is used in a context that is easy to understand.

■ Capture of Expertise—Development of an expert system allows for the capture and retention of perishable expertise. Capturing expertise reduces dependence on key employees and lessens a company's vulnerability from the loss of experts through retirement, reassignment, or other circumstances.

■ Training—An expert system is an excellent training tool. When using an expert system, trainees can create typical situations and see what rules are applied and the reasoning behind the application of those rules without an instructor or expert present. In addition, trainees can assimilate the expertise at whatever pace they are able, providing excellent flexibility in scheduling training and manpower utilization.

■ Cost Savings—An expert system has the potential to be designed to avoid unnecessary manufacturing costs by improving efficiencies in manufacturing processes particularly in areas of scheduling, machine fault diagnosis and real-time process control.

■ Extended Use of the Development Shell—The typical license that comes with an expert system shell allows development of several, if not unlimited, expert systems with the same shell. Thus, once the process is automated, the shell could be used to support development of additional expert systems.

■ Customer Relations—Every manufacturer is dependent upon its customer base for continued viability as a corporation. The loss of a customer who is frustrated with late orders or flawed products represents a devastating cost. Once a customer is frustrated they may become a lost customer.

Types of Expert Systems

Expert systems are used in a variety of ways. From mainframe based systems to stand-alone personal computer-based systems the technology can be applied in a variety of effective methods depending upon the problem needing to be solved.

- Stand-Alone Systems—There are more of these expert systems than any others. A stand-alone system most often contains fewer than 50 rules and is fully resident in a computer's memory and operates without ties to any other computer. Most diagnostic expert systems operate as stand-alones.

- Imbedded Systems—These expert systems function in tandem with conventional programs or other expert systems in a support role as a sub-routine performing certain tasks. Typically the main processing is performed algorithmically by a conventional program, manipulating numbers for example. An imbedded expert system might use its logic to make critical accountancy-based decisions based upon the results of the conventional program. During operation, the conventional program calls the expert system routine and after the expert system performs an internal consultation, the expert systems logic-based decisions are used to determine how the conventional program will continue processing. Imbedded systems are frequently used in accounting, fault tolerance situations and in critical defense related activities.

- Real-Time Expert Systems—These systems are used to respond immediately to inputs from sensors, conventional programs, or other computer-based devices. Real-time systems make instantaneous knowledge-based decisions that are needed to respond to different environmental situations. These expert systems are seen in combat support roles in fighter and bomber aircraft and in commercial aircraft as pilot advisors. Control systems for chemical plants, power plants and manufacturing machining process control are other examples of real-time expert systems. Real-time systems most often combine the features of embedded and linked expert systems.

- Linked Expert Systems—These expert systems are tied to other expert systems or conventional systems to perform logic-based processing tasks. In many cases the expert system obtains information from another computer through the downloading of needed files. Using the expert system, the user performs a consultation and the results are then formatted into history-type files that can be uploaded to the other computer. Linked systems typically feature expert systems operating on personal computers that are linked to mainframe databases. These systems are used in just about every type of industry.

- Dedicated Expert Systems—Dedicated systems operate like stand-alone systems but operate in a mainframe environment. These systems most often are used to solve only a single type of problem like those associated with military weapons and manufacturing process-control systems.

Languages and Tools

Most artificial intelligence programs have, historically, been developed using symbolic languages such as LISP, PROLOG, and OPS. These languages are versatile and very effective when developing logic-based systems from scratch. Artificial intelligence purist speak of these languages with high acclaim touting that the languages are the only effective method for developing expert systems. Ten years ago there was little debate to this argument; today is different. Although symbolic languages are very powerful, they require programmers who have the skills to use and get the most out of the language. Unfortunately symbolic programmers are not as readily available as are their conventional language counterparts. For this reason most companies developing expert systems today leave the symbolic programmers with the academicians and use expert system tools to fill their needs.

An expert system tool (or shell, if you prefer) is simply a collection of programs that you can use to build an expert system without needing to know a symbolic language. As described earlier, an expert system tool operates much like spreadsheet software. The editor is there, the input mechanism is there, the inference engine is there, the user interface is there, and the output mechanisms are there. All the developer needs to do is add the knowledge! It's not quite this simple, but using a tool is a quicker way to develop expert systems. Most of today's tools feature sophisticated user interfaces, advanced graphic capabilities, support object oriented programming techniques, allow for complex mathematical functions, can tie into just about every type of conventional program or database and enable developers to use them quickly and efficiently with a minimum learning curve. Expert systems tools have all but supplanted the use of symbolic programming languages for use in developing expert systems in the business environment. There a number of companies that make and market expert system tools; many of these tools were used to build the manufacturing systems detailed in this book. In fact, without the expert system tool vendors, the majority of these systems never would have been developed and this book wouldn't exist. Tools are available for just about every computer and application imaginable.

How Expert Systems are Used

Expert systems are being used in almost every industry imaginable. From agricultural applications that assist in planting, forestry and animal husbandry to financial use for lending and tax planning to petroleum use in finding oil reserves and controlling refineries to health care for diagnosing ailments and recommending prescriptive actions; expert systems are proliferating. Books have been written that deal with a myriad of applications, explaining what these systems can do, down to the last bit and byte. I urge you to examine a few of them. You will be amazed at the different uses these systems have. Here, however, our focus is on manufacturing and our contributors have their own important stories to tell.

Conclusion

By now you should be so well versed about expert systems and artificial intelligence that it is obvious why these systems are important to manufacturers. From this point the emphasis of this book shifts from generic explanations of expert systems to the real issues of how your company can and should be using this exciting, productive technology. The foundation is prepared—its time for the real building to begin!

Author Biographical Data

Rex Maus is president of KnowledgeBase Group of Austin, a leading AI consultancy specializing in manufacturing applications. Co-author of *Robotics—A Manager's Guide* and *Expert Systems—Tools and Applications*, both published by John Wiley & Sons.

Prior to this, he was Director of Advanced Technologies for CAP GEMINI AMERICA. He is senior editorial advisor to McGraw-Hill book company.

Chapter

3

Some Facts
About Expert Systems

Jessica Keyes,
President, New Art, Inc.
New York City, New York

Introduction

The manufacturing community has embraced Artificial Intelligence as
a technology that brings with it competitive and strategic advantages.
Nearly 80% of the Fortune 500 has dabbled in and dallied over this
newfound art. For the most part, the branch of AI that has made these
dramatic inroads into the business world is expert systems.

In a nutshell, the crux of expert system development concentrates
on the art and science of unravelling what makes an expert tick. And
while this process bares close resemblance to systems analysis, the
techniques used are different. They're different because knowledge
engineers must dig down deep to uncover often subconscious levels of
experience. They must persevere to locate the tactical level of
knowledge just beyond the surface knowledge which systems
analysis generally deals with. And then they must deal with the
technology that allows them to encode and encapsulate the essence
of this knowledge into something a computer can understand and a
user can use. To this end, the management of the AI development
process can make the difference between success and failure. This
requires a wide array of skills from the development manager.

A 1989 Northeastern University study[1] probed the management techniques used in successful expert system development. The goal of this study was to identify those management attributes and attempt to develop a usable methodology.

Determining the Level of Expert System Complexity

The first major attribute leading to success that they looked at was the complexity to be encoded into the system while the second was complexity of the chosen technology. Other factors examined were level of expertise as determined by education or experience level of expert, amount and certainty of information, duration of typical session and accuracy required of system's conclusion. Other attributes factored into this equation included number of different computers or operating systems that the expert system will operate and types of databases.

The Northeastern team looked at 12 expert systems in these terms and found that they really boiled down to four different types.

A *Knowledge-intensive system* is knowledge bound but uses a simple computing environment and usually acts in an advisory capacity. The other extreme is the *Technology-intensive system* which contains limited knowledge, or knowledge in a limited domain, but requires advanced computing prowess. This sort of system can usually be found in areas where improvement in organizational productivity is envisioned. The most exotic sort of expert system is a *Strategic-impact system* where not only complex knowledge is encoded but the system is technically complex. On the low-end are *Personal-productivity system* where limited amounts of knowledge as well as simple technology are the characteristics. The high end is exemplified by a batch plant scheduling expert system where advanced workstations are used, massive data feeds are integrated and complex rules of knowledge are encoded. The low end is exemplified by an expert system which is in use today by a dental company to forecast when manufactured false teeth will be ready for customer use.

Knowing how a potential candidate for expert systematizing is classified permits the manager to prepare for the development task. Careful preparation is the hallmark of good management.

---

1   Meyer, M.H., and K.F. Curley. 1989."Expert System Success Models." Datamation (September): 35-38.

Why Expert Systems Fail

An oft-touted statistic is that only ten percent of expert systems of a medium to large size are actually successful. The reasons for failure are many and diverse and are oftentimes dependent upon individual personalities and cultures of organizations. But in sifting through the multitude of failed cases we can come up with red-flags that we can watch for the next time we build a system.

The first red-flag is perhaps the most obvious: lack of an available and willing expert. Shearson American Express found this out the hard way in 1984 with their aborted attempt to implement an interest rate swapping expert system. The prototype had been quite successful in assisting the traders in winnowing down the large volumes of data to pick out possible swap partners. During the pilot the system appeared to have earned Shearson at least a million dollars, but this cash cow soon dried up as the real experts—the traders—bowed out. The system relied on the cooperation of the traders who were required to enter information into the system. What the manager of this effort did not understand was the psychology of the Wall Street trader and his penchant for independence and non-sharing.

AI managers occasionally err on the side of omission. After securing management support and locating a willing expert they often omit including the actual end-user of the product into the equation. One classic example of this is in the US Post Office. When the IS department automated the clerks who serve at the windows, they met with enormous amounts of resistance, even sabotage, which delayed the implementation of this system for years. Although this was not an AI system the reason for this wall of non-cooperation was due to lack of involvement of the users—up front—where it matters. The results of exclusion of the end-users in expert system development could be even more calamitous.

One of the biggest reasons for failure of expert systems is the inability to fulfill operational requirements. This is especially true in manufacturing expert systems that often rely on real-time input of instrumentation data. Expert Systems need timely connections to this real data. They need to keep up the pace with the real problem. Good management of an expert system development effort recognizes that a "smart" but very slow system is as good as no system at all. From this perspective several operational items should be included as tasks on the development checklist.

- **Rule firing speed.** If the system is built using a programming language, this can be controlled at the development level. However, use of shells preclude control over this item. It's not something that is commonly discussed, but rule firing speed statistics are available from the various vendors. During the development task of selecting a tool, the vendors should all be taken to task to prove the speed of their inference engines.

■   **Performance of the system.** There's a full repertoire of activities that management should include on the development task list that concerns performance of the system. How fast is the system? What mode of interface works best (e.g., text-based or graphical)?

■   **Database activity.** All systems require connection to corporate data. In what form is this stored? Is it stored in a hierarchical mode when a relational view is required for real-time expert system consultation? Is it even accessible to the expert system? For example, at one organization data was resident on an IBM IMS database accessible under the CICS transaction monitor. Unfortunately, the expert system did not run under CICS. And CICS does not like to share data. This forced the manager of the project to take a step back and create a redundant file for access to the expert system.

■   **Networking.** This is the age of distributed architectures. The manager must probe to discover requirements for multiple platform support of the expert system.

We should never minimize the importance of the heuristics we all picked up from years and years of traditional software development. Bypassing an expert system "development life-cycle" is foolhardy indeed. The most crucial difference between expert systems development methodologies and traditional data processing methodologies is the all important capture of knowledge — which is the essence or "raison d'etre" of the system. Conventional systems contain minimal knowledge and volumes of data while expert systems process volumes of knowledge and oftentimes volumes of data as well. Knowledge capture is, in and of itself, a fuzzy process. There are many ways of doing it and no one best way. And the road to success is often paved with obstacles. Take the case of the Vermilion dam. In 1986 Southern Cal Edison initiated a $300,000 project with Texas Instruments to replicate a civil engineer's knowledge gleaned from years of experience with the dam. Both sides grew frustrated right away. Time was ticking away and the knowledge engineers felt that Tom Kelly, the civil engineer, was not revealing enough. This was a complex task, relying on many rules of thumb. How did Kelly know that a small stream of water on one side of the dam meant a blocked drain at its base? And Kelly was unhappy too. He felt that the two TI knowledge engineers weren't picking up on what he was saying fast enough. Too ambitious a scope, coupled with lack of communication between knowledge engineers and expert crippled the progress of this system. Today, Project

Kelly has wound down. A good prototype was built, but not a deployable one. Southern Cal estimated that another $100,000 would have to be pumped into the Vermilion dam project to permit it to become a generic dam troubleshooting expert system.

Lack of communication between developers and expert stem from many sources. It might be due to untrained knowledge engineers or lack of an articulate expert. It can even stem from lack of recognition that the expert's knowledge lacks depth, that it's really just commonsense. No expert system deals satisfactorily with common sense. But a little commonsense is a plus. There have been some notable cases of omissions of the most trivial commonsense from a system. During beta testing of a Ford Motor Company expert system designed to do credit analysis of car loans, the system failed to notice that one twenty-year old applicant listed as ten the answer to the question of number of years of driving experience.

But perhaps the biggest reason for failure is the modus operandi of the manager himself. Failure here can come in many directions. Obviously, the case histories presented demonstrate failures that could have been avoided by scrupulous management. To manage these types of projects successfully belies a thorough founding in not only traditional development life-cycle techniques, but also in the subtle differences between AI projects and conventional projects. Managers of AI projects must be schooled in AI. This is obvious, but it's interesting to note how many projects are headed by people less than experienced in this sort of technology.

Eight Steps to a Managed Project

Management of AI projects requires a stepped and organized approach to the process. The manager must not only understand the technology, but should be familiar with the problem area and must be directly involved in the effort.

This section demonstrates a management methodology that has been used with success by many managers of successful AI projects.

Step One—Creating the Expert System Team

The first sign of good management is care in selection of an appropriate development team. Similarly to traditional systems development it's a good idea to use a team approach. There are two components of the expert systems development project that require careful planning. Even in the most high-tech systems, there will always be some traditional systems work that must be done. The majority of expert systems have components that require access to corporate databases. The added complexity of networking may be required to get at this data. For example, an expert system being built on a workstation may need to access data on the mainframe. The appropriate development staff needs to perform the programming tasks required to get this functionality. These tasks are data-driven and require a certain skill-set.

The flip side of the team are the knowledge engineers.

They generally possess quite a different skill-set, less data-intensive and more inclined to go after the source of knowledge. The issue is that these two skill-sets are quite different. While it is possible to find a unique talent that can do it all, it is recommended that two separate people perform these two very different functions.

A good knowledge engineer will exhibit some special global characteristics. Perhaps the most important characteristic is the ability to converse with the expert in the language of the expert. This characteristic is becoming increasingly important in the traditional development world. The trend towards distributing computing power to the department level has had the effect of distributing analytical and programming expertise right into the heart of user territory. Where, in the past, development staff could get away with being expert programmers with a vague sense of the user's domain, this lack of user knowledge is fast changing. This requirement for depth of knowledge about a user's domain is even more pronounced in AI-projects where the goal is not to program a report but to siphon the expert's knowledge and place it in a repository for general use. This person should also have the ability and the inclination to perform thorough research in the area chosen to be expert systematized. And finally, the selected knowledge engineer must have the literary capabilities necessary to document the results. What we have here is a rather unusual combination of skills that few systems people seem to possess.

Step Two—Picking the Problem

Not all applications lend themselves to expert system technology just as not all applications lend themselves to being on a PC or being on-line or being database applications. It is necessary for the manager of the project to not only sift through the plethora of ideas that are suggested by various personnel, but he must also be able to locate sources of good ideas as well.

There are many sources of ideas for expert system applications. Senior management is usually a good source of ideas. In this area you might find some good candidates in the realm of reducing organizational complexity. Elsewhere, other sources of good ideas are the company's customers, a task force specifically set up to come up with ideas, the research and development department, marketing, operations staff, the knowledge engineering staff and last but certainly not least, the expert.

The most suitable problems for expert systems development require application of knowledge rather than data, decision making over number crunching. For example, developing an expert system that advises, monitors and controls a factory is far better than one that just monitors a factory floor. The expert system gathers data giving supervisors dynamic access to the shop floor as well as performing analyses on this data. But, these two systems are not diametrically opposed. The conventional Factory Monitoring System could be tied together with the expert system Factory Advisor. This is what is known as an embedded expert system.

Once a problem is selected that looks ripe for expert system development, the idea needs to be scrutinized in more detail. The manager needs to ensure that a problem was picked where there is knowledge readily available. The availability of knowledge presumes there is an expert available. The manager must make sure that expertise in solving this problem actually exists within the company. At times, the manager will come to the realization that the knowledge to solve the selected problem is not readily available in the company. At this point, either the project must be terminated or the manager must go out on a search for outside expertise. This is not as uncommon as it sounds. Many systems have been built with reliance on expertise garnered from the academic community, the Big Six, as well as industry experts.

Of course, if the problem selected is one in which it takes the expert a few days to solve, then the problem scope may be too broad. For those project managers who are not expert in the subject area it becomes necessary to limit the scope of the project to ensure success. Therefore, if the scope of the problem appears too large it is best to scale it down to a more manageable level.

On the other side of the coin is a problem that takes the expert an extremely short time to solve. If this is the case, then the problem is too trivial. In these instances, the manager must try to widen the problem or replace the problem altogether. The best problems to work with exhibit what is known as depth in a narrow domain.

Step Three—Problem Identification

Upon successful selection of a problem to expert systematize, the next step is to define it. The manager must ensure that the problem is formally defined at the first working meeting of the knowledge engineering team and the experts. The definition should be short, to the point and extremely clear. There must be no confusion as to the goal and expected outcomes of the project. For example, the problem definition of the Factory Advisor might be written as follows:

The goal of this project is to develop an automated Factory Advisor (FA) that will exhibit the following features:
   a. Dynamically gather shop floor information
   b. Display this gathered data to supervisors
   c. Perform analyses of data with which:
      1. the supervisor can react more quickly to situational problems
      2. resultant conclusion of analysis provides automated direction to automated shopfloor equipment (i.e. real-time shop floor control).
   d. Act as a trainer in the Corporate Training Department in the area of training new staff in supervising shop floor processes.

At this point it is imperative that this problem definition be distributed to all those involved. It would be poor management not to solicit feedback from the ultimate end-users of the system as well as management.

In traditional data processing shops, the idea of AI technology presents some difficulties. With a large investment in traditional software many computer shops have become entrenched in conventional methodologies. Therefore attempting to move them onto a higher plane becomes difficult, and if not managed properly, impossible.

Most computer shops are a bit nervous about anything new. New software means new skills to learn and new programs to support.

Most operations types fear what they do not understand. All new software has the potential of degrading system performance to unacceptable levels. For some reason expert systems have garnered a bad reputation as a resource hog. Since expert systems can run on the mainframe this is an understandable fear. Unfounded, but understandable. Expert systems tools do have a lot of features. And any tool with a lot of features, that is not well-used, has the potential of degrading systems performance. This is where project management and systems testing come in. A well-planned and executed system should have minimal impact in the production environment. In fact, you can probably do more damage to the system with a wild Cobol program than you can do with an expert system too.

To counterattack these fears, it is recommended that the manager form a committee composed of development, systems and end-user personnel. The charter of this group would be to address, in an unbiased manner, the needs of the organization versus possible implementation tools for this type of technology.

Step Four—Conceptualizing the Problem

The manager must now concentrate on the details of the problem. A "grass-roots" understanding of the problem and all of its vagaries is absolutely necessary. In order to do this the team should concentrate on the WHO, WHAT, WHERE, WHEN and HOW of investigative analysis. This is the beginning of the knowledge engineering process. The end-goal is to develop a comprehensive set of what is called conceptual documentation. This is composed of several items.

- **Who is the end user?** This must be known because the consultation, which is the dialogue between the user and the system, should be tailored specifically for the level of audience using the system. A system geared for low-level clerks would look different from one designed for shop floor staff would look different from one designed for supervisors would look different from one designed for engineers. A good example of this was IBM's decision to provide a pictorial user interface to an expert system used by security guards to repair plant security doors that were malfunctioning. IBM's Sterling Forest facility is rather large and has many secured locations. When a security door malfunctioned a technician was called in for repair. Unfortunately there was only one technician who didn't take kindly to being called in at all hours of the night. To solve this dilemma, without having to hire additional staff, IBM created an expert system that would permit the security guard to troubleshoot and repair doors on his own. Instead of providing textual instructions, the IBM expert system's team decided that a graphical interface would be much more apropos and much more understandable to the security guard. The expert system displayed, in full color, pictures of all the steps necessary to track down the malfunctioning component and replace it.

- **Where does the problem fit in relation to the environment?** The problem selected is generally in an area that has stringent ties to other parts of the department and to areas outside of the department and even outside of the company. It is important that these relationships be analyzed and dealt with. An expert system monitoring a dam interfaces to three different levels of the environment. Level one is the interface to the dam and surrounding environs itself. Here sensor gathered data can be used for monitoring. Level two is the interface with the humans who work the dam. Level three is the relationship to the outside world: weather reports, reports from sites upstream and downstream, drought reports, and so on.

The *generalized domain flowchart* is a thumb-nail sketch of the world the process lives in. For the knowledge engineers on the team it serves the purpose of introducing them to the area the expert system will be situated in. For the expert, this sketch serves the purpose of acting as a reminder of the relationships between the process to be systematized and the rest of the world in which it sits. In this way the team ensures that major inputs and outputs are not overlooked.

- **Review the specifics of the chosen task.** A *problem area detailed flowchart*should show the specifics of the functionality of the job to be expert systematized. At a minimum it should show:

  - relationships to external sources, such as outside databases

  - data flows (input and output)

  - data manipulations to be performed

  - points of user interactions

  - any areas of uncertainty.

For the knowledge engineer this activity will serve to make the job functions clear and specific. For the expert it will serve as a check-list to make sure that all bases are covered during the development process.

- **Test case development.** The manager should understand that a thorough understanding of the problem is never enough. In order for a system to be successful, it most be testable. Quality of testing has always been a large issue in the area of traditional systems development. It is more so in expert system development. An expert system gives no black or white answers, it advises. Therefore it is absolutely necessary that steps be taken to carefully track the expert system's correctness. This can be done by developing a sampling of test cases that typify the process. There are several ways to achieve this end.

The wrong way is retrospective analysis. Asking an expert how a solution was arrived at after it happened will lead to fragmentary spurts of knowledge, at best. Several methodologies are effective.

- **Brainstorming.** The first meetings held with the experts serve to set the stage. These meetings are best when they are loosely structured so that the team may become comfortable with each other and issues may be raised spontaneously.

- **Structured interviews.** Once all relevant issues have been raised a series of more structured interviews is necessary to hone in on particular aspects of the problem domain. Here specific topics to be discussed are scheduled and prepared for.

- **Task analysis.** Since retrospective analysis provides such fragmentary insight into the expert's domain it is often necessary to analyze various tasks that make up the daily routine of the expert. The manager should ensure that not only does the knowledge engineering team spend considerable time observing the process but also spends time in questioning the reasoning behind the decisions made. It would be worthwhile to adopt an auditing approach to this process. An auditor performs the auditing function in three ways: questioning, observing and testing.

- **Think-out-loud.** Here the expert performs his task unencumbered save for the requirement that he "talk through" the process.

- **Constrained problem solving.** This activity provides a valuable insight into the logic processes of the expert. The expert is asked to do a task under some unreasonable constraint. For example, in one study two expert radar map interpreters were asked to interpret a radar map within a very short time span. Their reasoning short-cuts, as they tried to meet this deadline, provided the researchers with valuable information.

During this knowledge acquisition process it is extremely important that the manager ensure that adequate notes are taken. Notes may be in the form of paper logs or audio-visual records of the proceedings. It should be noted that an often quoted statistic is that it takes 10 hours of transcription to get the essence of one hour of video-tape.

These notes can then be used to tailor a set of test cases that will then be used during the remaining steps of expert system development.

Step Five—Formalizing the Problem

When the problem is well understood, a specification should be written and reviewed. This formalization should be clear and concise. It should include the following:

- A one-page description of the problem to be addressed, including all assumptions about the problem.

- A description of the user base
- A description of the design team
- The domain flowchart
- The problem area detail flowchart
- Test case documentation.

At this point a preliminary decision can be made about the hardware and software that would best serve the problem area.

Picking an appropriate hardware tool. Today expert systems run on many platforms—from PC to mini to workstation to mainframe. The decision of which platform to run on is more dependent upon such conventional considerations as corporate policy, where the data resides, experience level of staff and networking capabilities than on anything expert system related.

This will be the most difficult decision that the team will make and it is often the one that can make or break the system. The manager should be prepared to oversee this process and examine all of the criteria used to make this decision.

- Should the expert system be mainframe oriented? This is one of the more difficult questions. Even though there is a trend towards distributed processing, most corporate data resides on a corporate mainframe. It seems natural, therefore, to run the expert system where the data resides. The manager should address response time issues as well as cost considerations if a mainframe platform is contemplated.

- **Alternatives to mainframe processing**. With the introduction of distributed computing the development team has many more options. It is quite possible to create systems that run on PCs which access mainframe data. The manager should be aware, however, that this solution may or may not be successful depending upon the company's systems environment. One manufacturing company chose the option of networking workstations to the mainframe and downloading data as required. It seemed an elegant solution and, in fact, had been successful elsewhere. Unfortunately, in their systems environment it was less than successful. This company did not have a robust network in place. They were primarily a mainframe environment with some PCs attached as terminal emulators. While it's true that terminal emulation programs do permit transfer of data between mainframe and PC devices, this is not a sophisticated solution. The manager needs to be aware of the company's limitations and override the team when he believes that an extra layer of difficulty is not warranted.

INTERPRETATIVE	Inferring Situation Description from Data
PREDICTION	Inferring Likely Consequences
DIAGNOSIS	Inferring System Malfunctions
DESIGN	Configuring Objects under Constraints
PLANNING	Designing Actions
MONITORING	Comparing Observations to Outcomes
REPAIR	Executing Plans to Administer Remedies
INSTRUCTION	Altering Students Behavior and Knowledge
DEBUGGING	Prescribing Remedies for Malfunctions

Figure 3-1   General Categories of Problems

Picking An Appropriate Software Tool. Not all companies use solely one tool. Different tools are appropriate for different types of problems. Figure 3-1 shows the paradigms, or general categories of problems, into which you can place your particular problem. These words are important to know since they appear in the literature which describes the software tool being evaluated.

The software question is really two questions. Do you permit the team to build from scratch, i.e. write in Prolog, Lisp or C? Or do you select a pre-written shell and tailor it accordingly?

- **Programming languages.** This is the solution that will provide the system with the most flexibility and the highest degree of tailoring. On the negative side, staff with these programming skills may be hard to locate and keep, you just might be reinventing the wheel and it will most certainly take you much longer since the inferencing program must be written from scratch.

- **Expert system shells.** The majority of systems being written today are based upon one of the many shells flooding the market place. The majority of these products are written for the PC although there are contenders as well in the mini, workstation and mainframe marketplace. In going out to these vendors to evaluate their products, it's useful to go armed with a wish list

Vendor:

Cost: One Time: Maintenance:

Consultation Paradigm: Diagnosis Planning Design Other

Knowledge Presentation: Rules   Frame   Objects

Inference: Control: Backward Chaining
                                             Forward Chaining
                                             Depth First
                                             Breadth First

Implementation Language: C
                                             Pascal
                                             LISP
                                             PROLOG
                                             PL/I
                                             Other:

Interfaces: Databases: External Exits:
                                             Programming Languages:
User Interface: Instrumentation Interfaces:

Knowledge Engineering Interface: Editor:
                                  Debugging:
Documentation   Graphical: Support:

Training

Figure 3-2   Sample Expert System Tool Selection List

so that you can find out exactly what features their product
exhibits versus other products on your list. Figure 3-2 is one
such list. You can expand it or narrow it down to suit the
system's requirements.

Step 6—Prototyping

The manager must make certain that an iterative prototyping step is
planned for. The reason is simple. Expert run-time systems are

commonly called consultations. These consultations advise, assist and generally interact with the user. Due to this high-level of iteration and the nature of a consultative session, it is paramount that the user be included in a staged development of the system. In this way, the system can be built up gradually in much the same way that a human acquires knowledge. These expert users can add layer upon layer of knowledge to the system and at the same time comment on the interface between user and system.

It is sound advice to build a prototype for other reasons as well. There is the risk of making an enormous investment in time and money only to find that the problem is not suited to an expert system. Or you may find the knowledge base is faulty, the hardware is all wrong or the software is inappropriate. A prototype will highlight these problems before a large commitment of resources is made.

- **Pseudo-coding the system.** During the knowledge acquisition step of the formalization step, field notes were compiled which can now be turned into pseudo-code. Some developers may prefer to pseudo-code the knowledge base during the knowledge acquisition process itself. No matter when it is done, this is the step to create an english language set of production rules.

Even if the project turns out not to be successful, or is never pursued as an expert system, the creation of these pseudo-coded rules is valuable for the users. For what they've created is a procedure manual with that extra bit of expertise.

- **Prototyping rules of thumb.** This prototype should certainly contain no more than 50% of the ultimate system, and cut this number down more if possible. For example, if 400 rules were developed during the formalization phase, prototype only 200 of these rules. A good rule of thumb here is to further narrow the problem and encode those that make up this section of the system. In this way, a complete component at a time can be coded, tested and accepted by the users.

Most expert systems tools are different from each other. The syntax in one is different from the syntax in another, although similarities do exist. The syntax in an expert system tool is also very dissimilar to conventional software toolsets, they contain structures that can handle the nuances of knowledge. For example, the Texas Instruments' PC EASY product contains several structures to handle uncertainty and ambiguity: IS MIGHTBE, IS DEFNOT, IS DEFINITE, IS KNOWN, IS NOTKNOWN.

To do the prototype effectively all facets of the problem must be represented:

- Represent the major classifications of data and rules. Can the software chosen easily handle the requirements?

- Prototype the important database linkages, simulate the rest. Make certain that database access is tested.

- Prototype, do not simulate, the user interface in a manner that is useful and pleasing to the user.

Step 7—Reformalization or Reassessing the Problem

Once the prototype is complete it should be run to determine if the rules and objects are properly encoded and if, in fact, all of the rules are present that need to be present and if all the objects are present that need to be present. This phase is a reiteration of the initial formalization phase. The team must also make the determination whether or not the hardware and software chosen is indeed appropriate for the problem.

- **The confusion matrix**. It is important to run the test cases, developed in the conceptualization phase, against the prototype. Conclusions must be compared against those manually derived by the expert or experts. It is hoped that such care is taken during the development process that this stage will find the expert system generated solutions to be 100% correct. Unfortunately, this is unrealistic.

The manager should plan carefully for what is called confusion testing. In the field of psychology a confusion matrix is built to determine the ratio between correct responses and those responses that were close, but not quite correct. We can use this same technique to determine just how correct an expert system is. In Figure 3-3 we show a confusion matrix for a medical system. This expert system determines the type of disease the patient has based on what we say are our patient's symptoms. Our list of possible diseases is the FLU, a COLD and the BLACK PLAGUE. In a confusion matrix expert the expert system's responses are compared to the responses that a human expert would make. One would hope that these responses would be parallel. You can see from the confusion matrix that this didn't happen. Each slot or entry in the matrix represents the percentage of times that the expert system differed from the human expert—which really means the percentage of times that the expert system was confused. In the example, where the doctor diagnosed the FLU, the expert system recommended the BLACK PLAGUE 20% of the time and COLD 15% of the time. These erroneous conclusions totalled 35% of the time. This means that the expert system was correct only 65%

	black plague	flu	cold
black plague	n/a	35%	45%
flu	20%	n/a	15%
cold	10%	5%	n/a

Figure 3-3   A Confusion Matrix

of the time. The average of these error percentages will give us an error threshold for which each company must decide what is acceptable and what is not. In the case of our medical expert system the total error threshold turned out to be a low 57% (i.e. system was correct only 57% of the time) which would be unacceptable.

Since it is not possible to detail each and every facet of knowledge before undertaking this stage, running the prototype will uncover noticeable holes. Therefore it is important that reformalization be dealt with as an iterative step.

- **Hardware/Software Problems.** During this stage the team might find the expert system tool selected doesn't quite muster up to original expectations. This could be for several reasons:

  - the shell is not able to process the final number of rules in the rule base

  - the user interface is not adequate

  - difficulty in getting to the external data needed

  - too many "user exits" required to make the system run. When you exit the expert system shell it is usually to write code because the shell itself can't perform the functions that are needed. Too many of these is not a good sign.

  - PC used does not contain enough memory (1 MEG or more is recommended)

  - PC used is not fast enough (use a speed-up board)

- Mainframe performance is degraded (could be that you're not really at fault, recommend a systems tuning first)

It may also be necessary to upgrade to a completely different hardware device. If the PC is found to be inadequate, upgrading to a mainframe or specialized AI workstation may be warranted, although this is unrealistic in the era of the 386 and now 486 machines. If mainframe response is seriously degraded and the software tool is properly used then this may be a function of the "straw that broke the camel's back." Some mainframe production environments just can't suffer one more burden. In this case it will be necessary for the manager to make the decision to move the application to a mini, workstation or PC.

Step 8—Testing

An expert system needs to be tested as thoroughly as a conventional system.

Alpha testing. In alpha testing the system needs to pass muster with those people designated as experts in the user group. They should be given the opportunity to scrutinize the final system.

Beta testing. In Beta testing, which is probably the most important type of testing you can do, the system is sent out to the actual end-users.

These users need to be given adequate time to test the system by themselves. At the same time they should be given formal procedures for monitoring the session and noting any difficulties. A minimum of two to four end-users for this process.

The system should be set up to run in tandem with the current system or manual procedures that this expert system will replace or assist in. Staff should be made to feel free to comment on any difficulties, opinions and suggestions. A side benefit of this phase is to make the users "buy-in" to the system.

This phase should be called a pilot and should last at least one month.

Maintaining the Expert System

In some companies the AI team does all development and maintenance. In other companies the AI team just does the development and an operations group does the maintenance. Still in others, where no AI team exists, the development team does it all just like any other project.

The majority of systems being built today have not yet encountered this question, since the technology is still new. Good management of an expert system project will plan for this most certain eventuality. Since there is a movement to think of AI tools as just another tool in the developer's toolset, it is recommended that AI systems be maintained similarly to conventional systems.

A Well-Managed Project

Management of AI projects runs the gamut between too much management and no management at all. Following a stepped and well-planned approach, overseen by a manager with experience in the area, will lead to successful implementation of AI systems.

Author Biographical Data

Jessica Keyes is president of New Art Inc., a management and computer consultancy based in New York City. She is the publisher and editor of TECHINSIDER newsletter and organizer of TECHINSIDER training seminars and video training series. She is the author of the *The New Intelligence,* published by Harper Business and a columnist and correspondent for such publications as COMPUTERWORLD, Database Programming and Design, AI EXPERT, Expert Systems Journal, Software Magazine and many others.

4

Manufacturing Survey

Jessica Keyes
President, New Art, Inc.
New York City, New York

Introduction

A new buzzword has been added to the manufacturing vocabulary. Manufacturing managers have long realized that in order to be competitive in the marketplace the use of computerization was necessary. The term CIM, or Computer Integrated Manufacturing, sprang up in the not too distant past to describe a series of automated methodologies with which the manufacturing process could be finessed to the point of increasing productivity, quality and ultimately enhancing competitiveness. In the fervor to automate all things manufacturing, several other concepts (and buzzwords) were thrown around. CAD, or computer assisted design, and CAE, computer assisted engineering, is probably the most exotic of these automatic techniques. The use of a computer to simulate the design and engineering of an object such as a car brings important advantages. With CAD/CAE, for the first time an engineer can model and view multiple versions in a shorter period of time than had been formerly possible. Quality is a consideration as well for CAD/CAE software permits multiple angle viewing as well as other features with which an engineer can uncover design flaws at earlier stages in the process.

Manufacturing is the most difficult of all industries. Literally thousands of components, from plan to design to inventory control to actual development, must be taken into account for a firm to be able to compete successfully in the marketplace. The bottom line for most firms is to keep expenses down while turning out the maximum number of product. This has lead to the creation of the manufacturing buzzword of the 1980s. JIT (just in time) is a rather complicated notion. In broad brushstrokes, it is the concept of bringing a product to market using the resources made available specifically at the point they are required—hence, just in time. Rather than an inventory flush with parts that sit and gather dust, it is more practical,and more cost-effective, to plan to order these parts when and if they are needed. Computerization was added to the manufacturing equation when management realized that the process was just too complicated too handle. And for a time, in the mid 1980s, when CIM techniques reached a nascence of popularity, these methodologies proved to be adequate. But, there was a menace looming on the horizon. No longer could American manufacturers be content to do battle with each other for market share. The era of globalization was born with manufacturers across the globe scrambling to gain a competitive advantage. With this push across the globe came the realization that the techniques of the past were insufficient to meet the needs of the future. And so a new buzzword began to be heard around the halls of manufacturing companies. This buzzword was artificial intelligence. AI is no panacea; no computer technology is. AI won't conquer the competitive dragon by itself. But more and more companies that are using AI are finding that it lets them have ahead-start.

Bob Louch, of Ford Motors in Dearborn Michigan, says that nearly every manufacturer in the automobile industry is moving in the AI direction in order to survive, "it's not a question of wanting to, it's a question of if you expect to be here in ten years you have no choice. You have to reduce turn-around time from X number of years in half." Ford is typical of most automobile makers in the application of AI technology in diverse divisions within the organization. Louch's project is to explore ways of using AI in the CAD process. The thing most appealing about AI techniques, according to Louch, is the robust form of object-orientation of the toolsets he's reviewing.Object-orientation is a perfect fit to the design of an automobile and Louch thinks that use of it can pare down the time it takes to design and deliver automobiles dramatically.

Ford isn't the only manufacturer to use AI. There are hundreds of other companies and hundreds of other AI applications in the manufacturing arena—from process planning in multiwire cable manufacturing to a system that tells shop personnel what tools and materials to use in assembly of electrical connectors to a system for dispatch of work in progress to assisting in creating layout designs for a factory floor to non-destructive testing of bridge designs. The remainder of this article will give the reader an overview, in survey form, of some of the many approaches of AI in manufacturing.

Welding

If you were to look through a registry of manufacturing expert systems, one name would pop up continually. Gavin Finn, of Stone & Webster in Boston, Massachusetts, has quite a few expert systems under his belt. Stone & Webster,an engineering technology firm, has developed extensive experience in applying expert systems in the industry. Stone &Webster has developed a cadre of fielded systems that run the gamut from diagnosing processes in real-time, scheduling, troubleshooting machinery and designing products and facilities. The first I ever heard of this company was through their WELD DEFECT DIAGNOSIS SYSTEM. This expert system helps welding managers determine the cause of welding defects. The welding manager has another expert tool in Stone & Webster's WELDER QUALIFICATION TEST SELECTION SYSTEM. The manager is assisted by this expert system to choose appropriate qualification tests to select welders.

Welding is quite a popular task for AI infusion. Over the last several decades much progress has been made in this area with the design engineer having a menu of over one hundred processes, hundreds of fluxes and an equal array of electrode materials. The design engineer must select the optimum mix of these materials to produce the safest weld in the shortest amount of time. Not every design engineer is a welding specialist with the result of the possibility of overdesign of the weld.Expert systems can take the place of the welding specialist and assist in designing an optimum weld. Along with Stone &j Webster, there are many others in this field. Battelle Columbus Laboratories (Columbus, Ohio) has done this in the form of their WELDEX system which identifies faults by examining features in a welding radiograph. WELDEX's problem is to interpret radiograph features such as shape and size,location and contrast in search of faults. A LOOPS-based system with an astonishing degree of accuracy, WELDEX's goal is to model the knowledge, reasoning and judgmental skills of an expert welding radiograph interpreter in identifying welding faults.

Perhaps the strongest player in the welding arena is the American Welding Institute (Knoxville, Tennessee) which has quite a few projects under development in conjunction with the Colorado School of Mines. The W.I.N. (Welding Information Network)goal is to provide a national information resource to help in making American firms more competitive in the global marketplace.The American Welding Institute has already fielded expert systems in this area such as WELDHARD, a system for hardfacing and weld cladding; WELDHEAT, a system for determining pre-heating and post-heating; WELDSELECTOR, which acts as a welding specialist; WELDSTRESS, a system for determining residual stress and WELDTRACK which can be used to track welders.

Real-time Process Diagnostics

The role of the plant operator is to diagnose and interpret data from different strategic areas within the plant.His goal is to regulate plant parameters to achieve a smooth and trouble-free, operation. This, of course, is not as simple as it sounds. Manufacturing plants are large, replete with diverse equipment. Trouble can occur literally anywhere—on the assembly line, within a group of equipment, within a single piece of equipment. Real-time expert systems advise operators on problems within this process. These problems take the form of alarms, equipment checks, process upsets, and instrument failures. In addition, the plant operator makes note of trends and looks for ways to optimize production. To add to this already complex mix is the requirement of integration of diverse computers and data acquisition systems.

Stone & Webster considered each of these requirements in turn and came up with an architecture for implementing these types of real-time expert systems. They refer to this architecture as the Blackboard approach. The Blackboard is proprietary software that serves to integrate information from a multiplicity of expert system shells, data communications, graphic interfaces,simulation models and statistical models.

The Stone & Webster Blackboard architecture is used to advantage in their POLYMER PRODUCTION EXPERT SYSTEM. This is a real-time system that advises chemical company plant operators on polymer production. The Blackboard architecture permits the linkage to a distributed control system, a laboratory database and a commercial graphics package enabling operators to get expert help in diagnosing over 30 different process problems. Along the sesame lines, Stone & Webster's FIRED HEATER EXPERT SYSTEM advises on such problems as tube leaks, false drafts, false instrument readings and control valve malfunctions.

Nelsons Acetate Ltd. is an English company in the plastics industry. They manufacture cellulose acetate which is a material used in photographic film. Since acetate varies widely in quality, Nelsons has developed a series of techniques and processes for ensuring consistently high quality. The Nelson steam is composed of very experienced staff members and management. Nelsons realized that when key members of the team reached retirement age, the firm ran the risk of losing invaluable experience. It was for this reason that management agreed to the development of the CHEMICAL PROCESS CONTROL EXPERT SYSTEM. Delivered in only five weeks, the expert system consisted of 37 primary and 42 secondary rules in four knowledge bases. Additional rules were added to a fifth knowledge base for control purposes. The system provides automatic interface to plant instrumentation and a simplified user interface for plant operators. This was accomplished by using the KES expert system shell (Software A&E is located in Arlington, Virginia) embedded in a Microsoft C language program.

Con Edison of New York is a manufacturer of another sort. Probably one of the largest producers of electricity, it produces power for the country's largest city. Using the ART expert system tool (Inference Corporation located in Los Angeles, California), Steve Silverman's team created the SYSTEM OPERATION COMPUTER CONTROL SYSTEM. SOCCS' function is to monitor and assist in the control of electrical power transmission and distribution in the New York City area. This expert system gathers data from multiple monitored points and compares them to expected normal values. If there is a difference it generates an alarm. An associated expert system is Con Edison's SOCCS ALARM ADVISOR whose function it is to assist operators by identifying and suppressing repeating or toggling alarms, by analyzing the network's status and by recommending restorative actions. General Motors heeded the words of a manufacturing consultant who indicated that manufacturing companies in the U.S. will lose $200billion in 1990 because of inadequate maintenance. As a result, GM decided to put a critical emphasis on improving maintenance operations. They wanted to improve electrician skills in the diagnosis and troubleshooting of complex automation devices.

Commercially available, artificially intelligent machine diagnostic systems began to appear in the mid 1980s. At the forefront were Cimflex/Teknowledge's Diagnostic Assistant, Carnegie Group's Test-Bench and Automated Reasoning's I-CAT.

For the expert system that needed to be built for GM's Bus and Truck Division's Flint Metal Fabrication Plant, I-CAT was chosen and installed on a portable Toshiba 5200 computer and made available to electricians on the plant floor. The problem area systematized was the manufacture of truck fenders. Sheet metal goes through a series of stamping operations to emerge as a fender part that subsequently undergoes welding operations beyond the line. Automation devices are responsible for moving the piece of metal through the presses in a systematic and controlled manner. These devices are controlled by three control panels, each having its own Data General NOVA computer. The TRADES system presents the electrician with a graphical diagnostic model of the manufacturing system with built-in test point information to guide him through a series of tests to isolate the faulty component(s) in the system. Even though TRADES may recommend a particular testpoint based on the system's stored heuristic knowledge, the electrician may still deviate from that recommendation. He simply selects the testpoint of his choice with a mouse. If the electrician desires to backtrack, he may ask the program to forget the results of the previous test and return to conduct a new test.

Intelligent Applications, based just outside of Edinburgh, Scotland (Livingston Village), has emerged as a leading company in providing expert system applications for manufacturing in the United

Kingdom. Headed by Dr. Robert Milne, who was formerly Chief Scientist of the US Army's AI Center in the Pentagon, Intelligent Applications has been busy in the area of process control. For British Steel Ravenscraig, a very large real-time expert system based on a VAX to monitor the steel-making process has been implemented. Its goal is to identify faults which will affect the overall process. For Exxon Chemicals, an on-line continuous monitoring system has been developed. This expert system monitors the vibration characteristics of the main pumps and compressors of the plant.Perhaps their biggest contribution to AI in manufacturing is the AMETHYST system. Developed using the Crystal expert system shell as its base, AMETHYST's function is being marketed worldwide as a maintenance diagnostic system. The principle behind this particular expert system is a concept called condition-based maintenance, that is, repairing the equipment only when the conditions require it. Some installations will open a pump and overhaul it every six months. This can be very wasteful, often involving the overhaul of a perfectly good pump. In addition, there is always the possibility of further damage or mistakes being introduced to the system needlessly. Vibration based condition monitoring has not been largely used because there was little knowledge about where to measure and what information to collect. Further, most companies simply do not have the expertise to diagnose the state of the machine based on the vibration signatures. The AMETHYST expert system provides an accurate diagnosis of the state of a machine with no skill or training. Using a hand-held vibration data acquisition device from IRD Mechanalysis, data is fed into the expert system where it is analyzed by the rulebase to diagnose faults in rotating machinery.

Allocation, Scheduling and Planning

One important issue in the quest to achieve success in the manufacturing arena is the selection of methodologies used to control, manage and schedule capacity for each resource needed to serve each task. This is the goal behind the manufacturing buzzword JIT (just in time). If a plant is capacity limited, then there is a great incentive to produce every incremental amount of capacity by reduction of the idle time and maximizing resource allocation. On the other hand, if a plant is over-capacity, then the plant must be managed to reduce production costs. In other words, we're back to the original caveat of maximize capacity and minimize costs.

This was the goal for PPG Industries Coatings and Resins Division (located in Delaware Ohio) when they choose the MERCURY KBE SYSTEM (Artificial Intelligence Technologies located in Hawthorne, New York) to solve the dilemma of unforecastable order volumes in the manufacture of automotive refinish paints.These capacity swings range from 200% down to 25% which is a virtual shutdown of plant operations.

Using Mercury's robust CLOS compliant object system on a VAX/VMS platform and operating with links to DECs Rdb relational database, the PPG SCHEDULER reasons on hundreds of thousands of objects and analyzes opportunities which ultimately lead to an improvement of 20% more orders on or ahead of due date without cost impact.

The AIT approach to the PPG SCHEDULER was to create a hybrid of generic modeling and scheduling tools. Using these technologies it is possible to completely represent sequencing. The route planner component uses a combination of AI, CPM (critical path method) and PERT operations research techniques. The Resource Allocator/Resource Simulator provides a knowledgebase of alternatives to be reasoned about and optimized while a constraint system describes the global network.

At PPG's Springdale, Pennsylvania site, the Mercury expert system shell has been used to field the INTELLIGENCE STATISTICAL MODELING System. Prior to expert systematizing, the process engineer would spend anywhere from two to three man-weeks in generating a valid statistical model. The expert system created was actually a cooperating expert system which unified a statistical master, a process knowledge base editor and a interactive form driver for model development. A form driven user interface was used to process the knowledge base editor, enabling the user to converse and discuss process relationships and confidence on those relationships. Using both objects and rules the process engineer, even one with no computer science or statistical background, can generate a valid model in a one-hour session.

Thirty miles north of Pittsburgh, the Ellwood City Forge makes open-die forgings. These are forgings that don't require costly closed dies that have to be machined to shape for each job. They handle big forgings, from a quarter ton to 25 tons, with parts that need greater strength than casting can produce. Typically these large forgings are required in lots of only two to three pieces. Some manufacturing engineers would call it a"snowflake" job shop since no two jobs are ever exactly alike. With revenues of over $50 million a year, the Forge's goal has always been to deliver parts quickly at an economical price.In spite of their reputation for quality and price points, they do have their off-shore competition.

2000 orders are usually open at any one time. The forging process takes place in four separate departments. Work starts at the Melt Shop, where under computer control, raw ingots and scrap steel are melted together with such elements as molybdenum and vanadium, creating an optimally composed steel alloy. At this point the ingot is heated to red or white heat. Giant forging presses then squeeze this molten mixture into an oversized version of the finished part. At this point the part goes to Heat Treatment where again computers control the cooling and tempering process. The process of austenization is used to quickly cool the part at a precise rate to preserve a precise fraction of the critical 1% or so of dissolved carbon that makes iron into steel and controls its degree of hardness. Tempering occurs when the part is re-heated to a much lowered and precisely regulated temperature (usually around 400 to 600 degrees Fahrenheit) and kept there for several days to replace brittleness with toughness. Finally, the part may go to Machining, where it is shaped into the final dimensions on a lathe or milling machine.The key component here is scheduling and the catch-22 is again speed versus cost and quality.

The Forge joined forces with the Carnegie Group in Pittsburgh to fund a joint venture to improve plant-wide scheduling. Stephen Gifford, a senior manager at Carnegie Group, makes this statement about human schedulers, "There are no experts on scheduling a total manufacturing facility. The decisions are simply too complex, and they depend upon having clear and realistic guiding priorities." The system that this joint venture produced was developed to implement clear-cut guiding priorities. Its primary objective was to reduce order turnaround while improving raw material yield; its secondary objective was to improve energy efficiency. The payoff to this system includes an improvement in material yield, reduced energy consumption, and a reduction in turnaround time from the former eight weeks to less than three weeks. As a side benefit, the company is also able to estimate price and delivery much more accurately before the job begins, which is a definite competitive advantage.

Meanwhile back at Stone & Webster, a PC-based PRODUCTION SCHEDULING ADVISOR was developed for general planning purposes.In the manufacturing arena, Union Carbide Chemical and Plastics in Charleston, West Virginia uses it to make the company more responsive to global export opportunities. According to Craig Castro, manager of strategic planning for Union Carbide, "We sometimes need to turn around a decision on whether to accept an order in a day, and this involves calculating what the impact of changing production runs will be." This scheduling expert system runs under Windows 3.0 and OS/2 Presentation Manager and combines graphics, spreadsheeting, linear programming and expert systems (Information Builders Level 5 was used) to help production schedulers effectively meet changing demand, as shown in Figure 4-1. The expert system component comes into play when production requirements cannot be met based on the data stored in the Excel spreadsheet. At this point the expert system module is invoked to help schedulers decide on which constraints to eliminate in order to meet demand.

Of the Big Six consulting firms, perhaps the one with the largest presence is Arthur Andersen. The name Joe Carter pops up whenever discussion of AI turns to successfully fielded systems. Andersen Consulting has been involved in numerous manufacturing projects. Using Gensym Corporation's G2 real-time knowledge-based system,the HEAT TREAT SCHEDULING AND MONITORING

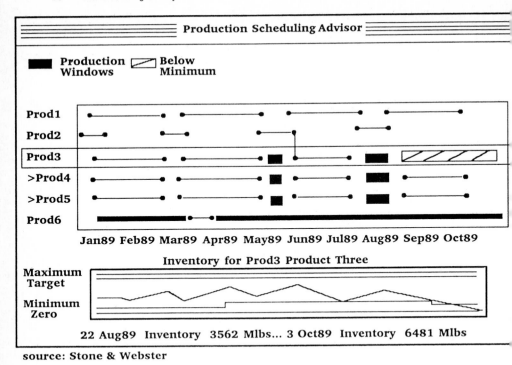

Figure 4-1   A simulation of the Production Scheduling Advisor

SYSTEM was built and installed at a large equipment manufacturer where one set of furnaces can process part sizes from 4 to 400 pounds. Heat treatment is one way of forcing a metal to become hard. After the part is heated it is cooled to a lower temperature, resulting in a modified molecular structure with greater hardness. Similarly to the Ellwood City Forge process, the next step would be tempering. Andersen Consulting was called to the scene when the manufacturer could not turn up an off-the-shelf solution to their real-time process data needs. Andersen selected the Gensym G2 system because of its ability to easily interface to equipment controllers in a real-time mode. The expert system component trapped expert knowledge in the area of scheduling simulations and process monitoring. The scheduling model executes a series of rules that determine the most optimal order in which to process the parts, and then simulates the actual processing. This simulation calculates when the order will complete and determines furnace utilization. These results are presented graphically through the use of color coded Gantt charts, as well as textually. This tool may also be used in a "what if" scenario since it's quick and easy to use. Schedulers often use it to determine the impact of high priority orders. The process monitoring component has three functions: equipment performance monitoring, alarm management and diagnostic help.

Equipment monitoring links to alarm management and expert knowledge in the form of diagnostic help. If the system suspects an equipment failure or a pending problem an alarm message is generated. If the furnace operator does not have the knowledge to answer the alarm the expert system leads with a series of questions prompting for the operator's response. This ultimately leads to a resolution.

Configuration

There is probably nothing more complicated to put together than a computer. Piecing together a personal computer is difficult enough so imagine the complexity of correctly configuring a computer that can have anywhere from 200 to 10,000 components. It was this configuration paradigm that first fostered the growth of the use of expert systems in the commercial arena. And this was due largely to Digital Equipment Corporation.

In the days before expert systems DEC, like all manufacturers, configured their products using technical writers. The problem was that humans make mistakes, especially when what is being configured is extraordinarily complex. In an association with Carnegie Mellon University a prototype of an expert system, dubbed R1, was developed that could handle configuration of complex objects efficiently and, most importantly, correctly. R1 was ported to the DEC AI labs and XCON was born. XCON, which stands for expert configurator, and DEC's sister products, XSEL (for sales) and XSITE (for installation), stand out as a models of applied artificial intelligence and paved the way for the generation of artificially intelligent computer systems that we're seeing today.

In 1990, expert configurators are still being built. The french computer manufacturer, BULL, has taken the lead from DEC and with Arthur Andersen has developed the NOEMIE expert system. NOEMIE configures not only hardware but software as well. BULL realized that the number of combinations in designing a complete system is enormous—virtually impossible for a human to configure correctly. NOEMIE is really four systems. NOEMIE/COM (commercial) assists the sales force in generating valid configurations. NOEMIE/TECH (technical) assists order administrators to validate and translate the orders placed by the sales representative into bills of material. NOEMIE/FAB (manufacturing) helps industrial personnel to automatically generate assembly or production specifications and NOEMIE/INS (installation) assists customer service engineers when installing the configurations on the client site. According to BULL the payback has been significant. There has been a decrease in the number of giveaways due to configuration problems, it's now easier for staff to master a new product offering, faster delivery is being provided to customers, the order handling has been automated which means faster and more reliable processing and finally higher accuracy of the final order.

Andersen consulting has also come out with an expert configurator available for sale. MAC-PAC's EXPERT CONFIGURATOR runs on the IBM AS/400 platform and provides expert system assistance in three areas: manufacturing, pricing and order entry. The Expert Configurator improves material and production planning by establishing that all-important link between Engineering, Sales and Manufacturing. It determines all components needed to make the final product, determines all labor and machine requirements for production and then automatically creates the work order to produce the item and backward schedules it based on the promised ship date. It also determines an expected production cost.

In the chemical industry the concept of design is often called formulation. The task of formulation is to select or modify recipes for rubber, plastics, chemicals or alloys to meet new performance or cost requirements. The difficulty is in determining how different proportions of each ingredient will effect properties of the compound. AI WARE, based in Cleveland, Ohio, found a solution for this complexity in their CAD/CHEM CUSTOM FORMULATION SYSTEM. CAD/CHEM is a neural net that provides several essential features to a formulation task. Each property can be assigned relative weights, a desirability function can be defined for each property enabling the user to express a relative preference for a property and constraints in costs and proportions of ingredients can be accommodated. Using a sophisticated graphical user interface the user enters relevant information and then proceeds to let the neural net system automatically configure itself. This it does iteratively. The user is given a suggested formula meeting stated preferences as closely as possible. CAD/CHEM can also support "what-if" analyses to explore design trade-offs and sensitivities between the many factors. The system also supports an expert system to handle constraint analysis. The user has the capability to enter a number of these variables such as mutual incompatibilities between ingredients, or ingredients to be avoided for environmental, cost or production reasons. This rule-based expert system works in tandem with the neural net to provide a strategy of diverse AI technologies.

Inventory Control

Steelcase Incorporated is well-known in the office furniture industry. This 1.5 billion dollar Grand Rapids based company decided that AI techniques were necessary as a strategic tool. Selecting AICORP's INTELLECT natural language processing system, one of the uses of this AI tool was to control Steelcase's vast inventory. Over 800 users access the 80,000 item manufacturing database. They can locate on-hand quantities of inventory and combine that data with customer order information to schedule materials and labor more effectively. Users can research up to the minute plant information without difficulty. A typical query might be, "list all the items in the chair plant that weigh over 50 pounds". As a result Steelcase has been able to cut its production lead time by two days and greatly reduce material and parts inventories giving a new meaning to JIT (just in time).

CAE/CAD

Computer assisted engineering and design is probably the most exotic of all computer-assisted tools. It becomes more so when AI enters the picture. In fact, some experts believe that this will become the primary area of application of AI technology. Stone & Webster's STONErule typifies the AI approach. Using AION Corporation's ADS expert system shell with commercial mainframe CAE/CAD systems, provides designers with advice based on expertise, judgement, standards, procedures and code specifications. Graphics information from CAD models are used as direct input to the expert system. The expert system then evaluates the design and advises the user as to modifications that might be made. Boeing, in Seattle, created the ENGINEERING MATERIALS ADVISOR to help a design engineer pick the correct structural materials. NCR Microelectronics uses its DESIGN ADVISOR to solicit advice in such areas as timing violations, synchronicity, critical path sensitivity and timing violations in it's standard cell products. Users can enter into dialogues with this expert system if they question its advice. Los Alamos National laboratory uses the INTELLIGENT OPTICAL DESIGN PROGRAM to design image forming optical systems.

On the automotive end GM Delco has created the MOTOR EXPERT which designs components of custom motors to conform with changing constraints such as space, weight and power. This expert system ultimately completes a production drawing complete except for signature. And if the user is confused, the MOTOR EXPERT will print out a written explanation summarizing its work.

Design

The design of complex products is a time-consuming and labor intensive process. In addition, multiple designs must be undertaken to ensure that the design selected is indeed the optimal choice. Ford

Motor Company built the TIES system (Technical Information Engineering System) to help achieve significant improvements in quality and production cycle time. What Ford wanted to do was to build a design tool that could cross functional product lines. The tool needed to store relevant engineering information, knowledge and heuristics and permit design teams to work separately but with consistency. TIES is based upon a structured development technique developed in Japan. The concept behind Quality Function Deployment is to provide asystematic methodology to ensure that the customer's desires are translated correctly into a technical specification. The framework that QFD uses is a chart called the House of Quality, which is a matrix filled with information describing the interrelationships between customer desires and technical know-how. Since more than one functional group may work on a particular automotive design (i.e. marketing, design engineers and manufacturing), there is a cascading effect of multiple houses of quality, as shown in Figure 4-2. An automated tool should have the capability of exploring alternative designs while maintaining the conceptual links. TIES uses a knowledge-based approach to accomplishing this end. Ford used SUN-4 workstations and Inference Corporation tools and staff to build a highly interactive graphical system. Through this graphics system user scan create visual representations of houses. These depictions can be refined and viewed in infinite variety.

The TIES architecture is quite complicated consisting of such information and knowledge representation schemes as contexts, frames, tree structures and rules. It is especially important to note that the Ford system is perhaps the first manufacturing company to automate

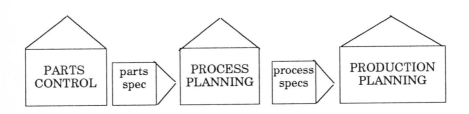

Figure 4-2   Automotive cascading QFD houses

the QFD concept and the only one that uses AI programming techniques for QFD.

Conclusion

There are few manufacturers that have not looked into the use of automated techniques to retain market share. Bally Engineering Structures Inc, a large maker of walk-in coolers, has looked to CIM (computer integrated manufacturing) to reduce manufacturing and inventory costs. Tom Pietrocini, the firm's president, stresses that CIM became a necessity for the company whose coolers and freezers must be custom-made. Bally's manufacturing process required 56 steps for each order and processed an average of 2.5 changes per order with a total weekly order base of 4000. Pietrocini estimates that even with a 99% accuracy that would still mean an unacceptable error rate.

In the coming global environment CIM, while a giant leap forward over than manual manufacturing process, won't be enough. In order to be competitive, and even to survive, will require large transfusions of expertise into the process. Dr. Ralph Gomery, president of the Alfred P. Sloan Foundation, apparently agrees. He heads a program that grants as much as $60 million dollars to universities that are studying the competitiveness of U.S. industries. And manufacturing is high up on that list. He hopes that U.S. universities will develop a more realistic view of manufacturing and subsequently to the tools and techniques that will make the United States regain its rightful premier place. In order to accomplish these lofty goals there are few technologies that offer more substance than artificial intelligence.

Author Biographical Data

Jessica Keyes is president of New Art Inc., a management and computer consultancy based in New York City. She is the publisher and editor of TECHINSIDER newsletter and organizer of TECHIN-SIDER training seminars and video training series. She is the author of the *The New Intelligence*, published by Harper Business and a columnist and correspondent for such publications as COMPUTERWORLD, Database Programming and Design, AI EXPERT, Expert Systems Journal, Software Magazine and many others.

Section

2

Integration of AI Into Manufacturing

5

Integrating Knowledge-Based Systems in the Manufacturing Environment

Janina M. Skorus, et al.
Andersen Consulting
Chicago, Illinois

Introduction

Knowledge-based systems have been used in manufacturing for a wide variety of applications. While stand-alone applications exist, many successful systems have been integrated into conventional production environments. These systems take advantage of existing data bases, hardware platforms and development tools to enhance their benefits to the organization. This chapter illustrates the benefits of integrated knowledge-based applications in the manufacturing environment through specific case examples.

Knowledge-based systems (KBS) have been used in manufacturing automation for a number of years. While some knowledge-based systems constructed in the early 1980s were available via standard timesharing systems,[1] they were rarely integrated with corporate data bases and processing environments. Initial benefits were often

---

1   Paul Harmon and David King, Expert Systems: Artificial Intelligence in Business, (New York: John Wiley & Sons, Inc., 1985.), p. 170.

recognized from these stand-alone applications. These "golden nuggets" developed by the research community or by early adopters of the technology provided some significant paybacks and created an initial enthusiasm for the technology.

However, like veins of gold, these types of applications were rarely found and, once discovered, were carefully guarded from the competition. The KBS market slowdown in the mid-1980s was attributed by some to the inability of early adopters to deploy applications.[2] Especially lacking were deployed systems which were integrated within conventional environments. Recognizing the need to deploy applications, advocates of knowledge-based systems began to view the technology as one part of an integrated solution, rather than "stand alone" applications.[3] To satisfy this role, moving away from the stand-alone system, developers of the late 1980s insisted that development software run on conventional platforms.[4]

The commercial market place began to respond to developer demands in 1987 with knowledge-based development tools providing:

- Delivery on the same systems that provide mainstream corporate applications (i.e., mainframe computers and minicomputers)

- Access to standard data bases

- Means to embed applications within existing systems[5]

Because of the availability of tools in more conventional computing environments, knowledge-based systems technology has become more acceptable to the mainstream data processing community.[6]

This chapter illustrates the trend toward knowledge-based applications which have been integrated into conventional data processing environments. Specific applications implemented at manufacturing clients include:

---

2   Stone, Jeffrey. "Commercial AI Trends Seen at AAAI-87." AI Magazine, Winter 1987, pp. 93.

3   Feigenbaum, Edward, Pamela McCorduck, and H. Penny Nii. The Rise of the Expert Company. New York: Times Books, 1988. p.259.

4   Feigenbaum, Edward, Pamela McCorduck, and H. Penny Nii. The Rise of the Expert Company. New York: Times Books, 1988. p.259.

5   Stone, Jeffrey. "Commercial AI Trends Seen at AAAI-87." AI Magazine, Winter 1987, pp. 93-96.

6   Johnson, Tim, Julian Hewett, Christine Guilfoyle and Judith Jeffcoate. Expert Systems Markets and Suppliers. London: Ovum, Ltd., 1988.

- NOEMIE: Computer Hardware and Software Configuration

- Product Manager's Workbench: Packaging Design

- MSDS: Material Safety Data Sheet Generation

- CHROMA: Process Monitoring and Diagnostics

The business problem, solution, approach and resulting benefits are described for each of these applications.

NOEMIE: Computer Hardware and Software Configuration

The authors wish to acknowledge the contributions made by J.C. La-Gache and F. Beauge of BULL[7] Worldwide Information Systems to the description of the NOEMIE application.

Computer system configuration is an exacting, and sometimes overwhelming, process. With a growing number of hardware and software options, determining the exact specifications and layouts of hardware and software to meet customer needs is getting more difficult. The NOEMIE solution was originally designed and implemented by BULL and Andersen Consulting in France, using knowledge-based systems (KBS) technology. Since then, BULL, one of the top ten computer manufacturers worldwide, has completely taken over development, extending NOEMIE's functionality and further demonstrating the unique advantages of KBS.

The Business Challenge

In designing a computer system, the number of possible combinations of components is enormous. For example, the BULL catalog covers the complete range of computer products, from personal computers to the largest mainframes. There is an ever-growing need to tailor each configuration exactly to customer specifications.

New components are continually introduced, old components are withdrawn from the market, marketing strategies and constraints change, and completely new technologies appear. These technologies can be very complex, as shown in Figure 5-1.

During the design of new products, only a few experts have a thorough knowledge of all the rules and constraints for configuration.

---

7   BULL, 20 Rue Dieumegard, 93406 St. Oven, France.

Given its value to business operations, however, this knowledge must be widely and rapidly disseminated. For example, the sales force needs up-to-date information in order to propose valid configurations to their clients. Once an order reaches the factory, order administration people need mastery of complete configuration knowledge to check orders and to translate them into industrial bills of material.

Systems developed with traditional techniques cannot easily handle the computer configuration problem. Considering the volume and pace of changes, experience indicates that the cost and time needed to maintain traditional systems in this domain are prohibitive. Yet the need for an automated tool remains.

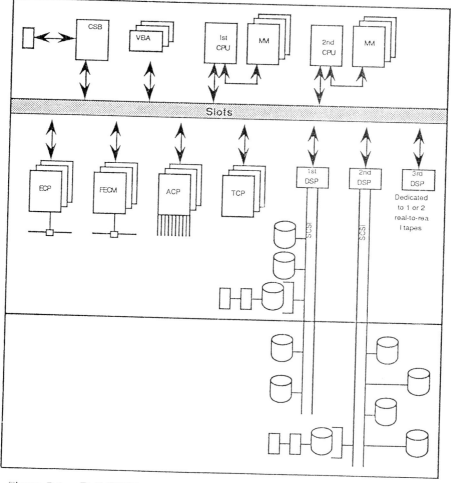

Figure 5-1   Bull DPX/2 320 Architecture

The Solution

Knowledge-based systems (KBS) techniques were chosen to handle the configuration knowledge, which is largely non-numeric and non-procedural. KBS techniques work well with evolving and incomplete knowledge since new knowledge can be incorporated by modifying or changing only some of the rules. Problems arising from the vast number of components, rules and constraints can thus be handled successfully.

The NOEMIE system, shown in Figure 5-2, consists of several applications for manipulating computer configurations:

- NOEMIE/COM (Commercial) assists the sales force in generating valid configurations

- NOEMIE/TEC (Technical) assists order administrators to validate and translate the orders placed by the sales representatives into industrial references (bills of material)

- NOEMIE/FAB (Fabrication) helps manufacturing personnel in automatically generating assembly sequences

- NOEMIE/INS (Installation) assists customer service engineers when installing the configurations on the client site

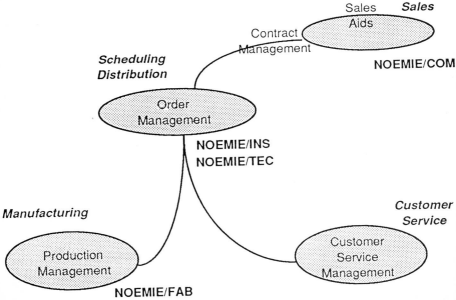

Figure 5-2   NOEMIE: A Set of Tools Integrated in the Order Cycle of Bull

NOEMIE/COM runs on personal computers and workstations, and is coupled to other sales systems. NOEMIE/COM covers most of the configured products of the BULL catalog, and is now operational in 25 countries (Europe, the U.S. and overseas), serving over 800 sales and marketing representatives. NOEMIE/COM is targeted for worldwide distribution to over 2500 users by the end of 1991.

NOEMIE/TEC is running in one industrial operation in France. It processes the daily orders coming in from different sales organizations. NOEMIE/FAB and NOEMIE/INS are still in the first phases of development.

The Approach

A hybrid architecture has been created for NOEMIE, using a relational database for the description of the products and KBS techniques for handling the configuration rules. This architecture, shown in Figure 5-3, led to a generic solution that can be used for different products and releases. Each NOEMIE tool is comprised of:

- A core configuration engine, independent of the product covered, that uses description data, fires rules, checks constraints and drives the interface modules

- An extract of the product description database for the product supported

- A set of rules and constraints specific to the product covered.

Results

NOEMIE/COM and NOEMIE/TEC have produced direct benefits in three domains: the commercial organization, the distribution units, and the manufacturing process.

In the commercial organization, the primary benefits are: a decrease in the number of products given away due to configuration problems; reduction in the time needed to understand and master a new product offering; reduction in the time needed to build configurations; and faster delivery (and billing) to the customer.

In distribution and scheduling, the gains come from the automation of order processing, which enables faster and more reliable handling. Consequently, orders are filled faster and there is less contact with the commercial organization to solve configuration problems. Overall, there is better information about the configuration and faster response to customer needs (especially during peak periods).

For manufacturing, the main benefits result from the higher accuracy of the purchase order placed by the distribution units. This decreases the number of orders that need to be recycled or rework, and saves time spent in data flow breaks and error resolution.

When full distribution is achieved, probably in 1991, the expected direct returns are estimated between 15 and 20 million U.S. dollars per year.

MSDS: Material Safety Data Sheet Generation

For some industries, such as the chemical process industry, accurate and timely information is critical. Occupational Safety and Health Administration (OSHA) regulations require that, for any chemical material used in the workplace, chemical companies must produce a Material Safety Data Sheet (MSDS). These data sheets must accurately describe characteristics of a product (e.g., how it interacts with other chemicals, how to treat improper exposure), so it can be used safely by employees and customers. For Lubrizol, a specialty chemical company, this means continually updating material safety data sheets for thousands of products.

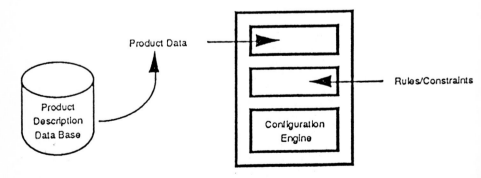

Figure 5-3   The NOEMIE Architecture

The Business Challenge

The manual MSDS generation process was highly iterative and time consuming. Lubrizol chemists researched formulary and compositional information for each product, and all intermediary and raw materials. Skilled individuals spent days or weeks completing a data sheet, which was then kept on file in thick binders in each one of Lubrizol's world wide facilities. Timely and accurate generation of these documents was difficult. Faced with a growing workload due to increased numbers and ranges of products, this paper-based system became cumbersome. Changing OSHA and international transportation regulations added to this burden. A better means of generating data sheets had to be found.

The business challenges that Lubrizol was facing when developing an automated MSDS system were:

- Keeping current with OSHA and international transportation regulations

- Being in compliance with those regulations, showing that their products were properly identified with MSDSs

- Being responsive to customers who often inquire about chemical properties of products to assure they are handled properly

- Being timely in generating MSDSs for new or modified chemicals.

The Solution

Conventional data processing techniques have been used elsewhere to provide automated assistance in MSDS generation and data management. The Lubrizol MSDS system is unique in that it combines conventional data processing with knowledge-based systems. The KBS component uses knowledge about computation of chemical compound interactions to determine information for the data sheets.

Using the new system shown in Figure 5-4, a chemist enters key information about a compound or raw material. Personnel in the Product Safety and Compliance Group, who have responsibility for ensuring that any Material Safety Data Sheet accompanying a product meets OSHA requirements, are then able to generate the appropriate MSDS overnight. Manufacturing operators within plants who deal with Lubrizol products on a daily basis and chemists responding to customer inquiries have access to this MSDS information through the on-line systems.

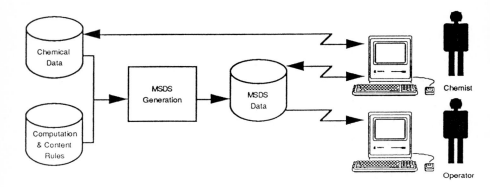

Figure 5-4   MSDS Architecture

The Approach

The MSDS system consists of the following components:

- A DB2 relational data base containing basic information on Lubrizol's chemical compounds

- A knowledge-based system which defines data sheet content and rules for computing the required chemical properties

- On-line support for update of chemical interaction and modification/generation of MSDSs

- Interfaces to Lubrizol's Bill of Materials and Order Entry systems.

The original knowledge-based system, built using ADS™ on a PC, took approximately six months to complete. The final system, running on an IBM 3090 mainframe, was completed in one year. Users throughout the organization have access to the resulting MSDS data via Lubrizol's world wide network.

The Results

The Lubrizol system illustrates many of the benefits expected of knowledge-based systems:

- *Knowledge Capture.* Regulatory and chemical expertise is captured and made available to personnel throughout the organization.

- *Job Enrichment and Leverage.* Repetitive tasks are leveraged to the system, freeing the expert for more challenging problems.

- *Accuracy.* MSDS monitors and records the calculations made in creating data sheets. Human experts may then verify the system's logic to eliminate errors or inconsistencies.

- *Consistency.* All MSDSs are written using the same rules and assumptions, providing consistent quality of documentation.

- *Safety and Customer Service.* The system's integration with Lubrizol's on-line systems ensures that customers and employees have access to the latest information about products. Response time for customer calls dropped from 1-2 days to, in many cases, immediate response.

The MSDS system not only assists employees in providing better customer service, it makes the lives of those associated with the products jobs easier and safer. This is an especially significant benefit; the Lubrizol application was recently awarded the *Computerworld Smithsonian Award* in Manufacturing, recognizing its benefits to society through the innovative use of information technology.

Product Manager's Workbench: Packaging Design

Packaging software products for the marketplace requires a collaborative effort across multiple departments in a software company. Yet, without an effective means of sharing product packaging knowledge, the prospects for productive cooperation are extremely limited. This problem was solved at Microsoft by creating a workbench which incorporates knowledge-based systems technology and distributed data access.

The Business Challenge

Packaging software products is essentially a collaborative process. Microsoft decided that greater sharing of knowledge would make the packaging process easier and more effective. Previously, word-of-mouth was the primary channel through which the planning engineering, purchasing, and product development departments communicated. No packing configuration standards were available to drive the product creation process. Product packaging designs were developed by the product managers, reviewed by the planning engineering department, and revised in an iterative process spanning several months, as shown in Figure 5-5 below.

Without interactive visibility into the entire product creation life cycle, product managers spent countless hours querying manufacturing about product designs, and checking with purchasing for component cost data. Further, these managers were limited to educated guesses about how factors, such as packaging, expected shipment quantities, and component costs would affect a product's bottom line. It was not the easiest way to make cost-effective packaging design decisions.

The Solution

Andersen Consulting teamed up with Microsoft to create a knowledge-based system (KBS) application to help product management teams create and design new products. The result is the Product Managers (PM) Workbench, which effectively bridges the communications gap between planning engineers, purchasing agents, and product managers, as shown in Figure 5-6. It not only helps product managers design and cost new products but also enforces company-wide design

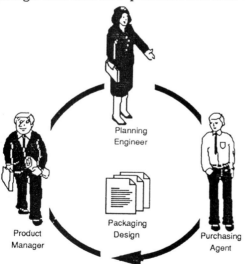

Figure 5-5   Original Product Design Process

and packaging guidelines. The PM knowledge bases, written in Nexpert Object,™ consist of:

- Corporate Guidelines. Microsoft's guidelines on packaging, such as the proper type and size of manuals or the number of colors to be printed on a carton.

- Cost Information. Packaging material costs supplied previously by purchasing.

- Configuration Rules. Domestic product standards developed by the Microsoft Applications Packaging Committee. These components are a permanent part of the PM Workbench.

While using the PM Workbench to enter information about a proposed product, the product manager is alerted to any corporate guidelines that have been violated. The PM workbench also lets users perform a what-if analysis on designs. This allows the product manager to quickly configure a product and estimate production cost for

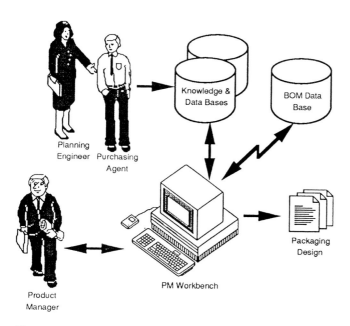

Figure 5-6   Revised Product Design Process

materials, labor and burden over the expected life of a product. The workbench immediately shows the bottom line impact for the product's cost of goods sold (COGS) in each scenario.

In addition to automatic cost of goods calculations on new product designs, the workbench also provides Microsoft with on-line viewing of CAD assembly drawings for existing products, and inquiry access to their minicomputer-based manufacturing package for the review of existing bills of materials.

Microsoft has 14 international subsidiaries around the world. In most cases, Microsoft products are "localized" into a number of different languages, operating environments, and even different cultures. It is not uncommon for a U.S. software product to be packaged into 8 or 9 different localized versions, each a separate product with its own configuration and bill of materials. Consequently, the value of a tool like the PM workbench becomes even more apparent to a product manager who creates 8 or 9 different localized designs for each U.S. version of a product.

The international version of the workbench will allow Localization Product Managers (LPM's) to take an existing product design and automatically create a default localized version through a set of rules contained in the workbench's knowledge base. The LPM's are then able to modify the defaults toward a specific market or language version and perform a cost analysis on the design to test the cost impact of design decisions. The scope of the design decisions may include the effect of such variables as the number of colors in a software manual or retail carton, the addition or deletion of pieces, or the possible combination of two pieces into one.

The Approach

From a purely technical standpoint, the Product Manager's Workbench represents one of the first highly integrated client-server architectures using SQL Server™ as a back end database access tool. On the front end, the Workbench combines a knowledge-based system, terminal emulation and CAD drawing visibility, all running under Microsoft Windows 3.0.™ Windows 3.0 provides the advanced operating environment which allows access to all available memory. Asymetrix' Toolbook™ and Microsoft Excel™ provide the graphical user interface.

The Results

The PM Workbench facilitates an integrated approach to product development and product life cycle management, and offers the following benefits to its domestic and international users:

- Higher quality product designs before they reach production, allowing manufacturing to concentrate on production effectiveness, as opposed to iterative design development and correction.

- A standardized set of rules and goals driving both the domestic and international design process, so product designs follow corporate guidelines.

- A standardized set of components used in product manufacturing, making the ordering process and inventory levels more predictable. Overall cost visibility for cost-effective decision making. A proactive approach to cost reduction.

The workbench is being used by about 30 Microsoft product managers to help determine the cost of packaging and marketing of a product.

CHROMA: Process Monitoring and Diagnostics

It is a common concern across today's competitive market that businesses utilize resources efficiently and minimize waste. In the fine paper making industry, even the slightest defect in the product can devour precious resources and seriously hurt a company's bottom line. One large firm enlisted Andersen Consulting to help build a knowledge-based systems (KBS) solution.

The Business Challenge

Manufacturing fine paper is an especially long and complex process. Production speed, pulp mix, and hundreds of other variables interact to determine the quality of the finished product. Even the slightest color deviation on a roll of paper, such as a light yellow tint, renders the paper unfit for delivery.

Only an expert who specializes in monitoring these variables can assess color quality on-line. However, even the most experienced operators have difficulty tracking and controlling color on a consistent basis. Furthermore, there is no real consensus on the best means to achieve high paper quality.

The Solution

Such problems pervade the process manufacturing environment, and in 1989 the CHROMA project set out to provide a solution for one large paper making company. The mission: to develop a real-time computer system that would diagnose color quality problems and recommend solutions to machine operators. The project had two key goals:

> To capture, within a knowledge-based system, the processes and procedures that would guide the operators and help them avoid, or quickly correct, quality problems as they developed.

> To define the interfaces needed to operate the system on-line, in real-time, within the existing mill-wide information system.

In gathering the information for CHROMA, the project team, consisting of consulting personnel and the client's scientific computing team, observed operators on duty for several weeks. Through observation and one-on-one interviews, the team learned how these operators reacted to certain events that might cause color to go out of specification (out-of-spec). Prototyping, iterative testing and revision of this knowledge base, using a variety of cases, led to the final system.

The Approach

The knowledge-based system works within a DEC-VAX™ based environment which uses on-line sensors to record events at each stage of the paper making process. CHROMA uses OPS 5,™ a rule-based expert system language, and has a custom-developed graphical user interface.

As paper is made, streams of raw data flow from a process control system into a database in real time. The information describes the optical brightener, additives, bales, grade specifications, flows, sheet brightness, and recycled paper percentage. The CHROMA system searches this database and extracts information influencing paper color quality.

The CHROMA system, shown in Figure 5-7, reads, evaluates, and interprets hundreds of pieces of data throughout the process cycle. "Intelligent alarming" prompts the operators if an "out-of-spec" situation occurs or, more importantly, if it appears that important quality factors are gradually deteriorating. Once the system detects either, it alerts operators to the real or potential problem and determines corrective action. If its advice fails to bring about the predicted outcome, the knowledge base can easily be corrected to include these new "experiences." As a result, the knowledge base becomes richer and its advice increasingly accurate.

CHROMA serves as an intelligent assistant, as it still allows the operators to make their own decisions. For example, if an operator is alerted to a color problem, the CHROMA screen may show that the blue dye exceeds the color limit. It advises him to decrease blue dye flow by a specific amount. The operator may agree with the advice, but conclude that the recommended change would be too drastic. He may decide instead to adjust the dye flow by more or less than the suggested amount, documenting his decision for potential system modification.

Results

The CHROMA system produces results in several areas:

- *Collaboration on Quality.* CHROMA does not replace the operators' jobs. Rather, it helps them do a better job by communicating standard procedures which let the operators make more accurate and informed decisions. It also makes the operators' judgment more predictable by providing consistent advice on similar problems.

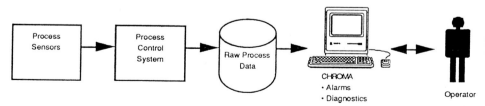

Figure 5-7 CHROMA Architecture

- *Continuous Improvement.* CHROMA provides an audit trail of color errors and actions taken. If an action does not produce the expected result, the experience will be incorporated into CHROMA's knowledge base. Using this corrected knowledge base, CHROMA will warn the operator the next time a similar situation arises. The longer the system remains in use, the higher its quality standards become.

- *Reduced Waste.* Most importantly, by sustaining consistency in the quality of a product, CHROMA significantly reduces the cost of waste both in operators' time and in raw materials.

The CHROMA system demonstrates that knowledge-based systems can enhance complex manufacturing processes such as making fine paper by capturing the know-how to effectively and efficiently operate those processes.

Conclusions

The examples described in this chapter demonstrate the important role that conventional systems integration serves in the overall success of knowledge-based systems in the manufacturing environment:

- A key benefit of the Lubrizol MSDS system, timely distribution of information, would not have been achievable without integration to conventional systems.

- At Bull, integration of the knowledge-based configuration system with order processing redefined the sales, technical support and marketing functions, allowing Bull to be more proactive in sales to the marketplace and, therefore, more competitive.

- Integration with manufacturing systems provides Microsoft managers accurate information for business decision making.

- CHROMA's integration with shop floor control data provides the necessary means to ensure production of consistently high quality paper.

The integration of knowledge-based systems within conventional data processing environments was instrumental in achieving the benefits of improved safety, improved quality and significantly reduced costs with these applications. Integration with conventional environments will continue to be a major factor in determining the success of deployed knowledge-based system applications.

Acknowledgements

The authors wish to acknowledge the following people for their contributions to the content and realization of this article: Mary Ellen McKee, Larry Buhl, Nicole Beauchamp. Thanks are also due to those who contributed to early versions: Joe Carter, Bob Caldwell, Gary Cole, Carolyn Hassel, Gezinus Hidding, Michelle Jensen, Barry McBride, Patrice Ruchon. Finally, sincere appreciation is extended to Bull Computer, Lubrizol, Microsoft, and other forward-looking companies who contributed to this chapter and gave us the opportunity to describe the applications mentioned here.

References

Feigenbaum, Edward, Pamela McCorduck, and H. Penny Nii. The Rise of the Expert Company. New York: Times Books, 1988.

Glitman, Russel. "Microsoft Puts Windows 3.0 to Work." PC Week, May 28, 1990, pp. 57-59.

Harmon , Paul and David King. Expert Systems: Artificial Intelligence in Business. New York: John Wiley & Sons, Inc., 1985.

Horwitt, Elisabeth. "Formula for an Expert Solution." Computerworld, 11 June 1990, p. 16.

Johnson, Tim, Julian Hewett, Christine Guilfoyle and Judith Jeffcoate. Expert Systems Markets and Suppliers. London: Ovum, Ltd., 1988.

KBS Applications. Andersen Consulting and Videosmith Productions. 1990.

KBS Makes Perfection Easier for Papermaker. Pamphlet. Chicago: Andersen Consulting, 1990.

Managers Put KBS to Work at Microsoft. Pamphlet. Chicago: Andersen Consulting, 1990.

Stone, Jeffrey. "Commercial AI Trends Seen at AAAI-87." AI Magazine, Winter 1987, pp. 93-96.

Systems Configuration Made Easy With KBS. Pamphlet. Chicago: Andersen Consulting, 1990.

Author Biographical Data

Janina M Skorus is a member of Andersen Consulting's Advanced Technology group. Andersen Consulting is a leader in the practical application of artificial intelligence to diverse industries.

6

Data-Based Systems and Knowledge-Based Systems

Synergy from Leveraging Resources

Paul Melcher
Caterpillar Inc.
Peoria, Illinois

Introduction

The advantages of expert or knowledge-based systems (KBS) technology has been lauded many times over during the past several years. Major among the advantages has been the ability to leverage existing human and machine resources to obtain a greater return with minimal additional investment. With interfaces between KBS software and data base management and other systems becoming more and more seamless, leveraging has become more obvious to the average data processor. Adding decision expertise to computer processes allows capturing a level of consistency in routine decision making not possible with conventional technology alone and this advantage is becoming more obvious to the average business user.

This paper deals with something less obvious—the computer based synergy potentially available through integrating knowledge-based systems that deal with similar problems in informationally dependent yet functionally diverse areas of a manufacturing company. Horizontally linking knowledge-based systems to reflect horizontal interdependence of business decision making processes provides a very high level of return to the enterprise in the form of streamlined business processes. The result is the need for fewer resources. Yes, we are talking about passing data from one function to another and that in itself is not unique. What is unique is the data being passed can embody levels of expertise that current systems cannot provide. The data passed is more than syntactically correct; it is "informatorily" correct.

Informatorily correct data means consistently applied expert knowledge went into determining its value. At the first level, many of the business processes centered on ensuring data accuracy are no longer needed or may be simplified to a great extent.

Some Background

Caterpillar Inc. became involved with artificial intelligence and in particular knowledge-based systems, during the mid 1980's as a research endeavor.

The company is multi-national, headquartered in Peoria, Illinois, and employs about 59,000 worldwide. Caterpillar designs, manufactures and markets earthmoving, construction, and materials handling machinery, as well as the engines used in them and in on-highway trucks and locomotives.

Caterpillar products are produced in 23 countries around the world and most of them are sold and serviced through a worldwide network of mostly independently owned dealers. The company enjoyed sales in 1989 of $11 billion.

Once the commercial viability of KBS technology was verified, Caterpillar searched for software that met its diverse business needs. The Aion Development System from Aion Corporation, Palo Alto, CA was chosen. ADS is a general purpose, multi-platformed knowledge-based system development and execution environment with varied knowledge representation and inferencing capabilities. It is approved for use by professional and end-user developers throughout the Caterpillar organization.

Data, Knowledge and Information

While transferring knowledge-based system technology from Research into Caterpillar's information processing mainstream it was necessary to differentiate between data, knowledge and information. Developers and users needed to understand how and why this technology was significantly different from existing capabilities. Making this differentiation helped. It is common practice to use the terms data and information synonymously in the information processing world. Knowledge is not mentioned at all as an application input or output. A closer look at these terms however, reveals the following:

- DATA is factual information (such as measurements or statistics) used as a basis for reasoning, discussion, or calculation.

- INFORMATION is the communication of knowledge or intelligence.

- And KNOWLEDGE is the fact or condition of knowing something with familiarity gained through experience or association.

Put knowledge with data and you get information. That is the dependency lacking in conventional systems technology. This relationship between data, knowledge and information, depicted in Chart 6-1, is the heart of the information generating process and the information management function.

	Is Kept In	(and) Requires	Using
DATA (is)	Databases	Data Management	Database Management Systems
PROCESSED (and used with)	Programs	Model Management	Callable Macro Code Libraries
KNOWLEDGE AND EXPERIENCE	Knowledge Bases	Knowledge Management	Knowledge-Base Management Systems
(renders) INFORMATION	CONCERNING PENDING OR EXECUTED DESISIONS AND ANTICIPATED OR ACTUAL RESULTS		

Chart 6-1   The relationship between data, knowledge and information

Data-Based vs Knowledge-Based Systems: The Impact of Decision Knowledge of Information Processing

Given these distinctions, a contrast is drawn between traditional computer systems called here data-based systems for simplicity, and knowledge-based systems:

A data-based system has the manipulation of data as its primary objective. Such systems use implicit knowledge about how data is to be manipulated or organized in such a way as to reveal problems or unusual situations. The programs used in the system are procedural and all possible results of the application are known ahead of time. The usual output is more data.

Information as it is defined above is not the output of traditional computer systems nor was it ever meant to be. More refined data is the common output. Obtaining real information is still a largely manual operation when data based systems are used.

In contrast, consider this definition of a knowledge-based system: A knowledge-based system has the manipulation of knowledge implied in a decision making process as its primary objective. It concentrates on applying user experience with many situations to a set of specific circumstances. Such systems utilize data and explicit knowledge about the relationships that exist among specific elements of data. The system considers the user's familiarity with a variety of previously experienced similar situations. These situations have been represented in a knowledge base. Using this knowledge along with given data, the system will infer new data, or a solution to a problem, or determine the significance of an unusual situation.

All results are not known beforehand as is the case with data-based systems. Recommendations and advice based on experience and expertise about what to do and how to do it, real information, are common outputs of knowledge-based systems.

Application Areas

Knowledge-based systems have two general areas of application. First, there is almost always a need to embed complex decision logic relating to data manipulation within data-based systems. Intelligent interfaces, embedded system tutoring/help and data monitoring/evaluation systems are examples.

The second place where knowledge-based systems are used is beyond the point at which data-based systems become too expensive to develop further. Conventional computer systems have been able to provide only minimal information because the cost to provide more is prohibitive. Software languages utilizing procedural programming algorithms impose limits, since all outcomes must be identified upfront, on the cost effectiveness of developing complex, information generating applications. Knowledge and experience are very difficult to define in absolute terms. Identifying all outcomes is simply not possible many times or too expensive to do so. The economics of data-based system development prevent many complex applications from becoming reality even though many of these applications are no more than extensions of existing data-based systems.

For example, have you ever seen an employee (maybe yourself) pouring over a report, taking a moment now and then to query a data base, to jot down a note? Ever wonder why the computer isn't doing that kind of thing? One can observe manual processes being carried on right in the heart of large scale data-based systems. Highly paid personnel using sophisticated computer capabilities to perform their jobs then refer to complex procedures written in multi-volumed books and manuals in order finish their work. This is an all too common occurrence. It's not so much that systems built with knowledge-based system technology cannot be built with traditional languages. Rather, knowledge-based systems are much more cost effective in development and on-going maintenance of complex applications.

Chart 6-2 demonstrates the relationship between data-based and knowledge-based systems in terms of complexity. Where there is a lot of routine, repetitive (algorithmic) procedure in the application, embed the knowledge in the program procedure. If there is little procedure, do the opposite.

The opportunity to leverage corporate resources with knowledge-based systems occurs at these two points -within data-based systems and beyond the limits of their capabilities. The resources involved are time, people and their expertise, and existing computer systems. This is direct leveraging. The synergistic qualities of knowledge-based systems lie in the ability to link systems performing similar tasks across diverse business functions. For example, providing the ability to link engineering expertise with manufacturing, marketing, distribution, service and administrative expertise concerning the same topical domain.

Data Based Systems	Knowledge Based Systems
————————COMPLEXITY————————	
Very algorithmic with little knowledge required by the application	Very heuristic with lots of knowledge required by the application
Embed knowledge in the programs	Embed procedures in the knowledge base

Chart 6-2   The relationship between Data Based and Knowledge Based Systems

An Application Example—EDNA

Each of the following Caterpillar applications has in both explicit and implicit ways leveraged human and machine resources. None of these applications are exceptionally creative or involve deep reasoning. You do not have to "go deep" to realize significant benefit from a knowledge-based system application.

The Engineering Drawing Notice Assistant (EDNA) embodies procedures established by Standards Engineering that are used by design engineers and detailers to complete the paper work associated with releasing a newly designed part or one where the design has been changed. The engineering drawing notice (which is similar to an engineering change order) is the basis of a great deal of computerized and manual activity throughout the company so the accuracy and integrity of the data appearing on the notice is critical.

The first thing the engineer does is tell EDNA the type of notice being prepared and from then on, EDNA prompts the engineer for only what is required by the particular notice under consideration. Figure 6-1 depicts the EDNA prompts for disposing of stock when the notice deals with an engineering change. Notes and details required during the completion of the notice are asked for at the appropriate times. EDNA automatically keeps track of them. Much of the data on the notice is inferred from data already given EDNA which relieves the engineer from the tedium of entering the inferred data. Since EDNA knows what data is required for parts of any given type of notice, there is no need for elaborate checking schemes.

Without EDNA, design engineers refer to printed procedures that cross-reference themselves making them very difficult to use. An elaborate system involving a number of people to manually check the notice for accuracy and correctness is necessary. Still a significant number of errors occur. EDNA eliminates these errors as well as the backtracking and cross-referencing. The system dramatically reduces the need for manual checking.

The ability to relieve staff of routine activity such as checking is a strength and attraction of conventional systems. While EDNA could be programmed using conventional procedural languages, the ease of maintaining the rules implied in the standards made a Knowledge-based system the system of choice.

EDNA is a phased project. The first phase requires design engineers to manually key in answers to questions posed by EDNA. It runs in PC emulation on Apollo workstations. The second phase will interface drawing data existing in Apollo and/or mainframe files. EDNA will then prompt design engineers for only information it can not infer from the data retrieved from the files. While phase 1 still requires more work of its users than phase 2, the level of accuracy and the ability to automatically infer other information from that already given has caused EDNA's users to insist it be made available as soon as possible. Even though the system is not complete, it is

Edna - Asking About Effectivity   20-Sep-90 11:24am

Please choose one of the following concnerning the effectivity of the
Change of IPI239.

> Urgent
> When Practical
> First Production
> Control Drawing No. & Change Level
> Other Effectivity Terms

> Please choose one of the above effectivity terms if this is the
> complete effectivity terms to be shown in the "effective with"
> portion of the notice. If additional effectivity is required, select
> "other Effectivity Terms" for additional listings.

Edna - Asking About Stock Disposition   20-Sep-90 11:24am
    Part Number   Version   Change   Bolt
        1P1239   -   1

> DUE TO THE EFFECTIVITY OF When Practical YOU HAVE
> DETERMINED FOR THIS DRAWING, YOU HAVE THE FOL-
> LOWING STOCK DISPOSITION OPTIONS TO CHOOSE
> FROM. PLEASE INSERT AN "X" IN THE APPROPRIATE
> BOX/S.

	ROUGH	FINISHED	PARTS DEPT
REWORK			
SCRAP			
OK	☐	☐	☐
EXHAUST	☐	☐	
RETURN TO PROD			☐

Figure 6-1   Caterpillar EDNA System

useful to the user community and the lack of additional knowledge
and data file interfaces are not detrimental to EDNA's effectiveness
concerning what knowledge it does contain. It is common to have a
knowledge base only partially developed when it is implemented
Iterative development is one of the benefits of the technology. It allows realizing a return on investment while additional capabilities
are being developed.

The described benefits are typical of knowledge-based systems. More subtle benefits are ones that involve the ease with which engineers complete the drawing notice. But the major benefit is the change in the process by which drawing notices for newly designed product and changes to existing products are generated. Ensuring the accuracy of the notice through embedding the expertise represented by the engineering standards governing notice creation at the time the notice is created has strong synergistic effects "downstream" of engineering.

Material Codes

Almost all corporate system codes having to do with continued processing of a product by engineering, manufacturing, marketing, service, distribution and the Caterpillar dealer organization rely either directly or indirectly on the data communicated by the engineering drawing notice. With fewer notice errors systems using corporate codes based on notice data provided by EDNA will run better and output more accurate data.

An example is an application called MATERIAL CODER which deals with a knowledge base that generates a 10 digit Production Material Commodity Code used by many areas in the company to conduct their business. Each of the digits represents a value, e.g., dimensions, or material type etc. Engineering data, most of it appearing on the engineering drawing notice, is used to generate the code as shown in Figure 6-2. Prior to MATERIAL CODER, the material codes were created manually using guidelines published in expansive volumes of documentation kept on a text data base. It was very easy to make mistakes. MATERIAL CODER has eliminated these mistakes for the codes it currently generates.

MATERIAL CODER runs on the IBM mainframe under TSO and uses objects (Object-oriented programming) for knowledge representation. MATERIAL CODER uses object-oriented programming and IFMATCH rules for knowledge representation which provides simple, straight forward system maintenance when new codes are required or existing ones change.

It is a shallow knowledge base. Ease of maintenance was again the main reason for choosing KBS technology.

Interpreting engineering data and translating it into the material code requires identifying a large number of relationships that take the form of pattern matching rules. System maintenance is done very quickly. Once again, less obvious is the benefit enjoyed by all the corporate systems that use the Production Material Commodity Code. The expertise used to generate the codes is now embedded in the commodity code itself and reflected through its accuracy.

```
ADS/MVS  Consultation Monitor          16-Aug-90 9:03am v5.11


  Message
  Central Purchasing Unformed Steel

    The material code derived by the expert system during the con-
  sultation was as follows:
                            0310209913
  Category:              Flat Bar Steel                 03

  Matl Specification:    1E0040                         10

  Type of Finish:        Hot rolled - Dry Pickled       2

  Thickness:             0000.2500                      09

  Width:                 0000.8800                      913

  Length 1:              0192.0000   Length 2:
```

Figure 6-2   Caterpillar Material Coder

Computer-Based Synergy

Caterpillar, like most other companies, grew its computerized systems which are usually driven by codes of some sort, out of independent efforts of individual application areas. Little attention was given integration and one result was a proliferation of "corporate" codes. These application systems continue in use and different codes representing the same product characteristic continue to exist. It is too expensive to revamp all those systems just to support a common corporate coding scheme. A knowledge-based system approach offers a highly cost effective alternative.

The experience gained with EDNA and MATERIAL CODER may be applied to all corporate coding schemes that are based directly or indirectly on engineering data. Doing so sets the stage for the short step to a Corporate Code Generator. Such a system would generate exactly the same codes used by the various areas of the company but from the same single or at least limited, very accurate data source. Examples include marketing codes used in marketing information systems which are based on interpreted material codes from CODER and engineering data from EDNA. Distribution codes that are based on marketing codes. And service management codes used to monitor service activity and warranty claims are based on engineering, distribution and other data.

All these corporate codes may be handled by a single knowledge base. Furthermore, because of the ease of maintenance and the independence of objects within the knowledge base, each area originating codes would be responsible for independently maintaining its own area of the knowledge base.

One can guess the value of having 95% or more accuracy in all the coding schemes used in production and end-user programs and systems throughout the company. Without question, the business processes with which these codes are generated today would be dramatically reduced in complexity and the number of people required to carry on the processes.

Someday the number of codes in use will be reduced. The logical time to eliminate a code is when major upgrades of a system or its replacement occur. The Corporate Coder would greatly assist in code conversions and elimination of codes no longer needed.

This is computer-based synergy. No systems currently in use need be modified or changed in any way to take full advantage of the Corporate Coder. Mundane jobs such as assigning a material code or filling out an excruciatingly detailed form or checking the work of someone else are reduced or eliminated. All other systems present and future will benefit from the results of simply putting the decision logic involved in creating any code into an easily maintainable knowledge-based system. Making it a corporate system ensures centralized logic and consistency in code generation methodology.

Other Applications

Corporate coding schemes based on common data and expert knowledge is one of the more obvious synergistic applications. Others also exist as exemplified by a knowledge-based system currently in prototype at Caterpillar's Mapleton foundry. The application deals with quality assurance and defect analysis of sand castings.

METLAB allows routine quality checks to be performed to ensure production is remaining within engineering tolerances. When casting defects do occur, METLAB is used to diagnose the cause and suggest a remedy. Obviously there is overlap between these two functions. However, casting irregularities are not the only reason for defects to occur. There are also machining defects caused by improper machining technique. Machining of castings occurs in different manufacturing plants to which Mapleton is a supplier. Defects identified at the manufacturing plant are first attributed to casting rather then machining problems and diagnosis starts with the casting. Again, knowledge about the defect, whether casting or machining, overlap. It is to the company's advantage to put knowledge of both in one knowledge base. A defect identified at the manufacturing plant can be diagnosed initially as a machining defect until the system says otherwise.

Synergistic benefits for both the plants and the foundry are the results of this combination of expertise.

Conclusion

Knowledge-based systems technology allows automating the information generation process and is viewed at Caterpillar as evolutionary rather than revolutionary. It is a more cost effective way to automate business functions heretofore too costly, therefore impractical, to automate.

Knowledge-based systems allow us to take computer automation another step and make knowledge the next level of data to be explicitly processed by computer. In so doing, existing human and computer resources are leveraged to yield far greater returns then anticipated. Computer-based synergy causes dramatic changes in the business processes yielding greater quality product at less cost. This is at the heart of the competitive advantage promised by knowledge-based system technology.

Author Biographical Data

Employed by Caterpillar Inc for 24 years, Paul Melcher has been working with knowledge-based (expert) systems since 1984. As a member of Corporate Information Services, Melcher has participated in KBS software evaluations and selection.

Melcher continues to be deeply involved in transferring KBS technology into mainstream information processing at Caterpillar. To that end, he has developed and continues to conduct seminars, executive briefings and classes in basics, advanced KBS topics and object-oriented programming.

Prior to 1984, Melcher supervised an Information Management group in the Parts Pricing Division. The group was responsible for user support of large scale development and maintenance of IMS systems used to establish and publish replacement parts prices.

Melcher's Professional affiliations include: Vice president of the Aion Midwest User Group and member of the steering committee, International Aion User Group, Member of AAAI. He has participated in the 1989 Aion Midwest User Group Conference, Chicago, and The International Aion User Group, Santa Clara, speaking on the topic of 'Technology Transfer, Bringing KBS Technology Into The Mainstream.'

Melcher also participated in the 1990 International Aion User Group, speaking on the topic on 'Knowledge-based Systems and Data-Based Systems: Synergy from Leveraging Resources.'

Workstation Integration of Expert Systems, Neural Networks, Image Processing, and Controls

James R. Blaha
Basic Industry Research Laboratory of Northwestern University
Evanston, Illinois

Introduction

BIRL, Northwestern University's Industrial Research Laboratory, applies the most advanced technologies to help industrial firms, associations, and government agencies solve problems and capitalize on emerging opportunities. The specific areas for research and development depend upon the needs of the client, and the Laboratory is staffed and equipped to be responsive to those needs.

BIRL is the anchor facility of the 24-acre Northwestern University/Evanston Research Park, which was conceived to accelerate technology transfer from the laboratory to the marketplace in a research environment combining the resources of a major university, a progressive community, and private industry. The City of Evanston, Illinois and Northwestern University are partners in this $400 million project. Northwestern is committing substantial physical, financial, and human resources to achieve the Park's primary objective: the application of advanced technologies for basic, high-tech, and manufacturing industries. These technologies include expert systems, neural networks, image processing, and control.

Presently, BIRL's three major business thrusts include materials for wear and corrosion applications, optical and electronic applications, and structural or packaging applications. Projects include applied research and analyses directed toward new concepts, systems, materials, devices, and processes, as well as providing support for fundamental studies seeking to develop critical new knowledge. In addition, BIRL has the flexibility to offer badly needed and cost-effective services such as process scale-up and production system evaluation. Since opening its doors three years ago, BIRL staff have been managing and conducting projects involving physical vapor deposition (PVD), chemical vapor deposition (CVD), electroplating, thermal spray technology, optoelectronics, semiconductor technology, composites fabrication, metallurgy, ceramics, and materials characterization. This case study of technology integration documents a unique personal computer workstation that is being used on several projects at BIRL.

Technology Background

Technologies usually associated with AI, namely expert systems, neural networks, image processing, and adaptive control, have come from very different fields and have been developed in distinct ways. In the 1970's, academic researchers of cognitive processes that seemed to underlie observable intelligent behaviors, created a computational model called a production system that is the core idea of rule-based expert systems. The rule bases and their inference method, the inference engines, embodied the well known predicate calculus theorem of modus ponens, i.e., "if a and b then conclude c." Other knowledge representation methods such as frames, objects, scripts, etc., have been integrated with this concept to enhance the production system idea into sophisticated systems that can model expert behavior in solving certain applications very well. A completely different style of representation that mimics the biological mechanisms of a neurological system, i.e. the neurons and synapses, resulted in neural networks, more properly called artificial neural systems (ANS). [1] Arising from research in cognitive science and control in the 1950's, ANS's have evolved toward specific application areas, in contrast to expert system shells that were developed for generic applications. The com-

---

1   Rumelhart, D., McClelland, J., (eds), Parallel distributed processing. Cambridge, MA: MIT Press, 1986.

putational method used to interact with the artificial neurons, or neurodes, and the links between them, have many different implementations,[2] some of them created by researchers specifically to solve a particular application. Thus, expert systems and neural networks have radically different roots and development requirements, yet both can be used to solve some of the same problems. It is easy to see why the 'AI community' and the 'neural network community' are quite distinct—having few researchers that develop or use both technologies.

Image processing, a form of 2 or 3 dimensional signal processing, is yet another technology which is not often developed (but frequently used) by the AI and neural network communities. AI researchers use image processing as a step to image understanding,[3] a high level inference process that builds a model of the observable environment from video images. ANS developers often use image processing algorithms to reduce a video image to a list of features which can then be used by an ANS for image classification to visually identify bad parts, defects, threat targets, etc. Integration of expert systems, neural networks, image processing, and control is desirable for a number of reasons. In research, it is not often known which technique can most effectively be applied to a problem. Open-ended research problems such as machine learning, image understanding, adaptive robotic control, and automated target recognition are but a few of the challenges facing researchers in these fields. Having a suite of cooperative tools is essential for such research. Invariably, the successful application of research is strongly dependent on having a relatively simple system configuration that can be packaged up and deployed in a customer's facility. A solution that requires multiple processors connected over a network is difficult to sell. Also, the customer ultimately wants the best solution, regardless of which technologies are used. Thus, having a single workstation to explore alternatives is a great motivation for this development effort.

Applications Require Technology Integration

Any application in a manufacturing or other dynamic environment with sensors and controls requires several technologies to participate in the solution. Frequently the integration of sensor processing, control programming, and AI technology is done very loosely through file transfers between software components or it may be simply left up

---

2   Carpenter, G. A., Neural network models for pattern recognition and associative memory. Neural Networks, Vol. 2, 1989, pp. 243-257.

3   Winston, P.H., Artificial Intelligence, 2nd ed., Reading, MA: Addison-Wesley, 1984. Ch 11.

to the operator to perform manually. Neither approach is particularly conducive to applications development for time-sensitive problems. The development of an interface between computer systems and their components is a major cost of many installations and an application developer should avoid it if possible. An integrated workstation solves this problem ideally by implementing communication through more sophisticated and responsive mechanisms.

A Comparison of Neural Networks and Expert System Technology

It is important to compare expert system and neural network approaches to the same problem. The development scenario is vastly different, and the tools have not often been used together on the same problem. A hybrid approach of expert networks is becoming popular[4], but the applications are still far from commercial implementation. Once the alternative AI technologies are examined, possible application areas can be considered.

Although neural networks are still in their infancy in terms of their number and range of applications, it is possible to compare neural network technology with expert systems as shown in Figure 7-1. The application scope of neural networks and expert systems augment each other to form a more complete system. From an application developer's standpoint, the neural network tools are primitive in comparison, yet the computational horsepower of a commercial neural network shell (and its coprocessor) is sophisticated. No commercial expert system shells utilize coprocessors directly to accelerate the recognize-act cycle. Little research has investigated how to execute the inference process in a massively parallel way. On the other hand, neural networks are massively parallel both in concept and implementation. Thus, one significant advantage of neural networks over expert systems is the advanced hardware support and the fact that the basic neural paradigm translates well into today's parallel hardware.

A more important consideration is the conceptual match of the technologies to the problem at hand. Table 7-1 summarizes some of these distinctions and illustrates where the technologies overlap in their respective approaches. Neural networks have fault tolerant characteristics because they are composed of many similar processing elements (PE or node), thus reducing the importance of any one PE to the overall behavior. On the other hand, expert systems follow a critical decision path, by sequentially chaining through rules toward a goal.

---

4   Caudill, M., "Hybrid expert networks as decision tools", Proceedings of Neural Networks: Opportunities and Applications in Manufacturing, April 3-4, 1990, Society of Manufacturing Engineers.

SCOPE OF NEURAL NETS AND EXPERT SYSTEMS IN AI APPLICATIONS

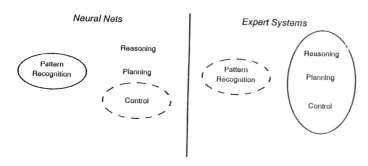

Figure 7-1   Scope of Neural Nets and Expert Systems

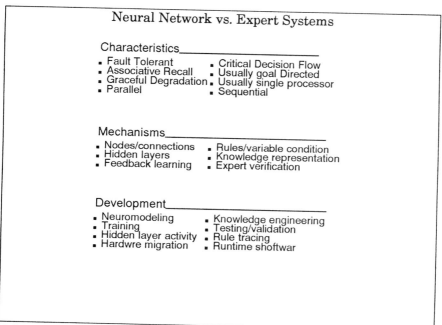

Table 7-1   Neural Net, Expert System Comparison

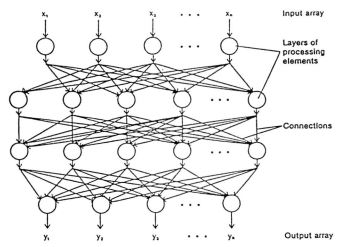

Figure 7-2   A Neural Network

The mechanisms of a neural network are nodes and weighted links that are subjected to an activation function that evaluates the weight or strength of connections entering a node, and passes on the result of the computation to other nodes in the network by setting the strength of its output connection. This is shown in Figure 7-2, adapted from [5]. Knowledge is never explicitly represented in a neural model; it is developed in the internal (hidden) layers through repeated training, which seeks to stabilize the network and cause the development of an internal model that effectively maps an input vector of data to an output class vector. An ANS developer is satisfied when the network performs its task to some specified level of accuracy, which is determined by comparing it to 'gold plated' data and results. An expert system developer is satisfied when the domain expert judges the program response as correct, or when trial cases continue to yield similar results to expert judgement.

Applications development is also dissimilar for the two technologies. Neural network development requires data manipulation and statistical processing as a preliminary to building a neuromodel. The model itself may be chosen from some three dozen available commercially or in the literature, or developed from scratch if the problem requires. Much of the time is spent tuning the model to achieve the proper classification or recall behavior and accuracy. This

5   Hecht-Nielson, R., "Neurocomputing: picking the human brain", IEEE spectrum. March, 1988. pp. 36-41.

requires building training and testing data sets, and submitting them to a working neuromodel. The actual training cycle can range from minutes to many hours, depending on the problem size and the hardware support. The cost of neural network tools range from public domain C code to $25,000 systems with coprocessor support[6] Expert system shells range similarly, but offer no real coprocessor support.

Artificial Intelligence Laboratory

Artificial intelligence is used at BIRL to solve problems in optoelectronic systems control, material design and fabrication, inspection, and manufacturing automation. BIRL's Electro-Optic System Development Laboratory shown in Figure 7-3, creates expert systems, neural networks, image processing algorithms, and advanced control strategies. These advanced computing technologies can be applied individually or in combination to develop solutions to challenging problems.

This laboratory configuration is capable of developing applications in the following areas:

- Integrated optoelectronic systems in visible, infrared, and ultraviolet spectral ranges

- Automated target recognition software and systems

- Sensing and signal processing systems

- Computer controlled prototypes

- Complex process models

- Expert systems for offline data analysis and online manufacturing control

- Neural networks for data classification, signal processing, and control.

---

6   Obermeier, K. K., Barron, J. J., "Time to get fired up", Byte magazine. August, 1989. pp. 217-224, 244-245.

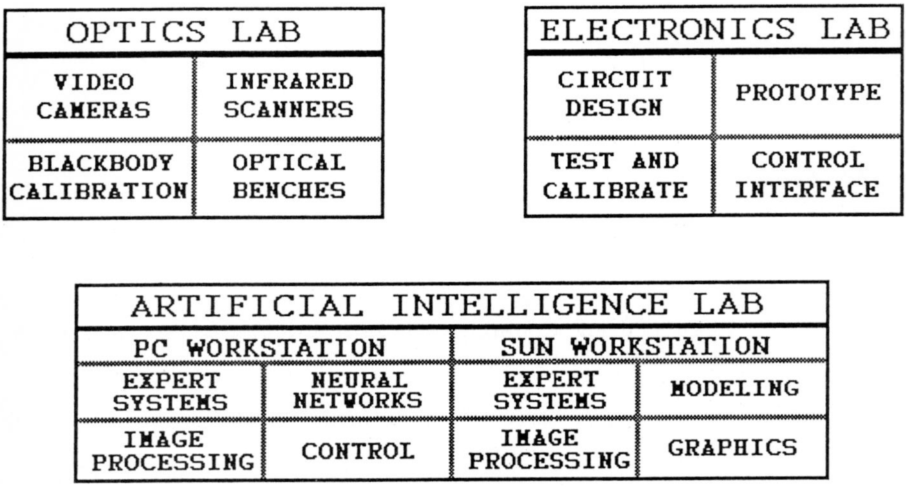

OPTICS LAB	
VIDEO CAMERAS	INFRARED SCANNERS
BLACKBODY CALIBRATION	OPTICAL BENCHES

ELECTRONICS LAB	
CIRCUIT DESIGN	PROTOTYPE
TEST AND CALIBRATE	CONTROL INTERFACE

ARTIFICIAL INTELLIGENCE LAB			
PC WORKSTATION		SUN WORKSTATION	
EXPERT SYSTEMS	NEURAL NETWORKS	EXPERT SYSTEMS	MODELING
IMAGE PROCESSING	CONTROL	IMAGE PROCESSING	GRAPHICS

Figure 7-3   BIRL'S Electro-Optic Laboratory

Integration Considerations

It is important for the computational aspects of a problem to be separated from the reasoning aspects, thus, a division of labor is required in the software. It makes little sense for an expert system to perform the computations necessary for finite element analysis or image processing, for example. An expert system is ideally suited to weighing alternatives and choosing a solution path. Neural networks are more suited to handling the classification aspects of a problem as a precursor to reasoning, and they have been shown useful in massaging data at a low level as well. The final link to the real world through a sensing or control element usually requires a hand-written program that can monitor equipment operation and maintain status information for interpretation by a higher level program. Commercial programmable controllers offer comprehensive capabilities for selected sensors and controls, yet a research environment tends toward customized sensors, controls, and interfaces. It is important that all the elements of an integrated workstation can link to custom source code for those reasons. As will be explained later, this can be through several methods listed in order of decreasing preference: subroutine calls, interrupt service routines, and file transfer. When configuring an integrated workstation, the most immediate concern is whether the hardware elements can physically function on the same processor bus. Often this must be determined through experimentation involving exhaustive testing of a matrix of configuration alternatives. One can not depend on a vendor to supply a comprehensive list of compatible host computers, let alone compatible software products. Some points to watch for are:

- Physical slot limitations (size, bus speed, power requirements)

- Board address range conflicts

- Interrupt request line conflicts

- I/O port conflicts.

In some cases, a board may be installed at different addresses, with different interrupt lines. Sometimes this can be selected by the software driver, making compatibility testing simpler; in other cases, DIP switches on the board may need to be changed. Before purchasing a board or auxiliary chassis, it pays to get this question answered by a technical person, not a sales representative. If available, the purchase of an on-site installation agreement is advisable, because the vendor becomes better acquainted with your problem. The operating system running on the host processor constrains performance of the solution, either because of speed limitations, memory address range, or multi-tasking requirements. Since both DOS and Unix are available for 80386 processors, i.e. IBM compatibles, this is an attractive platform. Of course the Motorola processors, i.e. 68000 series, handle memory more efficiently, but this is not a major concern for an application integrator. A great many quality software products and boards are available for DOS machines, and an application integrator must be able to select among several vendors for each system component. Although Unix is often characterized as a good tool for research, long acquaintance with DOS shows it to be equally viable. DOS is an easily accessible operating system, with countless books on it for the serious programmer[7]. The method of interrupt handling and file access is consistent among commercial products, and this is required knowledge for the system integrator. Over the years, DOS system programmers have discovered ways to beat speed limitations, (avoiding BIOS and speeding up the system clock), address range

---

7   Norton, P., Wilton, R., Programmer's guide to the IBM PC and PS/2. Redmond, WA: Microsoft Press, 1988.

limitations (using protected memory), and multi-tasking (through interrupts synchronized to the system clock). Microsoft Windows is multi-tasking, but Windows programming tools are only supported by a few software vendors. Windows 3.0 may offer a good alternative to DOS, but the support base is currently lacking [8]. Because of the vast availability of products, the standardization, and simplicity, standard DOS provides a good basis for the integration of expert system, neural network, signal processing, and control components into a workstation.

Potential Manufacturing Applications

The present integration strategy is based on earlier work on fusion welding control and spot welding control. The prototype robotic fusion welding system required several processors running expert systems, image processing, and process control loops to operate cooperatively on a network. A single integrated host computer would have been a better solution, had it been available. Many applications are attractive candidates for this integrated workstation, among them automated inspection, adaptive process control, robotic motion control, machine tool monitoring, metals processing, furnace control, and laser processing. The ideal match is a problem requiring signal processing and knowledge based or adaptive techniques. This system has unique capabilities to experiment with alternative AI technologies, and to investigate cooperation between software technologies.

Integrated AI Workstation

Integration of the software components requires the development of:

- Interprocess communication techniques

- Master/slave relationships between modules

- Flexible strategies to design and test an application.

The hardware components of the prototype integrated PC workstation are shown in Figure 7-4. This configuration was developed through

---

8   Hall, W. S., "Windows 3.0", PC magazine, Vol. 9, No. 15, (September 11, 1990), pp. 181-239.

careful review of competing vendor product specifications, and experimental methods described in the preceding section.

A key decision in software integration is how to structure the overall interface logic. If an expert system is to be the front end for the integrated system, then the expert system shell must have a full complement of interface capabilities:

- Subroutine calls with data transfer functions

- Interrupt service support

- File system support, and

- Control over memory allocation.

These capabilities can be found in shells ranging from complex (G2 by Gensym) to basic (CLIPS by NASA). Goldworks II by GoldHill, is midrange in price, speed, and power, and it also offers these interface capabilities. CLIPS is delivered in C source code [9], so it is very flexible

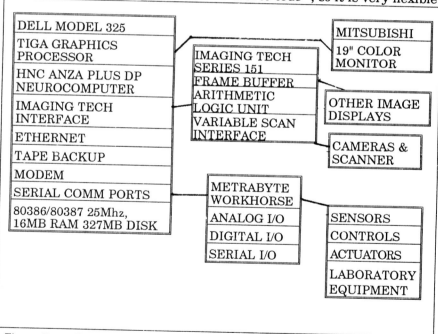

Figure 7-4   Integrated AI Workstation

---

9   Giarratano, J. C., CLIPS user's guide: Version 4.3. Lyndon B. Johnson Space Center. Distributed by COSMIC, University of Georgia, program number MSC-21208. 1989.

and can be compiled on virtually any machine. Since Goldworks II is implemented in LISP, it has equal flexibility because a full set of subroutine call and interrupt service support is available. It is run as a Windows 286 application, which complicates some matters while simplifying others. Most (if not all) image processing shells are not Windows applications, so the developer has to implement an interface from that shell to Windows, and link everything together. To do this however, the source code for the image processing shell has be available, which it often is not.

An alternate strategy is to use the LISP interface of Goldworks II to make subroutine calls to a custom C program which has been linked to an image processing library. If subroutine calls are not possible (perhaps because the image processing library is in Pascal or Fortran), a software interrupt can be generated in LISP to invoke an image processing function that has been previously installed as an interrupt service routine (ISR). Very little data can be passed directly through the registers during an interrupt request however, so one technique is to pass the physical address of the program's data buffers in one of the registers. One would use this technique in cases where portions of the image needed to be transferred from the ISR into LISP, for example. A better use would be to calculate a feature vector in the ISR (or subroutine) and pass it to LISP. In the other direction, LISP could pass command strings to the image processing ISR (or subroutine) to control the image analysis.

If interrupt service is not supported by a language, but physical memory addressing is, such as a 'peek' and 'poke' instruction,the Intra-Application Communication Area (ICA) of DOS provides a small area to get data transferred between programs. Located at memory address 0040:00F0, a 16 byte area is reserved by DOS for this purpose. There is very little documentation on it[10], and some commercial programs may use it without properly saving and restoring its contents. It can be thought of as a small blackboard, just large enough to hold some binary flags and a few physical memory addresses to serve a similar purpose as the registers in an interrupt call as discussed above.

When all these techniques fail, the developer may have to resort to communication by file. Nearly any program has the capability to read and write files. When one program needs to command another, the master writes a file of commands, temporarily exits (or shells out to a new process), then the slave is invoked and requested to read and process the command file. A results file is then generated, the slave terminates, and control is returned to the master which finally reads the results file. This sounds easier than it is, particularly if one wants automatic operation. In practice, the user would probably have to manually start and stop each program, and force the reading and writing of the command and results files. It may be possible to use a 'super-batch' type program or a program that sequentially

---

10   Norton, 1988, p. 62.

'feeds' keystrokes to the DOS keyboard buffer, thus mimicking the presence of the user. Unfortunately, some programs on the market (particularly wordprocessors) capture keystrokes directly, circumventing the keyboard buffer entirely. Experimentation is the only way to find out for sure.

In any case, it is desirable to have one module acting as a master to direct the processing of all other modules. For example, the expert system shell could serve as a front end, which calls a custom administrative program that communicates with the neural network, image processing, and application routines as shown in Figure 7-5. The custom routine would allocate buffers for temporary data storage, and thus would act as a master blackboard for all interprocess communication. This is probably easier than trying to get the expert system shell to serve this role by itself. If the expert system shell itself can be called as a subroutine, then the custom administrative program can reside at the top (user) level.

Once the integration mechanics are established, alternative application strategies can be evaluated. For example, one can attempt a machine vision application directly in the image processing modules. As an option, one could use a neural network to assist with the classification of calculated image features into different classes of objects or defects. Another option could be explored using the expert system shell to develop rules that associate image features with objects, or to develop a consistent labeling of objects in an image toward image understanding. It has been suggested that a neural network could write expert system rules[11], thereby shortening the time to create a rule based system, or making a rule based approach

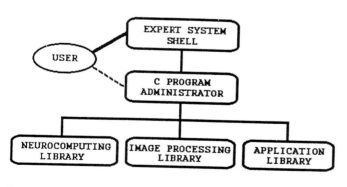

Figure 7-5   Expert System front-end

---

11   Caudill, M., "Hybrid expert networks as decision tools," Proceedings of Neural Networks: Opportunities and Applications in Manufacturing, April 3-4, 1990, Society of Manufacturing Engineers.

feasible even in lieu of a domain expert. With a control interface, one could investigate adaptive robotic motion control using such a scene analysis. Instead, one could develop a neural network that would map the image, part geometry, and robot geometry into a trajectory space for assembly type problems. Obviously, the range of strategies and applications one could tackle with this system are the best feature of this integrated workstation.

The Prototype Applications

The integrated workstation is presently being used to develop four different applications at BIRL. Not all of the available technologies are used on each one at this point, but as a benefit, the workstation configuration does not have to be altered when changing between applications. After a certain stage in development, the required system components can be cloned to a dedicated platform for installation in other laboratories. This is a very cost-effective solution for applied research and development. An application to the manufacturing process of sputter coating will be described in detail.

Sputter Coating

Hard materials have many uses in high-wear environments. Heat-treating is one process used to harden metals, but it is energy intensive and may physically deform the material. Twenty years ago hardened drill bits were created by bonding a carbide insert to the cutting edges of high-speed steel. Today it is possible to coat one material such as high-speed steel with a completely different material such as titanium nitride (TiN). The sputtering process is one way to create such a coating in a vacuum chamber. In essence, the material to be coated is placed in a chamber and the atmosphere evacuated as shown in Figure 7-6. A gas such as Argon is backfilled into the chamber, and an electric potential is created between the target (titanium) and the substrate material. A glow discharge plasma is formed in front of the target and positive ions from the plasma are attracted to the target. The ion's collision with the target ejects titanium atoms which are propelled toward the substrate to be deposited on the desired material.

Excellent adhesion results, and often only a very thin coating is needed to achieve wear-resistant properties. A variation of this process called reactive sputtering makes it possible to create a compound coating, without the necessity of having a compound target material. This process injects an additional gas into the chamber, for

SPUTTERING

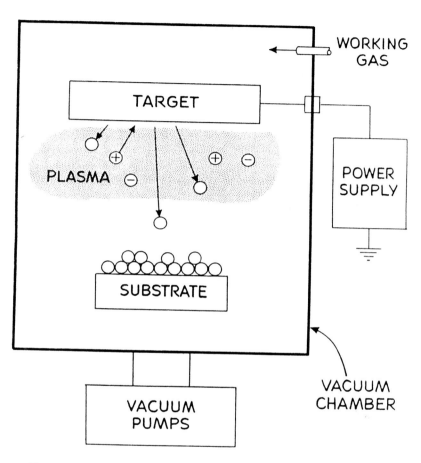

Figure 7-6   Sputtering

example nitrogen, to create TiN coatings. Key to the successful deposition is control over gas flows, voltages, currents, and vacuum pressure.

BIRL's Physical Vapor Deposition Laboratory contains advanced facilities for sputter coating. A full range of services are offered from coating samples through product development and scale-up to limited production. Special systems can be designed and built to a client's specifications as well.

Two large high-rate reactive sputtering systems are available. One of the sputtering systems consists of an in-line sputtering machine with both DC and RF magnetron cathodes, RF etch, and DC bias capabilities. Closed-loop feedback control systems provide high-rate reactive sputtering capability. Typically with these control systems operating, it is possible to reactively sputter nitrides of titanium, zirconium, and hafnium at essentially the same rates as those for the metals using a patented high-rate reactive sputtering process[12, 13]. For example, TiN can be reactively sputtered at 5000 Angstroms/min, and hafnium nitride can be deposited at 6000 Angstroms/min. With its DC and RF capabilities, this system can sputter all materials from metals to semiconductors to insulators.

Sputter deposition can also be carried out in a new opposed magnetron system [14], which is being interfaced to the computer control system discussed below. Two vertical 5 x 15 inch inset planar magnetron cathodes are separated by 10 inches, and the biased substrate holder, which rotates, is located equidistant from each target. Substrates can be placed on the holder and rotated while different materials are sputtered from each side of the targets.The reactive gas is introduced into the chamber through two separately controlled gas manifolds that are located next to each target. A baffle can be placed between the two targets close to the substrate to minimize target material cross contamination.

To ensure against contamination from the gas phase, the sputtering system can be evacuated by two sets of pumps. One set is a 6 inch diffusion pump that is backed by an 80 CFM mechanical pump; the second set is a 1500 liter/sec turbomolecular pump backed by a blower/mechanical pump combination with a pumping speed of 145 CFM. In addition to the high pumping speed of the system that allows fast initial pumpdowns plus high gas throughputs during deposition, the whole chamber is constructed with a water jacket around it. When the chamber is open and during pumpdown, hot water flows through the jacket to reduce the amount of water vapor contamination. During deposition, coldwater flows through the water jacket to ensure that no substantial amount of water vapor is released from the chamber walls.

The auxiliary equipment on both of these sputtering systems supports their high rate capabilities. Sputtering pressure is maintained with capacitance manometers, and gas flow is controlled with mass flow controllers. Both systems are continuously monitored during

---

12   Sproul, W. D., "High rate reactive sputtering process control", Surface and coatings technology. Vol 33 (1987), pp 77-81.

13   Sproul, W. D., "Reactively sputtered hard coatings show great promise for wear resistant applications", BIRL Newsletter, January, 1988.

14   Sproul, W. D., "New sputtering system enhances PVD capabilities", BIRL Newsletter, August, 1989.

deposition with mass spectrometers, which are instrumental in the control of the reactive sputtering processes.

The present system requires intelligent computer control because it is extremely complex in operation, having been prototyped from different commercial systems and custom-made hardware and controllers. A skilled operator must walk the process through every step using front panel switches and dials. The cycle time of the machine is about 4 hours, only 0.5 hours of which is spent coating the part to a thickness of 2.5 microns. The bulk of the time is spent in cycling various vacuum systems, and setting up proper operating parameters. Various feedback controls have been installed to adjust gas flow based on the output of mass spectrometers. Feedback control is also used on the target and bias supply (not shown in the figure). Vacuum pressures are under separate feedback control and serve as interlocks to the process sequence. Clearly, the operator needs almost the expertise of a coatings specialist to run the equipment.

The computer control system is being implemented in three phases. The first phase of Data Acquisition has set up the basic system architecture and established data logging of process variables. The I/O points are gathered by a Metrabyte Workhorse external chassis, mounted inside the sputter coating machine. Thirty two analog inputs and 32 digital inputs are available. Sixteen analog outputs and 16 digital outputs are also provided. Communication with the host computer is through an RS232 interface operating at 19.2 kilobaud. The system also uses three other RS232 links to the bias power supply, and two mass spectrometers. The serial communication software was developed using Blaise Computing's Asynch Manager library. Some of the menu interface and display screens use Blaise Computing's Power Screen software.

A single 'Serial_Interface' module performs the transmission and reception of messages to the bias supply and mass spectrometer units, even though the protocols are different. A high level message format was created to make it easy for the other modules to originate messages, and interpret responses. The other modules in the system are thus unaware of the particulars of the communication method. This has made it easy to simulate the communication link to individual sensors, controls, and actuators in the sputter coating machine.

The second phase is Computer Control, which establishes remote operation of nearly all of the sputter coating machine functionality. A relay panel installed in the machine operates as an intermediary between the front panel dials and switches, and the computer commands. Both the front panel switches and the computer process sequencing commands cycle through the respective relays, with front panel or computer activation selected by a master switch on the front panel for local or remote operation. This switch acts as a computer defeat, to allow regular operation of the machine through the front panel, whenever desired. This was required because the unit must remain functional for materials processing as the computer system is installed and debugged.

The main logic for the control is being developed in the Goldworks II expert system shell. Communication of computer commands and data values is through a custom written C module, 'Communications', which then calls the Serial_Interface module as already described. The expert system shell is responsible for determining the process plan and monitoring execution of the coating process. Machine diagnostics area also being developed as a separate knowledge base.

Phase three, Adaptive Control, will explore approaches to modifying the process plan on the fly, according to sensory data obtained from the gas analyzers and other sensors mounted inside the chamber. We expect that a neural network will help reduce the noise content of some of these signals, and improve the ability to detect significant process events during the coating operation. The image processor can be used to monitor the plasma surrounding the part being coated through visible light (CCD camera). Since the process generates heat, infrared imaging thermographers could be used as well.

Conclusions and Recommendations

The integrated AI workstation can serve the needs of the application developer very well by greatly reducing the effort expended to create hardware and software interfaces between technologies. The cost of a delivered system is lessened, and the development time shortened. Competing technologies can be evaluated for a problem, and used in optimal combinations and configurations.

Basic research can also benefit from this tool. Some very open-ended problems can be investigated when the right technologies are made accessible to each other. Some of these areas have been mentioned, but research is needed to explore the very nature of cooperative technologies. Neural network learning laws could perhaps take the form of an expert system rule. On the other hand, the expert system matching algorithms might be performed by a neural network classifier. Image understanding (and automated target recognition) will require close cooperation between the signal processing elements, models of the world, and knowledge of the current situation contained in expert systems. Adaptive robotic motion control will require similar elements, with the added complexity of actually moving metal. The integrated AI workstation offers extreme flexibility in pursuing these targets, at a relatively low cost, yet with all of the power available from commercial products.

Author Biographical Data

Mr. Blaha has a B.A. in Computer Science from the Illinois Institute of Technology and over 6 years of experience in industrial and government R&D in the areas of materials processing, manufacturing, computer systems, image processing, and artificial intelligence. He is an accomplished programmer of many computer systems in languages including C, Pascal, LISP, FORTRAN, PROLOG, and more. Prior to joining the Basic Industry Research Laboratory, Mr. Blaha was Principal Investigator of Intelligent Control Systems at IIT Research Institute where he was responsible for conducting programs in artificial intelligence and signal processing in a diversity of industrial and government applications. A patent was granted for his development of an expert system shell and image processing algorithms for real-time control of manufacturing operations. Additionally, Mr. Blaha's electrical design experience includes the prototype and testing of high voltage, modulated power supplies for the microwave industry.

Knowledge Engineering in a Manufacturing Environment

8

Knowledge Acquisition Issues in Manufacturing

Kim B. Noderer
Inland Steel Flat Products Co.
East Chicago, Indiana

Introduction

Knowledge acquisition is the process of understanding an expert's decision making process and transferring it into a formal representation or model that can be viewed and validated by others. This model may be on paper or in the computer.

Knowledge acquisition has been called the "bottleneck" in expert systems development the development step that is most prone to errors and most time consuming. Existing successful methodologies for this area are limited in use. Knowledge engineers often develop systems that are incomplete or incorrect.

There are at least four basic issues that make knowledge acquisition a slow and error prone step in expert systems development:

1. **Human decision making is not well understood.**
   We are complex, creative, intuitive beings and we've only just begun to understand all the ways we can solve problems. When the expert himself doesn't understand how he reached a decision, it may be a Herculean task for a knowledge engineer to try model that decision making

process. Science has a long road to travel before we can say that human problem solving is understood.

2. **The current set of knowledge representation techniques is inadequate to fully model human expertise.**
   The reasoning a computer system uses, generally backward chaining and forward chaining, often doesn't match the expert's reasoning process. Yet the knowledge engineer must create a model of the expert's knowledge using these strategies. The Expert Systems equivalents of data flow diagrams, HIPO charts and other analysis tools are not complete enough. Decision making is modeled only in bits and pieces and its difficult to get the complete process conceptualized using current paper modeling techniques. More powerful ways of representing knowledge are needed.

3. **The actual process of eliciting knowledge from the expert is difficult.**
   Experts may have long forgotten how and why they make a certain decision they just know that it works. A good knowledge engineer must be able to help the expert introspect and understand what the real decision making process is. If the expert is non-analytical this task can be very difficult and frustrating. This aspect of knowledge engineering, knowledge elicitation, is currently considered more art than science.

4. **Organizational issues may extend the knowledge acquisition step.**
   If the expert is not available, no knowledge acquisition can take place. Additionally, multiple experts may give conflicting information -policies should be set up front to handle these kinds of organizational issues.

These four issues are some of the major underlying causes of the knowledge acquisition bottleneck. Are these issues being addressed at all by current technology? What is the state of the art?

Research efforts over the past few years have yielded manual techniques (e.g.: case analysis, protocol analysis) that nibble at the edges of knowledge acquisition, but so far there are no robust bundled knowledge acquisition methodologies. Today's knowledge engineer must pick and choose from several dozen techniques, such as structured interviews and building a prototype in order to capture and formalize knowledge. The Yourdan of Expert Systems has yet to emerge.

Automated knowledge acquisition tools are also very primitive, still three to seven years away from being commercially robust. The current crop of commercial tools are narrow in focus and limited in application. For synthesis problems such as planning, scheduling and design not even the most primitive automated tool is available.

Given the current state of affairs, what can the expert systems manager do to ensure successful knowledge acquisition?

The first and perhaps most important key for success is properly selected and trained knowledge engineers.

Many of the traits that make a good systems analyst also apply to a knowledge engineer. Keep traits such as an analytical mind, attention to detail and good interpersonal skills in mind when choosing your developers.

Knowledge engineers should be familiar with a wide variety of interviewing and knowledge analysis techniques. A good interviewing class, one that describes various types of interviews and questioning strategies, is probably the most important. They should also know about case analysis, observation, role-reversal and other techniques. Although a few AI vendors offer some sort of knowledge acquisition training, these methods tend be taught under the communication skills banner rather than AI. Some of these techniques are mentioned only in books, journals and magazines. Through experimentation and experience the knowledge engineers will learn when the various techniques are best used.

A caution about a very popular knowledge acquisition and representation technique: Prototyping makes a good analysis and representation tool when the problem area is complex or vague, when other methods reach a dead end, or when the expert is just more comfortable working with a prototype. But don't expect to grow your prototype into a well designed, maintainable (by other than the developer) system. Chances are the prototype will need to be rewritten when the full scope of the problem is comprehended.

A second key point is to use intermediate knowledge representation.

Use paper based models or representations of decision making process before you build the system. Using paper modeling techniques such as flow diagrams translates into a having more correct and maintainable system down the road. Resist the urge to jump immediately into prototyping. Paper based models, even if high level, can save debugging time. The knowledge engineer can review the paper based model with the expert before coding starts, validating the engineer's understanding of the problem and avoiding major design mistakes.

If you are dealing with classification or diagnostic problems, decision trees are a common way to represent knowledge. Other representation techniques can be found in trade magazine, books and through vendor training and consulting firms.

For synthesis problems that lend themselves to frame-based representation, traditional techniques such as system flow charts, data flow diagrams and variants of HIPO charts can help the project get off to the right start. The understanding of the basic inputs, outputs and processes of the problems will help your knowledge engineers design a better object-oriented system.

Another key point is to attend to organizational issues.

The expert must be available or no knowledge acquisition can take place. This may sound self evident, but its critical to get the expert's time commitment upfront. A good policy is to require that the expert will devote 25-50% of his time for the duration of the project. It's highly preferable that he spends that time away from his work site (and interruptions.)

All the usual caveats about experts also apply in knowledge acquisition: make sure your expert really is an expert; when you have multiple experts, either choose one as the chief expert or be prepared for many long argumentive meetings; if you have a choice, choose the expert that has an analytical bent (engineers for example.)

Finally, understand the knowledge acquisition is an iterative and on going process through out the expert systems development life-cycle.

No single technique will get the knowledge out of the expert and into the expert system. Your knowledge engineers will need to use a combination of prototyping and other techniques over time to fully model the decision making process. Your expert may have taken 30 years to reach the level and skill he has today in problem solving. It's no surprise to learn that modeling and transferring this expertise is a time consuming and inexact process more akin to art than science.

Conclusion

Practical knowledge acquisition methodologies can best be characterized as being in a primitive state. Current methodology tends to be most available and useful for simple diagnostic systems. For more complex problems there is a paucity of techniques that will allow the knowledge engineer to wholly capture and verify the experts' knowledge.

The situation will improve over time as research continues, and as a body of practical experience builds up. A periodic review of publications on this subject is worthwhile as new books and articles appear on an ongoing basis. In addition, one should carefully select and train knowledge engineers, ensure the right experts are involved, use intermediate paper based models, and finally, adjust your project plans to incorporate the interative nature of knowledge acquisition. These strategies will help ensure success.

Author Biographical Data

Kim B. Noderer has worked in the Information Technology field for about 7 years. For the past few years she has been involved in the development and technology transfer of knowledge-based systems technology at Inland Steel. She received her B.S. in Data Processing and Information Systems from Indiana University Northwest, and also has a Post Baccalaureate Certificate in Artificial Intelligence from DePaul University.

Organization of Knowledge for an Expert System

The Step Between Acquisition and Representation of Knowledge

Joan B. Stoddard, Ph.D.
President, Stoddard Productivity Systems, Inc.
Canoga Park, California

Introduction

The purpose of this article is to present an efficient method for organizing knowledge, acquired for the purpose of developing an expert system. Because of the unique qualities of the proposed method, and the opportunity it offers for interviewing more than one person simultaneously, some aspects of knowledge acquisition will also be touched upon.

Knowledge gathered for use in an expert system may originate in interviews, in direct observation of job performance, or from a study of brochures, specifications, training guides, technical literature and price lists. The problem faced by knowledge engineers is how to combine knowledge gained from such diverse sources into one coherent Expert System knowledge base.

It has been my experience that conventional methods of organizing facts and ideas do not serve very well for the planning and development of expert systems. Outlines, flowcharts, and decision trees tend to favor simple straight-line analysis. What is needed for expert system development is something more flexible and imaginative in approach.

The Crawford Slip Method (CSM)

The approach to knowledge organization presented here is the Crawford Slip Method developed by Dr. C.C. Crawford, founder of the University of Southern California Productivity Network.

The basic design of the CSM is to transfer information taken from a wide variety of sources to a common, conveniently handled medium that can then be sorted and categorized into logical groups or sequences.

The Crawford Slip Method can be used at four specific points in the development of expert systems:

A) Planning and scoping a project

B) Acquiring information from interviews

C) Acquiring information from publications

D) Organizing information and planning the expert system.

HOW TO BEGIN

Take a ream of inexpensive 20 lb. bond paper to a print shop and have it cut into slips, 2 3/4" X 4 1/4" in size (see Figure 9-1). Don't use an office paper cutter for this job because these are not designed to cut stacks of paper and will allow your paper to skew. Slips of irregular size and shape will reduce the effectiveness of their later use for neat "shingle-style" layouts (see Figure 9-2).

Become acquainted with the use of slips by taking notes using information you have gathered for a current or recent project. Set a generous stack of slips in front of you and do the following:

1. Use only black or blue ball point pens to write your notes. Pencils and felt tip pens produce slips that are difficult to read.

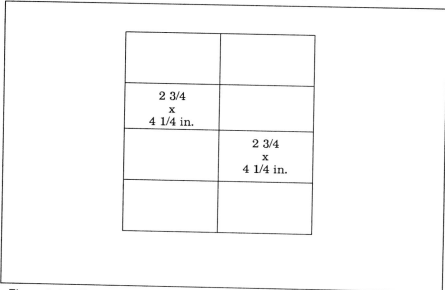

Figure 9-1   Pattern for cutting slips from 8 1/2 x 11 inch paper

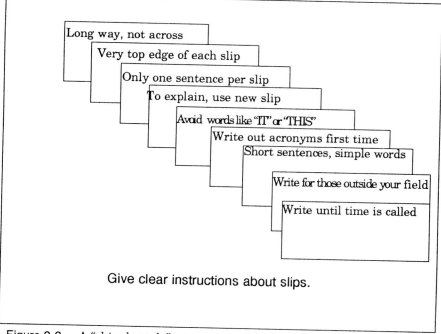

Figure 9-2   A "shingle-style" array of slips illustrating how contents of
such slips can be read at a glance. This set of slips
summarizes directions for writing CSM slips.

2. Write along the length of the slip—for those familiar with the term—you will be using the "landscape" format. Write close to the edge of the slip. Do not leave a "top margin".

3. Write no more than one sentence per slip. Try to keep that sentence short—about 10 words is best.

4. Each sentence must be totally independent of any other. You may, in some cases, write a series of slips that you want to be kept in sequence. For example, you might be taking notes from a chapter in a book, or you might be detailing the sequence of actions necessary for performing a specific task. When you have such a sequence, take a slip and write the subject of the slip sequence on it. Gather those slips, put the "title" slip on top and clip them together.

5. Do not number the slips in the pack. Numbering the slips tends to fix a sequence of slips in your mind and acts to prevent intrusion of information from other sources. One of the strengths of this system is its loose-leaf feature. Whenever you discover some piece of information that relates to the subject of one of your slip packs, you can insert it in the pack as you would put a page in a loose leaf book.

6. Avoid the use of a pronoun (e.g. "it" or "this") on one slip that refers to a noun or action on a previous slip. If you use pronouns relating one slip to another, the sense of the second slip can be lost when other slips are inserted into the pack and break up the original sequence.

7. If you use an acronym, be sure that on the first slip where it appears, it is accompanied by the complete name which it represents. Do the same for abbreviations.

Planning and Scoping the Proposed Expert System

Call a meeting of management personnel. It is strongly recommended that the person who authorizes the project be involved. Your goal in this meeting is to obtain a collection of slips that will provide the basic outline and scope of your system.

Put a stack of slips in front of each person and ask them to write responses to your questions. It is important that they do not speak to you or to each other. In this way the responses you get will be independent and not influenced by the remarks of any one person in the group.

Begin with questions aimed at producing an overview of the target task of the expert system. Then ask people to write slips focused on problems with the way the task is being performed at the present time. Ask people to write slips about the problem as they see it, what facets of the task take the most time, what seems to cause the most trouble and results in the most complaints.

As the slips are written, pick them up and look for repetition. This will indicate to you the degree of irritation which particular problems are causing, and those which, if corrected, would be perceived as resulting in the greatest benefit to the department.

Watch for any slowing down in the number of slips written. On average, a participant in these sessions can write ten slips per prompt. Try to set a pace that will keep most of the attendees occupied and involved.

Choose several of the most frequently named problems and ask participants to focus on remedies. Ask "How could we improve what we are doing?" or "How can we remedy this problem?" Ask the participants to answer in simple, declarative statements. What you are looking for is the possibility that a problem requires a less complex remedy than an expert system. The desire to build an expert system doesn't justify reinventing the wheel.

Continue to look at slips as you pick up them up, and use them as guidelines for further prompts. Let the slips help you determine where an expert system is needed and what it should be expected to accomplish.

The slips you gather at this meeting should be sorted into two stacks, the slips that show you what people expect the system to do, and the slips that you can use to set up your first outline of the system. Keep the first group to give you a check list for the success of the system in meeting its goals. Lay the others out on a table top to do a modified version of a HIPO analysis of the target task.

CSM AND HIPO

One of the areas researched for a new approach to information management was the HIPO form of system analysis. HIPO stands for Hierarchy and Input-Process-Output analysis of systems. Having studied several explanations of the method, it became clear that the principle could be applied to represent the functioning of an expert system. A HIPO analysis offers a means by which, for example, an expert system for specifying a complex instrument, could be planned from the point where information (e.g. specification of an instrument) was input into the system, through the processing of that information (e.g. selecting an instrument) to the point where the modified information was output (e.g. and order or quote was generated).

To use the HIPO technique with CSM slips, take three slips and mark them "Beginning", "Middle" and "End". Lay them in the center of the table, and then sort the slips to those headings by asking yourself "To what phase of the task does the information on this slip (or slip pack) refer?". If you have some slips that you cannot categorize immediately, put them in a "Miscellaneous" group to one side. Later in this sorting session, or after a consultation with your experts, you can assign these slips to their proper place.

This will be a "rough sort" just to establish your three main task divisions, so do not try to match slips up into any sequences. If you happen to see some slips that have a common theme, you can put a "cover" slip on them and clip them together but don't look for detailed relationships at this point.

Keep shuffling slips until they are in one of the four groups. Gather up the Miscellaneous group and two of the task division groups and put rubber bands around them. Take the remaining group of slips and sort them again. Now you are looking for divisions within the first phase of the task.

Let us examine the sorting procedure with reference to an expert system that will be used as an advisor for selecting flow meters. The management team may have come up with a number of slips that could be divided into the following phases of a task:

What size meter is needed?

What style and price?

What accessories are required?

At this point you can begin to envision the shape of your system. Let us say that you decide to do your second level sort on the question "What size meter is needed?" Look over the slips that you have assembled into this group. What questions do they suggest to you? What slips do you have that will tell you how to determine the size of the meter?

- Get information from the customer specifying a size

- Ask information on the viscosity of his product

- Ask information on the flow rate of the product

- Check brochure for information about the size of meters available.

By looking at these slips you have learned what items you will need to include in your system.

You are now ready to go down to a deeper level. Sort your slips into groups relative to these points and any others you may have identified as necessary in determining the size of a meter. You might, for example, take the slips on viscosity and sort those. You may find, or the slips may suggest to you, questions that should be asked about viscosity.

- What is the standard unit for measuring viscosity favored by the company: centipoise, centistokes?

- In what units are the viscosity tables in their brochures written?

- Do they consider pressure drop in relation to product viscosity and flow rate?

You will by this time have created on the table top an array of slips that will guide you in planning the shape of your expert system—in effect, a HIPO diagram made up of slips.

Take the main heading slips and tape them across the top of a large sheet of paper.

Take another large sheet of paper and tape down a representative set of slips from the different levels of the first phase and indicate the movement of information among them with arrows. The idea here is to go beyond generalities to the point where you can see the connections between elements, and begin to visualize the loops and straight-line structures of the system, but to stay one step above the level in which you will actually work when you write your system.

Now take each of the other two groups of task division slips and carry out the same procedure.

In addition to guiding you through the first planning stage of your expert system, the summary sheets will also be helpful in locating a place in the system for additional information to be added as the system grows.

The CSM and the Story Board

Another new approach to knowledge organization for expert systems is the story board technique used by cartoonists and motion picture script writers. Story boards are made up of a series of pictures depicting scenes in a movie. During a story conference, the pictures are moved about until a satisfactory and coherent story line has been worked out.

Creating a Prototype

The summary sheets of your HIPO analysis will provide a basic design for your prototype. As you write it, keep a pad of 2" square Post-it notes beside you, and write each question on one Post-it note at the same time you enter it into the prototype. When you are finished, arrange the notes on a large sheet of paper with arrows showing the order in which the questions are asked.

The HIPO analysis provides your guide to the inner structure of the system, but the story board, with the questions written on the Post-it notes, will give you a diagram of the system's user interface to display to your clients (or colleagues) during the demonstration of the prototype.

During the demonstration, have the story board on an easel so the attendees can follow the actions of the system. You might point out features such as, "If you choose this answer, the system will ask you for this information, but if you choose another answer you will next be asked that question." While you are discussing the system, you can move the Post-it notes around according to the comments and suggestions of the group. It will help you to clarify changes that may be requested. You can take two notes and say, "Would you rather have this question come before that one?", or if an objection is raised to one of the questions, you could say, "We'll leave that out." and remove the note from the page.

Using the story board technique is especially helpful for consultants who are unfamiliar with the terminology or standard industry procedures of their clients. When you leave a meeting in which you have used this tool, you will take with you a concrete diagram of exactly what questions should be asked, and in what order to insure your system's acceptance by its intended users.

CSM and Interviews with Personnel

As outlined above, you can interview one person, or several simultaneously, by holding a CSM meeting. If that is inconvenient, or if the interviewee is not receptive to writing slips, you can record the interview and make your own slips from the tape later.

Before you begin, go over your HIPO sheets and your story board. Outline the points you want to cover and give those points identifying numbers. Begin your meeting as you did the one with management, starting by letting people write their personal view of troubles involved in the target task of the system. Then move on to remedies. As before, collect slips and let them guide you in framing your questions. Keep the "trouble" and "remedy" slips together for later sorting.

Then move on to a more structured pattern of questions based on your list. The goal of this workshop is to get a step-by-step record of how a task is done so that you can integrate that knowledge and experience into your expert system.

Your first prompt in this part of your interview session should focus on the general purpose of the expert system. Ask a question such as "What do you do when an order comes in?" "How do you go about testing a piece of equipment?" "What information do you ask for in order to determine the kind of help or advice required by a customer who can't get his computer to work?" Ask your interviewees to describe the procedure, writing one slip per step. Ask them, as they write, to lay the slips out in the correct sequence. Before asking the next question, have them clip that sequence of slips together, and mark it with its identifying number from your list. You will then have an easier time sorting the slips in preparation for writing the system.

After you have finished investigating the target task on this general level, take the participants through increasingly detailed reviews of each procedure that is to be covered in the expert system.

Note the time and money saving elements here. Several people can "answer" your questions simultaneously by writing slips during the same time period, thereby reducing the time required of the customer's experts. In addition, if the experts write the slips for you, there is no need for transcription, as would be needed to retrieve information obtained in taped interviews.

Working from Documentation

Before heading for the library, or digging into the customer's publications put at your disposal, it is well to get out the HIPO diagrams and your story board and your list of questions that you asked your experts, and think through what information you need to fill in the gaps and complete the project.

If you find it necessary to take notes from tapes, books, or other reference material, first make a set of reference slips like library catalog cards. On each of these slips identify the source you are going to use and give it an identifying code number. Put those numbers in the lower right hand corner of slips you will use for taking notes so that later, if necessary, you can always refer back to your original source.

Begin all notes with a word or code number from your list of questions. As with the codes on the slips from the meetings, this will be an aid to sorting. Also, as with the directions for interviews, if a sequence seems logical, keep it together with a cover slip. Group slips together where it is efficient, but don't let it slow you down.

One advantage that using slips has over ordinary note-taking is the ability to deal with random thoughts and ideas that may come to mind. If something you are reading suggests an addition for a different part of the system, you can jot it down, put the slip to one side, and integrate it in its proper place later when you have time.

Organizing Information

Using the CSM, you can quickly acquire a collection of slips during the planning and knowledge acquisition of the project. Some of the slips may be in slip-packs, such as procedures written out by your experts, or sequences of notes taken from documents. You may also have miscellaneous "trouble" and "remedy" slips that do not fit together in recognizable patterns. Your next step is to find a large table where you can begin sorting slips.

Take your HIPO sheets and lay them out on the table. (If space does not permit laying out everything at once, then concentrate on the first phase of your expert system, complete that, and go on to the next phase, until all the slips are sorted.) Arrange your slips and slip packs in accordance with the HIPO diagrams. This is where the importance of the size and uniform shape of the slips becomes apparent. Because of their size and format, and the fact that the information is written along the top edge of the slip, we can lay many slips in several rows in front of us, in easy reach and with perfect legibility.

If you have several slip packs with the same code number (i.e. slips relating to the actions necessary to carry out one part of the target task which were obtained from different experts) lay them out in vertical parallel columns in front of you. Examine them to see where they are similar and where they differ.

As the system developer, it is your responsibility to determine which of two or more methods of performing a task belongs in your expert system. Consider whether combining slips written by several people will give you a more complete record of how to do a task than by following one person's account. On the other hand, check to be sure that the people who are doing things differently were not working with incorrect information or directions. These are the kind of questions you may want to take back to one of your follow-up interviews. Being able to make a detailed comparison of how several people are doing a job will help you to offer users the best level of corporate knowledge through your expert system. (see Figure 9-3).

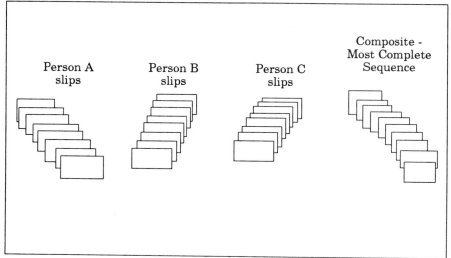

Figure 9-3   Arrays of slips on one topic contributed by three experts. The numbers represent descriptions of steps in a process. Column 4 represents the most complete representation of directions for performking the task.

Whenever slip packs on the same topic exist, repeat this procedure. Gradually winnow down the slips in your HIPO diagram array to the ones you will need for establishing structure and the ones that you will need for background information to be used in text, help and HyperText files.

When you have each section sorted and the duplicate or superceded slips removed, you will be ready to turn your prototype into a full-fledged expert system.

Other Sorting Methods using CSM Slips

Another way of sorting the slips is to lay them out in the form of a tree. Start with the point where you expect users to enter the system, and lay your slips out to resemble the branching of a "decision tree". When the "tree" threatens to grow off the table, make slip-packs of your first sequences of slips and continue to lay out slips representing "higher" branches.

If you are dealing with an expert system for selecting an instrument with many variations, so that you are not sure where to start, or what order to use for seeking and combining information, make a series of slips and try playing a form of solitaire. Sort the slips in different ways to see what features various groups have in common. Do it as though you were sorting a deck of cards, only instead of sorting for diamonds or spades, sort the slips for viscosity ranges, size, degrees of accuracy, relative price, capacity, or whatever features are salient to selection of the product.

If you are sorting the result of notes taken from documentation, for example a manual for repairing a piece of equipment, you may want to ignore the format of the manual with directions such as, "turn to page 23 for the rules on how to conduct a special kind of test", or a later request that the reader turn to the technical drawing in the appendix. Instead get the information you need to direct a repair procedure, and assemble the slips in the order in which the information will be used. Ask yourself where would you begin the task of repair, what action would you take first, what information would you need at that point, what options do you have, what do you do next, etc.?

Any of these methods can provide the insight which you can use to help you design your expert system. Working with the slips helps you to see how the information you have gathered fits together. It will also often indicate where information is missing, so that you can fill in the gaps.

Conclusion

The adaptation of methods of dealing with information developed by sources outside the AI community has made dealing with the "knowledge bottleneck" much less onerous. You no longer have to copy over (or edit in your computer) pages of written notes or shuffle stacks of file cards. All your sources can be reduced to slips which can be combined into one coherent package of information easily utilized for the development of expert systems.

Author Biographical Data

Joan B. Stoddard, Ph.D., is President of Stoddard Productivity Systems, Inc., a company specializing in the development of expert systems for the process control industry. Dr. Stoddard earned her B.A. in Psychology at the University of California at Los Angeles, and her M.A. and Ph. D in Psychology at the University of Southern California. She is an author of articles that have appeared in national and international AI Journals, and has authored a two year tutorial series on writing expert systems in Measurements and Control and Medical Electronics. Dr. Stoddard is a member of the American Association for Artificial Intelligence, the American Psychological Association and

Mensa. She is also a member of the University of Southern California Productivity Network.

Scheduling and Forecasting

Chapter

10

Knowledge-Based Dynamic Scheduling in a Steel Plant

Manesh J. Shah, Richard Damian
and Jonathan Silverman
IBM Corporation, San Jose, CA

Introduction

Applications of expert systems have been most popular in diagnostics and problem determination applications where extensive human/machine dialogue is common. In the application of process control the demands of execution speed for real-time response and the requirement of dialogue with asynchronous events occurring in a process, creates significantly different environment and challenge for expert-system tools currently available on process control computers.

In an earlier paper[1], we discussed an approach of applying expert systems to a process controlled by a hierarchical computer system.

The expert system is used for supervisory actions whereas lower level monitoring and control functions are performed by

1   Manesh Shah and Christopher Morley, Proceedings of ISA/88 International Conference and Exhibit, October 1988, Houston, Texas.

microprocessors and control computers. The application uses a rule-based system for dynamic scheduling of shared transporters, moving components, treated in sequential stations.

Numao and Morishita[2] described an expert system application to perform cooperative scheduling in a steel plant to allocate a variety of production resources with specific constraints to meet the requirements of customer orders. The main justification for the use of the expert system came from reducing the wait times between processing of charges and minimizing energy losses.

In this paper we present a somewhat different expert-system application in a steel plant for allocating individual resources capable of producing a variety of products in a hierarchically distributed control system encountered in a CIM environment[3] The dynamics of scheduling arise from the fact that the quality and specifications for the product may not be obtained from the scheduled run and the demands of the changing product specifications require constant modification of the equipment in real time. The information regarding the change in specifications is received from the upper level in CIM hierarchy, whereas the equipment-operational characteristics are constantly received from a lower level in the hierarchy which monitors and controls the production equipment. The requirement for an expert system solution rather than a traditional scheduling method in this environment arises when there is an extensive set of rules and restrictions which govern how orders can be sequenced through the equipment being scheduled.

The Steel Making Process

The steel manufacturing process consists of making pig iron from ore, producing molten steel in basic oxygen furnace, followed by casting into slabs, blooms or billets. The rolling step transfers slabs into plates, coils or sheets and billets into seamless pipes, using mills for each product. The scheduling problem arises in various steps of the steel making process. This includes the basic oxygen furnace, ladle refining (in some plants), and continuous casting. In addition, other steps of annealing, cold rolling, cutting etc. also require proper scheduling coordination to maintain optimum production, while servicing a variety of orders from customers.

---

2   Masayuki Numao and Shin-ichi Morishita, IEEE Transactions, June 1988.

3   T.J. Williams, "Recent developments in the application of plant wide computer control", Computers In Industry, Vol 8 No 2, pp 233-254, March 1987.

Today's competitive market environment requires timely manufacture of a variety of high quality products for customers. In large steel mills as many as fifty different products may have to be made in a day. With the constraints imposed by the equipment in the mills, the task of scheduling production becomes too complex for plant personnel. It is not uncommon to find a team of "expert" schedulers in a steel plant, examining the week's order requirements, previous week's unfilled orders as well as critical orders to be filled and solving the crossword puzzle by juggling the schedules of available production equipment to fit the order needs. Depending on the complexity of production equipment and operational constraints, it may require several hours of work to arrive at a reasonable schedule. The implementation of the schedule will be at the discretion of the plant operators, since the equipment is dynamically changing and may or may not meet constraints required for the product. When the product specifications are not met, it must be sent to a pool of inventory of available products with certain specification range. This pool then is used as an inventory for possible future orders whose specifications may match those available in the pool. In the mean time the expert scheduler must repeat the solution of his scheduling puzzle to meet the production requirements. In addition, the scenario may have to be repeated for each subsequent process since the planned schedules for those steps may also be affected.

The scheduling problem discussed above is not unique to steel production. It is typical in all manufacturing processes consisting of sequential steps to create a variety of products which require constant changes in the production equipment at each step as a result of changing product specifications.

The Problem

Process considerations

As stated earlier, the scheduling problem is defined as determination of what resources to allocate to each product and its accompanying process at what time. These allocations must satisfy the equipment constraints which are generally stated in the form of "rules". The rules are derived from operation experience with the equipment and are often changing as more operational experience is gathered. Thus, the scheduling problem becomes a constraint-satisfaction problem. While it may be desirable to solve the interesting problem of finding an optimum schedule, not enough information is available for an objective function required in most optimization algorithms. Also problems of this type lead to combinatorial explosions resulting in long compute times often using input data from production which may lack in accuracy. The goal then generally is to find a feasible schedule or schedules which meet the constraints with a flexibility for the expert scheduler to select one among multiple schedules when available.

The steel making process is fixed in that it is always sequential. The Basic Oxygen Furnace prepares a charge, which is processed through in fixed sequence of refining, casting, annealing and so forth. To maximize equipment utilization, the steps following the furnace are continuous. The scheduler must fully utilize the continuous processes without overlapping. Furthermore, provisions must be made for continuous maintenance changes in the equipment such as re-lining of furnaces, tundish changes and resting/setup times required for changeover from one product type to another. This does not include unscheduled outages due to equipment breakdowns and off specification products.

The overall problem

The scheduling algorithm/technique must meet the following requirements:

- Rule-based scheduling to be performed at each of the multiple points in the production line, such as continuous casters, finishing area, cold rolling etc.

- Each point has different set of rules, clearly because the production equipment is different.

- The rule set must provide flexibility, that is, it should be modifiable.

- The human scheduler must have a capability of violating certain rules in order to attain a feasible schedule or processing an expedited or special order at the risk of violating one or more constraints.

- When multiple feasible schedules are obtained, the scheduling operator must have the capability of determining cost benefits of one schedule over another. Also the capability must exist for "what if" scenarios for the operator to be able to override/modify the schedules and find the penalty in terms of rule violations and cost.

- In brief, an extensive man/machine interaction capability with a computer scheduler is required such that the operator may be able to start with a partially filled schedule and obtain feasible schedules or in the case when no feasible schedule is available, to ease the constraints until an acceptable schedule is obtained.

The rules

Since the system is rule based, the complexity of the problem is based on the number of rules at each point of production. The continuous caster presents the largest number of rules or constraints. It may vary from 25 to 100 depending on the definition of rules. Other units may have 10 to 15 rules and generally the constraints are not as severe as for the casters. It is important to remember that a single rule may comprise multiple constraints (in mathematical terms) but in a rule based expert system, multiple rules may be formulated with a single rule statement, as we will show later. By the same token, many of the rules are guidelines or soft, which make mathematical or algorithmic statements difficult. A typical soft rule may be stated as follows: try to avoid making low carbon steel as the first product after start up of the caster or a rule such as special alloy production should occur in batches of three in afternoon shift if possible. These rules come from the intuitive experience gained from the equipment by the operators.

Clearly, the task of "knowledge engineering" in this case is a formidable challenge in attempting to formulate the problem for any expert system tool.

The solution approach

It is not the scope of this paper to demonstrate the solution of the entire steel mill scheduling problem with all the rules at each production step. Also we will not delve into the problem of optimum selection of one of the multiple resources in the sequential production steps as shown in reference 2. We will concentrate on the problem of allocating a single equipment resource to create multiple product variety continuously within the equipment constraints as dictated by the rules set up and learned from experience. The procedure is to be dynamic in that the schedules may need to be continuously revised as the production goals are in the process of being fulfilled. While this is a subset of the entire scheduling problem, the purpose of demonstrating the use of expert system tools will be well served by this example. In addition, the techniques of reference 2 may be utilized to perform cooperative scheduling when multiple paths are available for an order leaving a production point.

It should be noted in this context that the ability to efficiently sequence jobs through a single resource, as illustrated in this paper, can provide important benefits in many applications. A significant number of process industry manufacturing lines contain a critical resource or a type of equipment for which the sequence constraints are particularly restrictive. In this case an expert system approach can be used to develop a satisfactory production sequence and schedule for this resource. This schedule can then be used to drive the standard scheduling calculations such as simulations, for the remaining resources in the manufacturing line. Some examples of applications which contain a critical type of equipment for production sequencing include continuous caster scheduling in a primary steel mill, loom scheduling and sequencing of fabric types or SKU's within a textile mill, packaging line sequencing dictating processing equipment schedules within a pharmaceutical plant and paper machine run sequences driving conversion process schedules within a paper mill.

The Simplified Problem

In Figure 10-1, we show a proposed CIM hierarchy for steel production. The lowest level in the hierarchy in our figure has the control computers which perform single or multiple loop controls and may also be programmable logic controllers (PLC's). These "cell level" computers provide the equipment status and other information to the area level systems which contain the plant management software (reference 3). The area level systems, communicate to the corporate level system, where planning, forecasting and perhaps company wide (multi-plant) production functions are performed. While many different configurations and levels of systems of CIM hierarchy are discussed in the current literature (see Reference 1 and 3), for our discussion we will limit ourselves to the three-level hierarchy.

In our case the corporate-host level determines production goals for each of the mills to satisfy the current backlog of orders and the forecasted mix of future demands based on planning, capacity, and status information for all the mills geographically distributed in the country. Whether optimization algorithms (for example using linear programming techniques) are employed or other planning tools are used is not relevant to our discussion. The production goals are then transmitted to the area-level system at the steel plant which has the responsibility of fulfilling the production goals for the mill, by month, by week and in general, meeting the orders for the customers. Note that the area level in our case also may maintain work-in-process inventory including the pool of off-spec products which could meet some of the orders with or without rework depending upon cost considerations.

The area level system, in our hierarchy also maintains the status of the production equipment, maintenance schedules and so forth. At this level a determination is made to keep a running schedule of orders or products to be made at each of the production points in order to meet the weekly/monthly orders to be shipped from the plant based on the information received from the corporate system and taking into account factors such as expected yields, cycle times and outfall percentages at each step in the manufacturing process. The goal of the rule based dynamic scheduler then is to arrange the schedule of production of different orders, each with distinct specification in such a way as to satisfy the constraints (be they hard or soft) imposed by the production equipment. The constraints are converted to rule based logic and expressed in as much operator language as possible to allow modifiability as well as turning rules on and off. As stated before, a highly interactive environment is required by the expert scheduler to work with the area level system to perform the dynamic scheduling function as he/she chooses.

An example

For illustration purposes let us select a weekly order set for the continuous caster to manufacture 30 to 40 products. The specifications also indicate the expected quantity of heats required to meet that order. This number is calculated by the area planning system based on the process yield data and outfall percentages of aim specifications. The order set is shown in Figure 10-2. Each product has specifications for carbon content, alloy type, special run, and width. For the sake of simplicity let us say that there are two turns (12 hour each) per day, designated as day and night turn and four products (heats) can be made per turn. Six different product types have to be made and they are classified as 101 to 601. Each product type is distinguished by carbon level, width and special or not (based on alloy type). The carbon level equal to or below 0.01 is called low carbon steel for the purpose of this illustration. Up to two turns are scheduled downtimes when the caster is shut down in addition to Saturday and Sunday. Also, the transition rules listed below are reset (reinitialized) after a down turn. With the assumptions above, the week's schedule is to be prepared for the orders on the continuous caster which is operated with constraints or rules listed below.

1.   Carbon levels must not decrease on successive heats.

2.   Width of slabs must not be increasing on successive heats.

3. Width decrease for the slabs on successive heats cannot be more than four inches.

4. A maximum of two 401 type products can be made in a turn.

5. A maximum of three 401 products can be made in a day.

6. The first heat in a turn cannot be a 401 product.

7. Only one run of specials can be made per day

8. Each run of specials has between two and three heats.

9. No specials to be made after the first three heats in a turn.

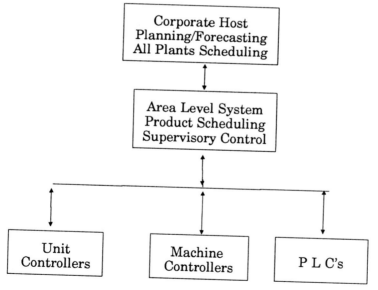

Figure 10-1   A Distributed CIM Hierarchy

10. Specials should be made in the day turn as much as possible.

11. After a down turn, start caster with low carbon for the first two heats.

Order #	Width	Quantity	Carbon	Special
101	90	2	5	Y
301	87	1	50	N
301	83	2	50	N
301	74	1	50	N
401	86	1	50	N
401	75	1	50	N
501	87	1	50	Y
601	82	1	150	N
101	89	2	5	Y
301	86	1	50	N
301	80	1	50	N
301	73	1	50	N
401	84	3	50	N
401	74	2	50	N
501	82	2	50	Y
201	90	2	10	N
301	84	2	50	N
301	75	1	50	N
401	88	1	50	N
401	80	1	50	N
501	88	1	50	Y
501	76	2	50	N

Figure 10-2   An example weekly order subset for the continuous caster

1	/0 *9*	/0 *17*	/0 *25*	/0 *33* XXX /0
2	/0 *10*	/0 *18*	/0 *26*	/0 *34* XXX /0
3	/0 *11*	/0 *19*	/0 *27*	/0 *35* XXX /0
4	/0 *12*	/0 *20*	/0 *28*	/0 *36* XXX /0
********	***************	**********	**********	*****************
5	/0 *13* XXX	/0 *21*	/0 *29*	/0 *37* /0
6	/0 *14* XXX	/0 *22*	/0 *30*	/0 *38* /0
7	/0 *15* XXX	/0 *23*	/0 *31*	/0 *39* /0
8	/0 *16* XXX	/0 *24*	/0 *32*	/0 *40* 301 /86

Figure 10-3 A typical weekly schedule table for continuous caster

The Human Approach

In normal operation the human schedulers receive the input list of orders from the ordering department office and try to find a feasible sequence of production of the orders in a table. A simplified form of the table is shown in Figure 10-3. The scheduler, proceeds by first blanking out portions of the table where the production equipment, in this case the caster, is known to be out of service for maintenance or other reasons. Next, the orders are first grouped and sequenced in a manner to meet the basic constraints or rules, for example, the rules one through four in the above list. In the second round of analysis the scheduler attempts to meet the secondary constraints (5-9 in the list above) and satisfy the production requirements. In the last round of analysis, the scheduler attempts to meet all the soft constraints such as those with "should" or "try" to complete the production schedule. When feasible sequencing is not possible, the constraints are eased somewhat to see whether orders can be scheduled with some penalty for violating the constraints. The orders from the following week may be also looked at to pull some of them for exchange with current week's orders to arrive at a satisfactory schedule at the same time altering the priorities of the delayed orders.

It should be clear from this description that a great deal of human intuition and judgement is utilized to arrive at a satisfactory schedule. Furthermore additional complications may arise as time progresses and experience is gathered from the operation of the equipment as follows.

As an example a modification of constraint 3 in the above list may be that the width decreases must be limited to every other heat. Another example may be that new products with different carbon levels or alloys and special characteristics are to be manufactured so that new rules have to be evolved and developed based on production experience with the caster.

The scheduling operation is thus continuously evolving and the "expert" must continuously modify rules/constraints in order to arrive at the basic sequences even without the problem of dynamic rescheduling required as a result of equipment malfunctions and off-spec products.

Many of the rules are based on quantities or variables which could be parameterized. For example the decreasing width between successive runs may be made as a parameter which can be modified prior to the start of a new schedule, thus facilitating a vehicle for relaxing of a constraint.

This application poses a formidable challenge for any software architecture whether based on procedural coding techniques or on expert system tools and languages.

The Expert System approach

Our approach was to include the human scheduler in the decision process as much as possible which presented the following requirements:

- Provide ability to observe the schedules as being filled in by the program especially as the constraints are being met.

- Allow manual intervention at any point so as to override the order sequencing and manipulate/exchange orders from next week's orders

- Allow on line easing of constraints when feasible schedules become difficult or impossible to obtain. Provide cost of violating the rules or easing of the rules to meet the order requirements.

- Allow the ability to activate or deactivate the rules as deemed necessary by the user.

- Allow the user to "seed" the schedules for the software to start from, provided that the seeds in the schedule follow all the rules.

The changing nature of the rule necessitated either a familiar programming environment if a totally programmerless system was not possible. Because of the nature of the problem a fully object oriented program was also deemed difficult to design. Also the plant environment required that the scheduling system be made to run on conventional computer systems available at the supervisory level in the plant CIM hierarchy. With these restrictions and based on our past experiences[4,5] we chose to formulate the problem using Knowledge Tool.[6] This tool provided high execution speed of a compiled language, efficiency of RETE algorithms, PL/I statements to provide procedural language for the sequencing logic and graphic interface required for user interaction as well as the rule formulation in PL/I like structure.

The Method of Attack

An examination of the table in Figure 10-3 will suggest that if there are no restrictions, the number of possible schedules to fit the orders in open slots will be very large indeed, the typical N-tuple numerical explosion. In actual practice because of a much larger number of orders and slots, the combinatorial explosion would be even greater than posed by our simplified subset under examination. The method of solution of this dilemma is to proceed as follows:

- Start with a one or more seeds in the slot(s) which are the beginning after transition. For example lowest carbon highest width heats on Monday morning and in the example of Figure 10-3, Friday morning, can be scheduled. As an alternative the expert scheduler may provide other seeds.

- With the seeded schedule determine children schedules for the next slot and match them against the constraints using the rule based algorithm. Discard the cases which violate the rules. Now the remaining partially filled table become the new parents from which the next slot may be filled using the remaining orders (heats). Again, the new set of children schedules are matched against the rules and only the valid ones are saved.

- This process is continued until all the slots are filled.

In this scheme the final completed schedule may have multiple feasible solutions or may have none, depending upon the order/constraint combination. The scheme however allows quantifying soft constraints (such as "try avoiding" and "should if possible") and operator/user relaxing them as the solution progresses. Also, note that as the rules/constraints are increased, the solution speed increases because

---

4   Manesh Shah and Christopher Morley.

5   Masayuki Numao and Shin-ichi Morishita.

6   IBM Knowledge Tool User's Guide and Reference. SH20-9251 IBM Corporation, P.O.Box 2328, Menlo Park, CA 94025.

the number of discarded children schedule increases thus reducing the number of grandchildren schedules. The scheme also allows user intervention with partially filled schedule and manipulating manually part of the schedule when feasible schedules are not possible.

Discussion of the Algorithm

The data representation and analysis scheme will now be discussed in some detail. The functional layout of the program is shown in Figure 10-4.

As orders are read in ("arrive to be filled") they are initiated in Knowledge Tool working memory. Orders have attributes of order number, ASTM alloy classification, width, special flag and others. As working memory elements, they are accessible by the inference engine and therefore may be referenced by the premises of the rules.

Schedules are represented as members of a Knowledge Tool working memory class. This is a PL/I-like data structure. The body of the schedules is a array of pointer-like operators, called selectors in Knowledge Tool. A particular schedule slot is assigned to an order by

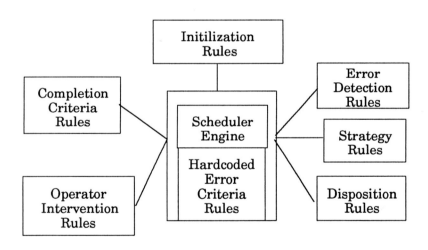

Figure 10-4   A function layout of the program

pointing the slots' selector to the order. The selector scheme avoids duplicating order information and reduces the amount of region occupied by the schedule. Note that not all slots in a schedule may be filled. The user is prompted to block out an arbitrary number of 4 slot regions as "downturns". These represent times during the work which is devoted to maintenance and no steel orders are processed. The schedule also contains a hash list of assigned order identifications. This permits rapid checking of a schedule for the list of assigned orders. Also present are a series of error counters associated with a particular error condition. For example, there is an error counter which maintains a count of the number of width errors present in the schedule. This feature allows use to separate tracking of errors from the actions or disposition of the schedule.

Functional aspects of the algorithm will now be presented in turn.

Initialization

After the orders have been read into working memory, a single parent schedule is also instantiated. There are a small number of rules which specify conditions that allow immediate placement of orders. A typical proscription is that orders for low carbon steel must be filled first at the beginning of an active work cycle. The rule-based seeding of the schedule results in less than 10% of the orders filled on the parent schedule, but it is important for program efficiency and does not impact its generality. Given the combinatorial explosion of possible schedules, it is highly desirable, and the authors feel justified, to seed the original schedule with orders.

Scheduler engine

There is a single rule whose responsibility it is to generate new instances of the schedule class in working memory. A single schedule is chosen by the rule. This schedule becomes the "parent". The list of orders is scanned and is compared with the parent schedules' hash list of orders which are already present in slots on this schedule. This results in a list of orders not yet on the schedule. The strategy rules have determined the single empty slot (target slot) on this parent that will be filled. Using the parent schedule as a template, "child" schedules are generated that look just like the parent, with one exception. The target slot on each child contains a different member of the list of unfilled orders. Since the information that the parent contains has been passed to its children, the parent schedule is deleted from working memory and its space is freed.

After the schedules have been generated, they are examined by the error detection rules. Since our scenario included several conditions that could never be violated, these conditions (3) were hard coded with the conclusion of the scheduling engine rule. One of the rules was that carbon level must never decrease through adjacent slots. These rules represent the invariant portion of scheduling knowledge. They could also have been implemented as a cluster of rules called from the right side of the scheduler engine. These rules prevent the generation of schedules that would otherwise be recognized as in error and would be immediately deleted by the disposition rules.

The scheduler engine is a low priority rule. The next time it is active, it will pick one of the "child" schedules just generated and use it as a "parent" to generate further children. This new "parent" schedule will then be deleted. In analogy to natural life, a parent is always deleted after consideration by the scheduler engine, whether it can produce children or not.

Error detection rules

The purpose to this medium priority rule cluster is to check newly generated schedules for error conditions. Since new rules may be added rather easily, this represents the area where transitory knowledge may be maintained. As a schedule is found to violate a particular condition, the rule counter for that schedule is updated.

Strategy rules

These rules decide the next slot to be filled. In general, they try to fill slots in ways that will most constrain the schedule for the next schedule generate cycle. It is much faster to kill a child early with an abundance of rules, than it is to pattern match that same rule set against the multitude of children that it may produce. Strategy rules also select "promote" schedules via a priority counter to the scheduler engine if they are almost complete.

Disposition rules

These rules have been broken out from the error detection rules so that both the user and the program had some control over "desirable" conditions. That is, the user, and later the program should be able to relax certain conditions or reject a schedule for a variety of not necessarily connected conditions. These high priority disposition rules are monitoring the error counters associated with each schedule. Thanks to the RETE algorithm, disposition rules pattern match only on schedules whose counters have changed since the last match-select-act cycle.

Operator intervention rules

This aspect of the program allows operator intervention when a schedule of a particular type has (or hasn't) been found. It allows the operator to reseed a schedule at any point in program generation and to specify it as the prototypical parent schedule from which to generate other schedules. This provision allows the operator to override the program and also to force processing of a new block of orders that may just have been presented to him.

Completion criteria rules

These rules recognize the first acceptable schedule and present it to the user. They could also be used to selectively reseed a new schedule if no acceptable schedule is found or to consider additional orders. Note that this program does not have an optimum schedule criteria, therefore it is picking the first acceptable schedule. This prevents the necessity of considering all possible combinations which could prove very costly. This situation is analogous to the problem of consistency and validity in logic.

Conclusion

In Figures 10-6 to 10-10, we show screens from the program run as the slots are being partially filled. Heats which violate the rules are highlighted (in color in plant displays) and discarded schedules are pointed out with rule violation reasons.

At the start of the program, the schedule operator has the capability of modifying certain parameters which adjust the rule set as well as the caster downtimes, and activation of some special rules. This is followed by either allowing the operator to seed the schedule or letting the system to start with a seed (see Figure 10-7). As the schedule is being built, at certain soft violations, the operator is informed and allowed to intervene to remove the order and enter other orders manually or let the system continue with removing the rule violating orders and creating other schedules. This cooperative rule

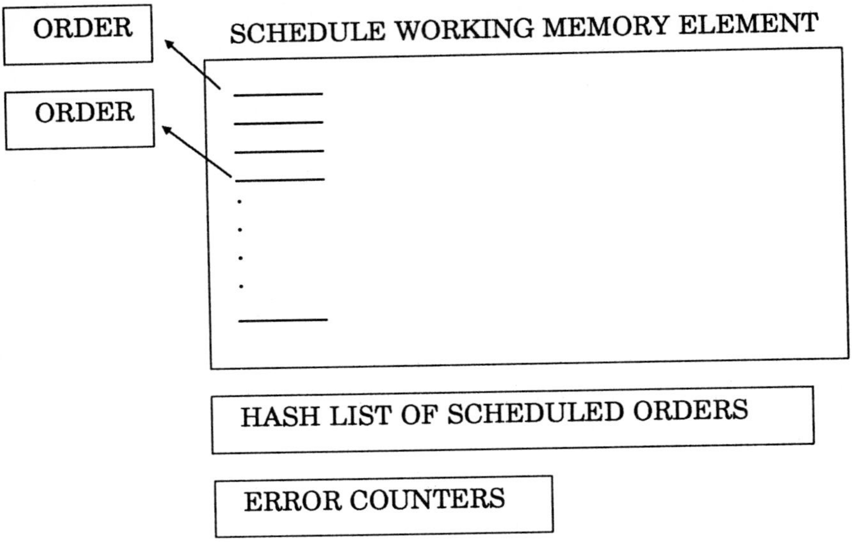

Figure 10-5   A functional layout of the memory during program run

Place an X at the downturn locations

Day	Night	
—	—	Monday
—	X	Tuesday
—	—	Wednesday
—	—	Thursday
X	—	Friday

What is the maximum acceptable width change?
Should the alternative width rule be activated? (Y/N)
Do you want to direct the output to a file? (Y/N)
Do you wish to print verbosely? (Y/N)
Maximum allowable night turn specials

Update Menu and then hit enter

Figure 10-6   User input for scheduler parameters at start

schedule building process continues till the first complete schedule which obeys the rule set is found or till no children can be generated to satisfy the complete rule set and the final incomplete schedule or schedule set is presented to the operator. At the conclusion of the scheduler run the operator may restart with new seeds, relaxed constraints or modify the computer generated schedules as they are built and make multiple runs with the scheduler until a satisfactory solution is found for implementation on the plant.

The seeding process allows the operator to rerun the scheduler in the middle of the week due to either a machine malfunction or arrival of other high priority orders. This facility then provides a comprehensive cooperative scheduling capability for the operator to run at any time and keep up with the dynamic conditions of the plant.

As the schedules are prepared they are then sent to the plant management program which has the responsibility of communicating

Monday	Tuesday	Wednesday	Thursday	Friday
1	9	17	25	33 XXXXXX
2	10	18	26	34 XXXXXX
3	11	19	27	35 XXXXXX
4	12	20	28	36 XXXXXX
5	13 XXXXXX	21	29	37
6	14 XXXXXX	22	30	38
7	15 XXXXXX	23	31	39
8	16 XXXXXX	24	32	40 301/86/050/N

Unassigned Orders

201/90/010/N	501/87/050/Y	401/74/050/N	301/76/050/N
201/90/010/N	601/82/150/Y	401/74/050/N	301/76/050/N
101/89/005/Y	101/90/005/Y	501/82/050/Y	
101/89/005/Y	101/90/050/N	501/82/050/Y	
301/87/050/N	301/86/050/N	301/84/050/N	
301/83/050/N	301/80/050/N	301/84/050/N	
301/83/050/N	401/84/050/N	401/88/050/N	
401/86/050/N	401/84/050/N	401/80/050/N	

Here is the initial schedule . . . . Hit Enter

Figure 10-7   The progressive filling of weekly schedule table, first phase

Monday	Tuesday	Wednesday	Thursday	Friday
1 101/89/005/Y	9	17 101/90/005/Y	25	33 XXXXXX
2 101/89/005/Y	10	18 101/90/005/Y	26	34 XXXXXX
3 501/88/050/Y	11	19	27	35 XXXXXX
4 501/87/050/Y	12 601/82/150/N	20	28	36 XXXXXX
5	13 XXXXXXXX	21	29	37 201/90/010/N
6	14 XXXXXXXX	22	30	38 201/90/010/N
7	15 XXXXXXXX	23	31	39
8	16 XXXXXXXX	24	32	40 301/86/050/N

Unassigned Orders

301/87/050/N	401/74/050/N	501/76/050/N
301/83/050/N	401/74/050/N	
301/83/050/N	501/82/050/Y	
401/74/050/N	301/84/050/N	
401/75/050/N	301/84/050/N	
301/80/050/N	301/75/050/N	
401/84/050/N	401/88/050/N	
401/84/050/N	401/80/050/N	
401/84/050/N	501/76/050/N	

Error - A run of specials must not exceed three heats
enter the number of the order to be removed

Figure 10-8   The progressive filling of weekly schedule table, second phase

the operational instructions to the plant operators or in an ideal situation directly to the unit level computers. By the same token, the plant management software continuously updates the actual status of the parts produced and sends the current pending orders to the based scheduler so that the schedule operator can run the scheduler at any time.

Monday	Tuesday	Wednesday	Thursday	Friday
1 101/89/005/Y	9 401/74/050/N	17 101/90/005/Y	25	33 XXXXXXXX
2 101/89/005/Y	10	18 101/90/005/Y	26	34 XXXXXXXX
3 501/88/050/Y	11	19	27	35 XXXXXXXX
4 501/87/050/Y	12 601/82/150/N	20	28	36 XXXXXXXX
5 301/84/050/N	13 XXXXXXXX	21	29	37 201/90/010/N
6 401/80/050/N	14 XXXXXXXX	22	30	38 201/90/010/N
7 501/76/050/N	15 XXXXXXXX	23	31	39
8 301/83/050/N	16 XXXXXXXX	24	32 401/73/050/N	40 301/86/050/N

Unassigned Orders

301/87/050/N	401/74/050/N
301/83/050/N	401/74/050/N
301/83/050/N	501/82/050/Y
301/74/050/N	301/84/050/N
401/86/050/N	301/84/050/N
401/75/050/N	
301/87/050/N	
301/84/050/N	
401/84/050/N	
401/84/050/N	

Error - A 401 order is the first heat of a turn
enter the number of the order to be removed

Figure 10-9   User interaction screen for order rearrangement

Monday	Tuesday	Wednesday	Thursday	Friday
1 101/89/005/Y	9 401/74/050/N	17 101/90/005/Y	25 501/76/050/N	33 XXXXXXXX
2 101/89/005/Y	10 501/82/050/N	18 101/90/005/Y	26 501/76/050/N	34 XXXXXXXX
3 501/88/050/Y	11 301/82/050/Y	19 501/87/050/Y	27 301/73/050/N	35 XXXXXXXX
4 501/87/050/Y	12 601/82/150/N	20 401/86/050/N	28 401/73/050/N	36 XXXXXXXX
5 301/84/050/N	13 XXXXXXXX	21 301/84/050/N	29 301/74/050/N	37 201/90/010/N
6 401/80/050/N	14 XXXXXXXX	22 401/84/050/N	30 401/74/050/N	38 201/90/010/N
7 501/76/050/N	15 XXXXXXXX	23 401/80/050/N	31 401/74/050/N	39 301/87/050/N
8 301/83/050/N	16 XXXXXXXX	24 301/80/050/N	32 401/73/050/N	40 301/86/050/N

Unassigned Orders

Program Complete

Figure 10-10 Final Schedule Screen

The scheme described here is in the process of implementation in a steel plant in U.S.A. Similar scheme has been implemented by the NKK/IBM team in Japan where the results have shown substantial economic benefits.

Author Biographical Data

Manesh Shah is a Senior Engineer with IBM Corporation in San Jose, California. After his undergraduate degrees from the University of Bombay, India, he obtained a M.S. in Chemical Engineering from University of Michigan and a Ph. D. in Chemical Engineering from the University of California in Berkeley in 1960. He has been with IBM Corporation since 1960 and has worked in areas of Process Control, Simulation, distributed control and application of Artificial Intelligence Techniques in Process Control. He was an adjunct professor at the University of Puerto Rico in 1978-79 and a lecturer at Stanford University in the years 1980 to 1986. He is the author of 35 papers and two books.

Richard Damian is an Advisory Programmer at IBM's Artificial Intelligence Support Center. He holds B.S. degrees in Chemistry and Mechanical Engineering from Michigan State University. Prior to his joining IBM he worked as a metallurgist with Ford Motor Corporation. He was awarded IBM Chairman's Outstanding Technical Award for his work on Monte Carlo Modelling of magnetic head error rates. He is the author of several papers on finite element modelling of laminated structures, image analysis and rule based scheduling.

Jonathan Silverman is a Senior Software Design Engineer with IBM's Industrial Sector Division in Santa Clara, California. He has over five year's experience in designing, developing and marketing CIM solutions. He is currently responsible for the architecture and design of planning and scheduling solutions for Process Industry Manufacturers.

Mr. Silverman has a B.S. degree in Mathematics and Statistics from the Hebrew University of Jerusalem and a M.S. degree in Operations Research from Stanford University. He is currently completing his Ph.D. dissertation in Operations Research at Stanford University.

11

Managing Complexity:
AI in
Batch Plant Scheduling

Herman F. Bozenhardt
Artificial Intelligence Technologies
Hawthorne, New York

Introduction

The automation of a batch chemical plant generally begins and ends with providing the real-time control and recipe handling. This level of automation replaces manpower in some basic functions and generates some incentives in the capacity area. However, a major area of opportunity in the automation of a chemical operation is the automation of the decision mechanism to select the products to be run in specific equipment. Currently, even with automated systems critical decisions impacting product cost such as product, scheduling, and equipment, allocations are left to operators. Secondly, process supervision makes allocations based on intuition rather than true cost calculations.

The degree of scheduling complexity increases in some specialty chemical and resin plants where production cycles are dependent on operator intervention and QC sampling/analytical testing. This non-deterministic nature of cycle time also provides a vast opportunity for improved capacity utilization. Plant scheduling via an automated and/or intelligent mechanism can be done using a blend of operations research techniques, heuristic methods and queuing theory. Prior to the development of some critical innovations in Artificial Intelligence and expert systems, blending several technologies and managing multiple solution techniques was not possible. However, with the advancements in hardware platforms, new expert system tools and AI techniques, an environment for implementation readily exists now. Specifically, in the AI techniques area, the advancements in cooperating expert systems, integrated AI, distributed AI, and time critical AI, the managing of complexity in a plant environment has been simplified.

Scheduling techniques to sort out campaigns, schedule recipe implementation and allocate process equipment have been implemented with success. These specialty chemical and resin plants have reaped benefits in energy, raw materials and capacity areas, plus the product quality area.

This paper will cover the economic incentives and the physical reasons behind the benefits. Secondly, it will also cover the concepts, technology and techniques and its practical application. The paper is geared for control engineers and operation's personnel. We believe the presentation will address a common industrial "nagging" problem and suggest a solution method.

Modern chemical processing today has begun to take advantage of micro-processor based control systems. These efforts have primarily fallen into two categories: improvement of operator interfaces and automation of regulatory control. This automation provides a basic foundation to keep the current operations philosophy on target and improve operations on a loop basis. However, a common danger in a low-level automation is contribution of the same method of product recipe handling, equipment assignment, campaign strategy, and allocation as were used in the non-automation time period.

With the installation of microprocessor-based control systems, the modern plant manager must consider what tools are available to solve the following scheduling allocation problems.

- Scheduling batch reactors, batch distillation columns, and other vessels requiring steam and other common resources in order to maintain the resources grid's performance and eliminate idle time;

- Scheduling batches consecutively to minimize equipment change-out;

- Scheduling CIP, and line flushing cycles for minimum disruption in product cycles;

- Optimization of scheduling to maximize the use of efficient equipment and idle inefficient equipment;

- Providing intelligent advice to the operator to assist them in making choices on product transfer, recipe handling and equipment allocations.

Techniques to accomplish these tasks, when implemented, have resulted in increased capacity, improved quality and reduced consumption. The static decision techniques fall into the linear programming/transportation algorithm category, while the dynamic network techniques are based on dynamic programming. AI techniques are used to emulate the human searching process for actual plant equipment configuration and scheduling.

This paper demonstrates that advanced "process level" techniques are of benefit to a plant environment. It presents techniques to solve commonly encountered problems and illustrates them with aid of a sample problem.

Multiple Variables

The manufacturing of resins or organic chemicals in a medium to large plant is a complex operation from a processing standpoint due to the many aspects of kinetics, thermodynamics, vapor-liquid equilibrium, solid-liquid separation, and material handling. The fact that it is a multi-step batch process increases the difficulty. Each step of a typical process from premix to reactors, crystallizers, filters, and centrifuges requires decisions as to when to process and in which vessel. Since dedicated equipment and dedicated streams rarely exists, operators, foreman and supervisory personnel must answer questions such as the following:

Should I drop the filter, recoat it, and begin a new run, or continue?

One of my reactors is at the end of its cycle. What crystallizer and centrifuge will I send it to?

I need to modify some cycle parameters. How much can I change, and how does it impact on my costs and schedule?

Why are batches always being held up because of the common manifold?

And always, where am I in capacity now?

The desire to use more efficient equipment or, more importantly, to use equipment that is currently operating most efficiently, is also part of the decision process. Ultimately, people make many key decisions, daily, primarily on consideration of one variable. In modern plants, the number of variables, overlaid with the number of products multiplied by the number of vessels is a multivariable problem best solved by a computer. To provide a computer based tool and solution method, an emulation of how the human makes his decision must be synthesized. The key elements of the human decision process is the knowledge of the process (model), searching and an iterative evaluation. The scheduling or decision support to help operators and engineers is a task which may require traditional OR tools but must use AI techniques.

Scheduling Incentives

The decision process, regardless of its validity, generates a measurable cost. Costs generally involve energy, capacity, manpower, raw material, and quality related factors. However, since quality is closely monitored and fixed and the raw materials for a recipe are fixed, actual variations in process efficiencies are energy (glycol cooling, steam, etc.) and capacity (gallons/day of product).

Capacity increases can be achieved in two ways. One via time-related activities and allowing more material to flow through (turn over more batches) in a given period of time; and by better material usage (higher product yield obtained). The material usage concept is less obvious, however a 1% greater yield in a 10 million pound/year organic specialty plant, would yield $200,000/year in revenue. This variability when not minimized, understood, or controlled becomes a hidden cost (opportunity cost) to the plant. Capacity has two aspects which allow a plant to achieve increased revenue:

(1) Dissipation of overhead and controllable costs. The higher capacity allows a plant to reduce its cost per pound during operation. Elimination of controllable costs associated with cooling towers, process overtime, and recurring maintenance is possible if the production rate is increased to allow shutdowns to lengthened (this assumes a fixed market).

(2) Increased revenue from incremental sales allows greater cash generation from the plant's fixed assets. This increases the return on net assets (RONA).

Capacity increases can be achieved by utilizing more idle time and prevention of hold up time. From an overview, the coordination of the tank farm, the reactor buildings, finishing area and warehouse provide the most potential for capacity incentives. Some examples in the plant are as follows:

- Scheduling filter runs and filter change-out in order to maximize average flow over a long time period;

- Staggering batch cycles so that the transfer manifold, often a bottleneck, is continuously being used in sequence with no batches are on hold;

Automating transfers when equipment is available. One potential problem is that human interaction which is based on availability or attention, can cause unnecessary delays or holds.

The other area of potential increased revenue, energy consumption, is equally important. In reactors, evaporators, or any boiling or heat exchanging systems, energy efficiency is very variable. It is related to many fixed and variable factors:

- Equipment selection due to vessel design geometry either promotes or deteriorates conductive and convective heat transfer, therefore newer equipment designs are more efficient.

- Scaling on the tube walls changes to U_o1 and decreases its effectiveness in heat transfer[1]. This is a function of a number of batches or stream time.

- Improvements in modern materials of construction, tolerances, and range of operating conditions have been continuous, yielding greater efficiencies compared to older vessels.

- Because of multi-user utility systems, vessel location in the manifold (physical distances) often produces different efficiencies due to the longer heat-up or increased cycle times for similar vessels.

These factors manifest themselves in plants where repeated usage of vessels over a period of time result in specific variations in lbs. of steam per lbs. of product. The key to scheduling is to maximize the use of the most efficient equipment and idle the inefficient equipment through careful selection. Typical variation of 1% to 5% which may be insignificant (noise) at first glance, easily becomes thousands of dollars over one year's operation.

A complimentary consideration is the trade-off between energy and capacity. As an example, a decision to begin a reaction in an older, less efficient reactor may be justified because of the revenue generated by the additional capacity. Another example may be to lengthen a cycle time on a mixer in order to schedule it into a different reactor and then a newer, efficient batch still. Each time a

---

1   1 U_o = Overall heat transfer coefficient (Btu/F-ft2-hr)

trade-off is made, one must consider the total picture and the cost of the incremental steam versus incremental revenue of that product.

The optimized scheduling capability provides several other benefits:

- Better utilization of manpower-instead of spending hours of frustrating batch balancing, a computer is far better at searching and evaluating on a NxM problem.

- Better turnaround/shutdown planning, to bring the process down slowly by restrictive scheduling, planning the start up, and planning single stream emergency start up to meet a market demand.

- Reduce Maintenance Cost via Planning - The scheduling system can run and illustrate the impact of shutting down various equipment (or combinations of equipment). Maintenance planning can be systematically planned and enhanced with this feature.

- Finally, a greater awareness of stream time and capacity is gained via the capability to search, allocate and synthesize current decision analysis and implementation of optimal scheduling.

Solution Systems Overview

Once the value of optimal or even better scheduling is recognized at the plant, several different problems can be addressed:

First, based on total demand and known capacities of equipment, what is the best allocation of production to specific equipment or groups of equipment? Linear programming based allocation provides the solution to a "macro" system or large overview problem. This planning exercise may be done on a shift, daily or weekly basis. The following is a discussion of the techniques required to go from a macro overview solution to a finite schedule.

LP Technology

The LP technology can assess a plant loading by producing a bulk allocation of resources to a task, or recipes to equipment. LP technology is confined by a systematic search within predefined global constraints directed by a global linear objective function. The relations between each recipe or pieces of equipment can only be modelled as a simple, linear and deterministic. The true nature of batch scheduling is either event driven or time driven and not a continuous function as an LP formulation may represent.

The transshipment algorithm which is a reduced formulation of an LP for supply/demand is an excellent method for recipe load analysis in a plant. This can produce an "infinite schedule". In addition for preventative maintenance, LPs can be used to balance required maintenance on machines with available mechanic time.

Dynamic Programming

Another important technology is dynamic programming. DP is a traditional operations research technique to solve networking problems. Because DP is an exhaustive calculation and search routine it is only appropriate for networks of limited size.

As an example flow path analysis between major pieces of a chemical plant equipment or between processing modules is an appropriate problem. The network modeling of all process hardware (pumps, values, line tees, etc.) would present a computational nightmare. Both the "backward-induction" method and the Hamilton-Jacobi method search through the network exhaustively to develop the most economical path.

An example of a proper application of DP is tank form management and distribution. The production lines are considered source nodes, tanks are nodes on a network and distribution points (dock lines, rail car loading "spots", tank wagon loading "arms") are the network terminal nodes. DP is an excellent mechanism to find the optimal path. The path costs are capacity related, manpower and pumping costs (electricity). These costs can be accumulated and be considered fixed. This is a computationally achievable application for DP.

Statistical Methods and Queuing Theory

A significant amount of the plant loading capacity, and resources are not deterministic nor can be modeled in an OR format. The nature of order entry volumes, recipe frequency, and manifold availability are critical supply, distribution and flow parameters. These are more appropriately expressed by probability density functions. The measurement accumulator and characterization (Gaussian, Poisson, etc.) of these distributors can provide expected values and bounds to calculate critical ratios, expected flows, expected average capacities, and arrival time analysis.

Although many organizations would openly characterize their quality attributes on products, few would discuss the distribution of time from product sample to information on the results in the hands of the operations personnel. This issue is a critical gating point in product delivery. It can be represented as a delay and should be characterized by a distribution. QC departments are a limited resource with a variable service time. By adequately recognizing this function as a distribution, this provides the first step to a better automated/improved procedural system.

Artificial Intelligence

The science of AI and the uses of expert systems provides a conceptually new approach to model the facility with objects. Object oriented programming allows the expression of relations and association. AI provides the mechanism to search, which is the essential human scheduling science. Intelligent search or directed search provides the approach to make process routings link work centers and assign specific time, resources and recipes. It is the intelligent search which human schedulers use. This is a search directed by years of experience in knowing the delays, bottlenecks, service time, distribution and feasible routings.

AI technology provides a modeling capability to construct and represent the plant. Expert systems can provide a rule based architecture to sort techniques and technology, and to apply the correct technology at the correct problem. Lastly heuristics and search can be combined to apply the best know how and methods to generate a finite schedule. This essentially emulates the best combination of science and human experience in scheduling.

Technique Comparison

After serious study of several batch facilities, a strategic decision was made to base a substantial part of the solution mechanisms on AI techniques. The following is a summary of that philosophy.

- Most batch facilities present a substantial number of attributes representative of a flexible manufacturing environment. It is the flexibility and the potential combinations of equipment needed for a product which makes modeling via LP difficult and often prohibitive in scope.

- Uncertainty of the product cycle times, and QA release of a product are not as easily represented in a LP. A localized statistical model which continuously is updated may be of more value when combined with logic that constantly assess operations alternatives in time. This alternative method is best generated by a simulation or intelligent search.

- Mathematical representations of capacities, utilizations and status simultaneously is again difficult in OR but a natural for an object oriented system.

- Lastly, the expert systems techniques are built for incorporation of plant heuristics,[2] experience, shop floor knowledge and general "know-how". OR techniques are not geared to retain that type of information. Experiences show in operations planning, and shop floor scheduling, "know how" is a critical factor. The "know how" (i.e. rules) that operation's personnel retain, embodies procedures, diagnostics and information that enable the operation to succeed. Without this "know-how", any scheduling system would fail.

The modeling of relations, the flexibility needed in representing and modifying constraints, and the representations of goals and objectives make the AI approach the most valuable from an overall technology management perspective.

The method chosen to solve the problem is the linear programming (LP) concept. The LP model allows the user to optimize a system of linear equations to a linearized cost function, all bounded by series of constraints. Traditional LP models can handle these problems, but their equation setup is complex and often not meaningful to the end user.

A simplification of the LP methodology was developed many years ago to routinely solve distribution networks and optimal warehouse stocking. It was called the "transportation algorithm," or "transshipment problem". This method serves the function of an LP in an all-constrained model (valid for batch plant scheduled on a time basis). The traditional LP solution via simplex algorithm is prohibitive; s shortcut method allows us to quickly transform our networking-type algorithm to a recipe/processing assignment solver.

The LP model of a system generally reduces to a tableau that facilitates solution via the simple method (Table 11-1). The transportation algorithm modifies the original tableau to a matrix of constraints and introduces new variables (U_1, V_j) to simplify away from the row/column matrix elimination technique (Table 11-2). Hiller and Lieberman explain this result in detail in their book, Operations Research. The LP concept is again an iterative search

---

2   heuristics—those intuitive "rules of thumb" which humans apply to solve problems. These are usually successful, although the complete analytics of why they are successful are not completely known.

technique which searches solutions in a feasible region. The key phrase in the LP is search which is the embryonic concept relative to AI.

To illustrate the campaign solving technique, the following example is given: the values used are whole numbers to simplify understanding.

A plant has four reactors. Reactor 1 has a weekly maximum capacity of 10K gallons. Reactor 2 has 20K, Reactor 3 has 30K, and Reactor 4 has 40K. We can assume that the plant has a cost history of receiving products A, B, C and D in each reactor. Given that headquarters has ordered 15K gallons or product A, 0K gallon is of B, 25K of C, and 40K of D, which is the campaign strategy for this given week.

With costs and constraints we need to iteratively search the combinations and permutation of this seemingly simple 4 x 4 problem for the optimal solution. The LP concept provides the best tool to iteratively search, evaluate and allocate resources.

The problem can be reduced to the tableau in Table 11-3, which contains the current cost information. The elements within the matrix are the associated costs of running recipe j in vessel i.

As an uncomputed approach, a human scheduler might estimate the allocation as shown in Table 11-4 ($550). The allocation guess starts at Recipe D and places product in the most efficient available vessel.

The initial feasible solution of a vessel/recipe campaign (i.e., it obeys all the constraints) uses the "Northwest Corner Method" (i.e., computer-like generation for an initial "seed"). Notice its cost is $580 for the week's production.

Next we calculate the incremental costs that are incurred by each allocation through the selection of both recipes and vessels. All C_{ij}'s (total costs) allocated (assigned) are equal to V_j (vessel component cost) plus U_i (recipe component cost). This is done first by assigning the row (vessel cost) U_1 that has the most number of assignments equal to zero. Now, knowing all the C_{ij}'s with $C_{ij} = V_j + U_1$, there are N equations with N-1 unknowns. This lets us generate all incremental costs, U_1 and V_j.

Figure 11-1 shows a typical concept of a batch plant and the information systems which must be unified for a comprehensive scheduler. Depending on the complexity, each area can be "solved", or provide relevant input to the master scheduler. Each system is different and may require LP, DP queuing, AI or traditional computing technologies. However, it is the selection of techniques to be applied in each area which can only be implemented by an object oriented and rule based expert system. The expert system must be able to

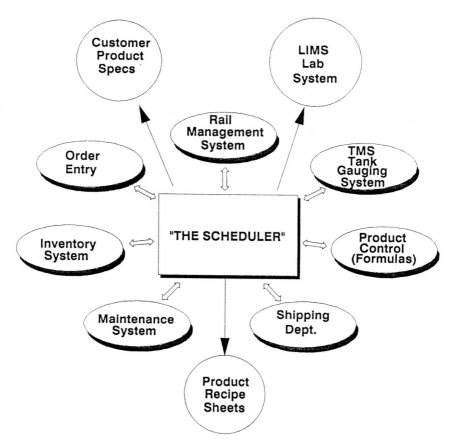

Figure 11-1   The Scheduler

combine strategies and implement the paradigm of cooperating expert systems.

Traditional OR Tools

The first area to be explored closely is the use of the DP. For this scheduling function, we will express our plant as a network model.

Using "Figure 11-2. Chemical Plant Network" as the model, we have assigned costs to each path (path plus downstream unit) in dollars per pound of production. The assigned costs are given in "Figure

Coefficients of equation							
	Basic Eq. Variable No.	Z	X1	X2..Xn	Z1..Zn		Right Hand Side
Objective Function	Z	0	-1	c_1	$-c_1..c_n$	0	0	0
	z_1	1	0	c_{12}	c_{12}	I	0	bI
	z_2		0				0	
	z_{n+j}	m+j	0				1	
Constraints		m+n	0					bm+n

Table 11-1   Coefficients of equation

Matrix of constraints								
				Destination				
		1	2	3	4	n	Supply	U_1
Source	1.	c_{11}	c_{12}			c_n	S_1	
	2.						S_2	
	3.						S_3	
	.	c_m				c_{mn}	S_m	
	M							
Demand	V_j	d_1	d_2	d_3	d_4	d_n		

Table 11-2   matrix of Constraints

Example of batch problem solving cost tableau

			Recipe			
Reactor	A	B	C	D	Capacity K gals
No. 1	4	7	3	10	10
No. 2	6	9	3	11	20
No. 3	3	4	3	6	30
No. 4	2	4	3	7	40
Demand	15	20	25	40	100

Costs in $/gallon of product for that recipe in that reactor

Table 11-3   Example of batch problem solving cost

Example of uncomputed approach

			Recipe			
Reactor	A	B	C	D	Capacity K gals
No. 1		10			10
No. 2	15	5			20
No. 3				30	30
No. 4		5	25	10	40
Demand	15	20	25	40	

Cost = 550 units

Table 11-4   Example of uncomputed approach

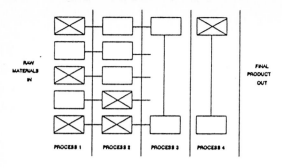

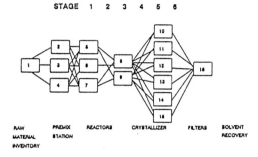

Figure 11-2   The Process Network

CURRENT COST* DATA

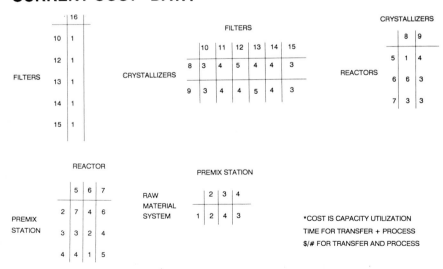

Figure 11-3   Current Cost* Data

	16	X5
10	1	16
11	1	16
12	1	16
13	1	16
14	1	16
15	1	16

	10	11	12	13	14	15	14	X4
8	(4)	5	6	5	5	(4)	5	10 or 15
9	(4)	5	5	6	5	(4)	4	10 or 15

	8	9	13	X3
5	(5)	8	5	8
6	10	(7)	7	9
7	7	7	7	8 or 9

	5	6	7	12	X2
2	12	(11)	13	11	6
3	(8)	9	11	8	5
4	9	(8)	12	8	6

	2	3	4	11	X1
1	13	12	(11)	11	4

X1—X2—X3—X4—X5
1—4—6—9—10—16

Figure 11-4   Induction Calculation

11-3. Current Cost Data". The actual costs in a plant would be very close to one another, and the optimal path would yield an improvement of 1-3% over the next alternative. However, in this example, the differences were exaggerated to illustrate the technique.

The matrices in "Figure 11-3. Current Cost Data" represent the costs of the path plus its destination node. So, in the upper right hand table, the transfer from crystallizer 8 to filter 15, plus the cost to run filter 15, is $3 per batch. The cost for the non-utility consumers included in a model could be represented as the revenue lost for a reduced production equal to the vessels potential, if used.

For the solution calculation ("Figure 11-4. Induction Calculation") we will use the dynamic programming technique (1) which employs a simple backward induction calculation. This contains the elements of choosing a starting point, search/evaluate and repeat the procedure. Assuming that we want to calculate the best path from the raw material system to the solvent recovery stills, we start at node 16*. Since all costs are equal, we chose the lowest numbered vessel, filter 10, at a cost of 1 unit for phase #5 of the process.

Node 16 is the last node; notice we work backwards—the backward induction algorithm.

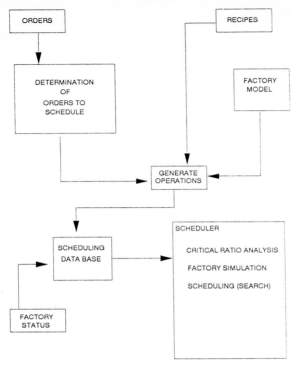

Figure 11-5

The next phase (looking backward) looks at: What is the lowest cost on phase #4 from filter 10? The paths from either crystallizer (8 or 9) to filter 10 will cost four units. The paths from the reactors (5, 6, 7) indicate that if crystallizer 8 is chosen, reactor 5 is the likely pairing while crystallizer 9 should be paired with reactor 6. The next phase (#2) and the last (#1) follow from reactor 6 to premix 4 and finally to the raw material area. The final solution shows that the best path (based on "real time" data) is now premix 4 to reactor 6 to crystallizer 9 to filter 10.

This concept, although viable for network analysis, would be best utilized in simple networks, large manifolds, and distribution systems where equipment is dedicated, options are few (or none) and costs are fixed.

The other area of scheduling concerns developing the best preselected schedule for a given time period. This scheduling involves finding the lowest cost allocation of a campaign or a bulk of product recipes. This method is designed at a level above the real-time operation and provides the "marching orders" for a week or a month. It could supply the first feasible allocation for a finite scheduler.

Based on these cost increments, we can calculate for all the unallocated areas Cij - Vj - U1. This shows the incremental cost (- for savings or + for additional cost) used to reallocate assignments. A savings of $5/gallon of product can be made if we reallocate part of the Recipe C campaign to Vessel 2. To reallocate capacity, we will begin by adjusting allocations in a closed-loop pattern of alternate additions and subtractions, centered around putting Recipe C in Vessel 2. Again, we recalculate the incremental cost based on the current allocation and recalculate all the cost changes of the unallocated spaces.

By reallocating the campaign assignment to the largest global cost savings, we improve the solution again. A total of four more iterations yields a true optimal solution for the LP model (i.e., no additional cost savings are achievable). Each iteration is a cost evaluation, search and trial solution. Note that the cost is greatly reduced by means of a non-obvious solution and the costs are an absolute low of $445 for a week's production.

As an extension to our solution, lets assume reactor #2 developed a severe crack in its glass lining. The reactor must now come out of service for a full week for inspection and repair (or replacement). This immediately reduces the capacity of the plant to 80K gallons of product per week, however, headquarters still has a full slate of product. The decision now must be made of, what products can not be made in the remaining reactors 1, 3 and 4, we will manufacture 80K gallons at (a base cost of $385). Superficially, the strategy maybe to continue produce on this allocation and simply not make 20K gallons of recipe C. Upon examination, however, since the plant is not at the same conditions as previously existed, a reanalysis is warranted and the LP should be solved again with the new conditions to search out a modified strategy.

Since capacity cannot be created, the reactor now should remain in the solution tableau as a "dummy capacity". This means any and all capacity assigned to it will not be run. In order to model what has happened, we want to assign a large cost (M) to all the cost values in row #2. This will force a reevaluation to schedule away/around vessel #2 as best as can be accomplished. This revaluation will rebalance the allocations.

Starting from the optimal solution, the costs are reactor 2 are replaced with the value M. With the knowledge, M is a large positive number we calculate our U1 and Vj values, and calculate our reallocation incentive. This immediately shows the system is in an non-optimal condition! After four iterations of evaluate, allocate and recalculate, the LP stabilizes, and shows no incentive to allocate. These iterations reveal a completely different solution. The cost for 80K gallons is $305, and the LP indicates to stop production of 20K gallons of recipe D (it is assigned to vessel #2). The solution allows a partial shipment of recipe D (20K gallons of the 40K gallons) and full production of A, B and C. Clearly from a production availability and a cost ($385 vs. $305), this is the best strategy.

The utility and flexibility of this simple method can be used for many types of production dilemma, with the use of modified costs, "dummy" capacities and "dummy demands". The method provides a searching and evaluating approach far more efficient and convergent than the trial method.

The "dummy" capacities and "dummy" demands can be used in a expanded (1 row and 1 column) to do any capacity balancing. A dummy vessel with cost of M will be used when an automated scheduling system tests the number and volume of capacity to be produced in a single time frame. Orders will be screened for due date and fixed number of the last 100 orders will be separated into recipes. The recipes will be bulk allocated and the algorithm will, because the cost (M), force fit the maximum volume into the real capacity. The allocation assigned into the dummy capacity is the unachievable capacity. Therefore, a routine would identify the recipes allocated to the dummy vessel and "shed" them based on priority and/or due date.

The most obvious aspect of scheduling by LP or DP is that the system is based on the model quality. Since LP's and DP's provide only for a fixed methodology, real production systems must be forced fitted into a model. The danger is always that the solution one generates is a solution to the model and not the reality of the plant. Many iterations on the model may be required to approach the plant representation, however, some complex situations never completely converge.

AI Methods

As we have discussed several methods for support functions of scheduling. The key area is once we know the bulk recipes to be scheduled, basic statistics and distribution methods, we must generate a finite time and/or event derived schedule. The determination of what specific batch and when it will be scheduled is our goal.

Modeling

The following section is a discussion of the methods and definition of the terms to be used in building an AI based model of the batch chemical facility for the purposes of scheduling.

- ResourcesResources are any physical entity that the system will schedule. Resources include machines, tanks, reactors, stills and some personnel.

- Capacity
Each resource has a capacity model. The capacity model reflects the time availability of a resource, and is used to model the plant's shifts. For equipment, it includes volume, operations performed, cleanout and maintenance. For personnel, this also includes information such as holidays, lunch, meetings, and distinction between regular hours and overtime.

- Work Centers
Work centers are aggregations of resources for reporting purposes. For instance, a premix blender and several reactors might be grouped as a work center.

- Process Routings
Routings will identify the sequence necessary to produce a specific product code. Routings are similar to the existing recipes, but routings have more information on the exact steps, times, and resources required to produce the product. The routings could be derived from recipes, process knowledge, and the relationships among resources, or simply entered manually. This area generally requires substantial study on plant and procedural specifics.

- Customer Orders
Orders represent what must be manufactured. Each order identifies what product is to be produced, how much, and the due (requested completion) date.

- Operations
Operations (SOP's) describe the steps necessary to produce a specific order, and are normally obtained from the routings. That is, when a new order enters the system, the operations will be created from the routing for the product code that is ordered. Each operation describes the required time to perform the operation, the resources needed, and the time phasing of the resources. For instance, a polymerization recipe might require a premix blender for 30 minutes, a reactor and associated cooling capacity for 9 hours and a monomer stripper for 2 hours. Operations can describe alternative sets of resources and times, and alternative sets of operations can be used to describe different methods of producing the product. Operations are also the method of entering machine maintenance requirements into the system.

- Goals
Goals will be used by the system to compare alternative partial schedules in order to generate an overall schedule that provides the best optimization among the goals. These goals will often be

in conflict with one another. For instance, shipping an order on time might require working overtime. Some of the goals in the system could be:

- Minimize "work-in-process" costs

- Ship by due date

- Minimize overtime costs

- Maximize resource utilization

- Minimize cost of total operations

The user has the ability to specify the importance of each goal relative to other goals.

- Constraints
  Constraints will be used for plant specific knowledge such as:
  Overtime policies, limitations, and preferences.
  Tank volume capacities
  Pre-assembly storage space
  Specific information to drive the
  automatic scheduler based on plant knowledge

- Resource Preferences
  Although multiple units might be available to perform some operation, some units may be preferred to others. The degree of preference can also be expressed based on efficiency or manifold location (as examples).

The AI methods used for the scheduler model provides a mechanism to model and represent the plant in a natural way. By this, we mean the plant can be represented by equipment, resources, people and services and the links/relations are represented as in reality. The primary method is for resources to be represented by objects and their slots (attributes). Objects are entities which contain capacity, link to other resources and process materials.

2. Scheduling
   This section describes how the program would actually schedule the plant. The result of the schedule will be resource "reservations" which will be stored in the database. The reservations identify the operation that is consuming a particular resource for a given time interval. These reservations can then be used to generate reports

that show schedules for resources, work centers, and orders. The major steps in the scheduling logic are listed below and depicted in Figure 11-5.

1. The system will review all operations for every order to be scheduled and perform a standard critical ratio analysis. For each operation, it will calculate the earliest possible start date (relative to resource availability, prior operations, and materials availability) and the latest date by which the operation can be finished and still meet the order due date. The critical ratio (operation time divided by the difference between the early start date and the late end date) will be calculated. This information will be used to guide the actual scheduling in latter phases.

2. A simple discrete event simulation of the plant will be run to generate an estimated load through time for every resource. The simulation will be based on plant specific scheduling knowledge and a dispatching rule. For instance, the dispatching rule could be to start the operation with the closest operation due date from the available queue of jobs.

3. The simulation results will be examined to determine the relative criticality of each resource. This information is then used in the step below.

4. Each operation will be scheduled. The logic will be similar to the simulation, but alternative resources, sequences, and overtime variation will be explored, or searched. For instance, when scheduling an identified critical (maybe bottleneck) centrifuge for an order in a specialty organics plant, the scheduling system would create a number of potential ways of scheduling the same operation by examining:
   -Use of different blenders, reactors and crystallizers
   -Different times or sequences of operations for the units above
   -Using overtime

5. The scheduler will evaluate these possible schedules based on plant adjustable goals for due dates, overtime policies and cost, preferred resources, a work in process cost

estimate, and such other identifiable and measurable goals that the plant may identify.

6. The system will use the information from the simulation and the critical ratio analysis, plus plant specific knowledge, to focus the scheduling on the critical orders, resources, and goals. The relative resource loading from the simulation step will force examination of more potential schedule alternatives for an operation that uses critical resources. Simultaneously, a combination of the critical ratio for each operation and the criticalness of the resources needed for the operation will force deeper exploration of the more critical operation and which of them can be scheduled and in what order. These mechanisms cause the scheduler to work more on the important scheduling problems (i.e., immediate due dates), and to quickly schedule the easier operations, thereby generating good schedules in a reasonable time.

The time required to schedule will be adjustable - that is, if more time is available, a wider search will be performed, which should result in a better schedule. However, statuses will be continuously monitored of changing plant floor activities. Time criticalness is foremost in each decision.

The system will allow for manual scheduling to override any automatic scheduling logic. Specifically, an operation may be forced to use certain resources, forced to be done during a specified time period, forced to precede/follow another operation on a resource, or any combination of these. This system will provide the capability to override the scheduler when a manual examination has discovered a more optimal partial schedule, and when factors must be considered that the scheduler does not model. When the automatic scheduling system is run, manually scheduled operations will be rescheduled within the bounds defined by the manual scheduling. Thus, if an operation is forced within some time period on a specific set of resources, the system will schedule that operation within the time period. If the time period is the same as the operation time, then in effect, the operation is completely manually scheduled.

The Flow of Information

Information used throughout the system will basically fall into **three** categories:

- Status - this specifically relates to resources, work centers and orders.

A resource is either available or unavailable (in use or scheduled to be in use). This is used in understanding its "inventory" of resources and if work centers can be used for new orders. Orders have a status of being processed, scheduled waiting for processing and unscheduled. In general being used or currently scheduled have the same status of already being allocated.

- Statistics - each model of resources and work center has utilization statistics to benchmark productivity and assess capacity.

- Costs - each order will have a cost to manufacture assigned to it based on its scheduled assignments. These cost values will be used in the decision process to find the lowest cost to process the order. In addition, final costs after processing are retained with each order as a measure of success.

Information flow is accomplished to and from the central database of the scheduling system. The scheduling model will need to acquire data frequently from the user's or their distributed control systems. The need to acquire data will be based on an interrupt driven basis. The need to acquire data will be based on an interrupt driven basis. The scheduler will assess 1) status changes of resources and work centers, 2) change in the order entry queue, 3) operations changes mandated by supervision or 4) change in goals or constraints.

Based on these changes the scheduling system will assess the need to reschedule. Rescheduling may require data acquisition and prompts to the scheduling personnel or QC personnel. The relational database will update all the appropriate statuses, statistics, costs and models. The data flow will typically be active order entry data, (marketing department), equipment status (production department), personnel list (production department) and raw materials (inventory) from terminals. Work center terminals will be the focus for Quality Assurance (QA) and production people, to receive these instructions, as well as to interact and answer questions posed by the scheduler. They will also input their status such as:

- step in the operational procedure

- status of batch

- QA release, test or fail

In summary, the system will rely heavily on a robust operator interface design integrated into the relational database. The scheduling system will scan and assess statuses for rescheduling and maintain a scorekeeping function for costs and statistics.

The AI based scheduling system schedules based on a realistic representation which is based on the users knowledge and not an OR scientist's forced fit model.

Uniting The Solutions

In order to unite OR, statistical and AI search techniques into an industrial solution and integrate into a CIM architecture a series of four paradigms must be implemented in the computing environment.

■ Cooperating Expert Systems - the concept of cooperating expert system allows each "bubble" in Figure 11-1 to exist as a separate entity. These entities can be LP, DP, AI, statistical, MRPII, LIMS, Fortran programs or a database. These can be developed independently with a common communications protocol. Conceptually the decomposition of the problem into separate expert systems allows for rapid contained development with each system operating in one of four modes.

■ Cooperation - where decision A from the expert maintenance system reports its information to the inventory system so it may produce a decision B (i.e., the pump is broke so the inventory is unaccessible). This may also be as simple as reacting to the intelligent reasoning cycle such as: 1) A says new orders are in the queue; 2) B says there is capacity unscheduled, then run the heuristic search again and schedule the next recipe.

■ Collaboration - where the decisions of A and B are in conflict and the two entities resolve the conflict by constrained manipulation.

■ Meta Mode - where ES#1 chose the decision A based on its objective and ES#2 chose B based on its objectives for the same problem. When the two expert systems have no common ground, the meta (the master scheduler) arbitrates the conflict, and to solve the conflict, chooses the best overall plant decision.

■ The User Mode - is imposed when a conflict cannot be resolved by the meta because it does not have a basis or rules. It must ask the human users to choose and it retains the data and decision for future application.

The cooperating expert systems provides an supervisory scheme to resolve problems which relate closer to human scheduling and human decisions. Ultimately, it provides a schedule which is physically and economically feasible and optimal or it will ask the human for assistance. A traditional OR mechanism would not provide the flexibility, and instead report a solution which the human must recognize as incorrect and then resolve by reconfiguration.

Integrated AI

This concept implements standard communication protocols integrated with standard relational databases. The goal is to unify all the methods on the cooperating systems regardless of the technology. Industrial strength AI environments must be used to develop and integrate the system.

Distributed AI

This concept unites the cooperating concept with the integration concept and allows each cooperating system plus a "meta module" to exist in each CPU used. This provides for fault tolerance and increased execution speed. The concept of transactional processing and the implementation of a network based client server paradigm is a critical design consideration for a plant level system.

Time Critical AI

This last concept imbeds in the meta and all calculation blocks the decision time or event time. This is necessary to control the length and time devoted to searches. This concept especially implements the philosophy of "how good can the schedule afford to be". The searches will have specific time limits to generate a solution. The searches will sequentially view simple feasibility and increase their beam search width and probe for better solutions. True optimality is often unachievable and generally prohibitively costly.

These methods provide for a time integrated solution system with data exchange/communications which allow constraints to be propagated between cooperating members. This mechanism provides for flexible environment and not one that is constrained by traditional modeling.

Scheduling a complex batch plant requires the use of two specific technologies—scheduling techniques and advanced computing technology.

Scheduling Techniques

The batch plant scheduling technique that will be implemented as the primary methodology is a flexible and adaptive search.

Secondly, AI paradigms are needed to monitor control, and use existing installed systems in a cooperative mode.

The intelligent search methods will be supported by:

- critical ratio analysis

- bottleneck analysis

- simulation

- predictive allocations and

- forecasting methods

These support methods will prioritize, structure and direct the search algorithm to develop the most optimal schedule in the short period of time. AIT believes the search strategy is one of the technical keys to success in scheduling.

Advanced Computing Technology

One of the most significant considerations in the development of an AI application for scheduling is performance. Performance of a system is its speed at which information is acquired, reasoned with, and action taken. This performance cycle can often limit the size of the system and scope of the problem. Until recently most AI tools were geared for demonstration systems, small systems and prototypes and they were limited by one of three factors:

- Separate or poorly integrated database, production engine, and communications if any; this generates a burdensome computing overhead which inhibits the transactions and limits the use of the reasoning engine.

- Traditional reasoning algorithms were not designed for a large number of objects in a large problem model. These algorithms required extensive convergence time especially on industrial standard hardware.

- Exotic hardware is the origination of many of the AI system tools today. The traditional reasoning algorithms ran more efficiently in the specialized (Lisp, Symbolics) machines, however, their connectivity to industrial hardware and multiple user interfaces was burdensome to the system or inadequate for the application.

Conclusion

One of the most difficult tasks in modern manufacturing is to plan the production campaigns of a highly integrated multi-purpose batch plant. The physical gathering of data and developing a timely feasible solution has been clearly the work of experts. To consider scheduling optimally was not a potential until recently, with the dawn of industrial scale artificial intelligence and expert systems.

Batch plants now have a unique opportunity to acquire a computing technology which it can integrate into its current automation and computerization. This technology can first link the many disciplines involved in scheduling. Then, upon acquiring the current state of the production and market demand it will produce a schedule using:

- the accumulation of human expertise

- high speed computing with operations research tools

- artificial intelligence to search, analyze and develop the lowest cost finite schedule based on detailed and substantial chronological history.

Because of the power of modern computing machinery, this system can accomplish in minutes what teams of planners could take days. An Intelligent Scheduling System can operate continuously to provide the plant a constant and infinite supply of scheduling talent, planning "know how" and plant production supervision. Specifically, some major areas of incentive are:

- Planning the campaign to optimize costs and to insure delivery schedules are met

- Planning preventative maintenance activities to coincide with and work harmoniously with production schedules;

- Provide alternatives and optimized equipment reallocation if a process failure occurs.

- Providing a constant analysis of the process via energy consumption, raw material utilization, process conversion and manpower utilization.

- Using this cost analysis to track equipment utilization and operations to minimize cost, maximize capacity objectives and develop a cost profile for each product.

These analysis techniques allow the unification of all the major operating departments data into a central relational database to share information in a time critical fashion for the benefit of the entire plant.

Ultimately scheduling, planning, inventory, manpower assignment and production are all so interdependent that an optimal solution cannot be considered without these as integral parts of the method. If the method to schedule truly includes these facets, then the problem is "combinatorially large" and does not allow (by sheer size) the ability for a human based solution. Only an AI/expert system of large scale is capable of solving this problem. Other computing solutions which utilize only traditional techniques (i.e., operations research only) cannot permit a complete solution exploration and optimization.

Author Biographical Data

Over the last 14 years, Mr. Bozenhardt has held various engineering and management positions in major corporations implementing integrated control engineering, artificial intelligence and operations research based solutions to operation plant environments. Mr. Bozenhardt's main contribution to the technical community is the demonstrated infusion of highly profitable technical solutions into operating plants. The result of these efforts are the subjects of his numerous papers which touch many industries. Mr. Bozenhardt is a recognized expert in the areas of control engineering, distributed control systems, artificial intelligence, and plant computing environments. He is a graduate of Polytechnic in New York and a member of AICHE, ISA and ISPE.

LA-Z-BOY Implements Manufacturing and Shipping Scheduler

Elizabeth Cholawsky, Ph.D.,
Customer Support Manager
and Don N. Harris,
Principal Knowledge Engineer,
AICorp
Waltham, Massachusetts

The authors would like to thank Frank Kolebuck, Jim Boughey and Stan Kirkwood of La-Z-Boy for their active participation in the development of this article.

Introduction

The La-Z-Boy Chair Company, based in Monroe, Michigan, is a leading furniture manufacturer with nearly $600 million in annual sales, and one of the world's best known brand names. For the first time in 1990, it joined the ranks of the Fortune 500. La-Z-Boy was not, however, an overnight success. It was a classic garage start-up when Edward Knabusch began making patio chairs with his cousin, Edwin Shoemaker, in the late 1920s.

smart," he would ask, "then why did we start the company in 1929 (on the verge of the Stock Market Crash and the Great Depression) and incorporate in 1941?" (only to hand over a brand new factory to the government for the war effort).

Coincidences of U.S. history aside, however, La-Z-Boy has a heritage of leadership and innovation. Most of the several hundred U.S. furniture makers are much smaller companies ($50 million or less); and La-Z-Boy itself was only at $1.5 million in sales in 1960. A few years later, in 1964, the company patented a mechanism for a recliner/rocker that provided the spark for dramatic growth. Football star Joe Namath became the company spokesman in an enormously popular advertising campaign of the early '70s, and La-Z-Boy has been synonymous with relaxing at home in comfortable, quality chair, ever since.

Although La-Z-Boy is known first for its recliners, it is also a leading manufacturer of sleep sofas. While furniture remains its only business, it has expanded into case goods (wood furniture for living rooms and dining rooms) and office furniture, including partitions and cubicle systems.

Upholstered furniture is still the company's mainstay, and La-Z-Boy officials believe the company is the largest user of upholstery fabric in the world, surpassing even General Motors for that distinction. Every week a staggering volume of fabric is stitched onto the more than 36,000 pieces of upholstered furniture produced in La-Z-Boy's manufacturing facilities.

It Starts with the Production Schedulers

Production schedulers at the busiest La-Z-Boy plants are deluged with more than 1,500 order tickets per day. Each ticket represents an order for a separate piece of furniture from La-Z-Boy's vast distribution network. The schedulers' job is to sort through the orders and determine the most effective schedule to produce, ship and deliver the merchandise behind every ticket. Efficient use of the production lines, coupled with a company goal of reducing lead-time on deliveries, makes the production schedulers' job complex.

At a glance, the production schedulers' area in the plants resembles a huge post office. This juggernaut is the area of La-Z-Boy operations where the potential need for an expert system was first identified. While the work is often repetitive and boring, it is important and requires careful consideration of many factors.

A prime consideration for the schedulers is making the most efficient use of the Gerber cutter, the huge computer-controlled machine that cuts fabric for upholstered furniture. But, the task is more complicated than simply sorting tickets and ordering fabric. Production schedulers must take into account fabric texture, nap, thickness, patterns and special requirements such as decorative buttons.

La-Z-Boy had done a good job of automating material requirements planning (MRP) in the late '70s and early '80s. What had remained cumbersome, if not overwhelming, was the human intervention required to physically sort the thousands of tickets, each representing an individual piece of furniture ordered.

The production scheduler has a myriad of other details to keep in mind. For instance, the Gerber cutter can handle as many as 45 layers of some fabrics in a single cutting, but far fewer of others. Also, the various production lines are dedicated to different groups of styles, so the volume of orders must be sufficient to keep a crew working productively. Sometimes, a ticket calls for a fabric that is out of stock, or a style that doesn't easily fit into a manufacturing group. La-Z-Boy's production schedulers did a great job under trying circumstances. Nonetheless, Frank Kolebuck, La-Z-Boy's Vice President of Management Information Systems, knew there had to be a better way.

Taking Advantage of the Mainframe

La-Z-Boy's corporate database has been DB2 since the mid-80's, running on an IBM 370-based machine with IBM 3270 terminals. The current system is an IBM 3090-200E. AICorp's INTELLECT, the first IBM mainframe-compatible natural language query system, had been implemented in the mid-80s to handle ad hoc queries and user-generated reports. It was also used to report certain shop-floor manufacturing operations, such as inventory requests and production volume.

While production scheduling was a prime area for automation, Kolebuck appreciated its complexity, and realized that a standard application package would not possess the necessary judgment. La-Z-Boy MIS managers Jim Boughey, CPIM, and Stan Kirkwood, CPIM, the primary designers and developers of the KBMS application at La-Z-Boy, wanted to address this issue without writing programs if at all possible. They knew that DB2 would help, but that programming would be long and arduous. They also had to consider that while information systems for La-Z-Boy manufacturing plants are centralized, there are important variations from site to site in the way the plants operate.

A production scheduling system would have to handle multiple levels of complexity, and take into account that a change in one part of the process would ripple through the entire system. Analysis of the scheduling bottleneck at La-Z-Boy revealed the classic characteristics for a good expert system. The task is repetitive, yet complex, and therefore prone to error. Also, considerable training goes into making an efficient production scheduler.

Kolebuck, Boughey and Kirkwood looked at 3rd and 4th generation tools available at the time. They thought if they could query the databases in an intelligent manner, they would be able to provide enough information for the schedulers to do the job better. Once the interface was supplied, the next thought was, "Why not have some smarts in the program to take that information and present a schedule, rather than have the schedulers summarize the information and then do it?"

Moving Toward the Knowledge Base

Because it was an application requiring judgment, it was logical to look at expert systems. Already familiar with AICorp because of INTELLECT, La-Z-Boy was intrigued by the potential of its new product, KBMS, the Knowledge Base Management System, when it became available in 1988. KBMS is a knowledge base application development tool for building large-scale expert systems. It is a multi-platform, integrated knowledge base system for IBM mainframe, PC and Digital Equipment VAX VMS platforms. It offers a robust set of reasoning methods — forward-chaining, backward-chaining and hypothetical reasoning, as well as object-oriented capabilities.

In particular, KBMS's forward-chaining inference engine is ideal for tackling La-Z-Boy's scheduling problems where data that includes countless variables must be quickly and efficiently processed according to complex rules. Other systems La-Z-Boy had considered were primarily backward-chaining.

When Kolebuck, Boughey and Kirkwood began to seriously consider KBMS, in order to minimize their risk, they established certain criteria for a functional knowledge base system. These included:

- Applicable to existing production process

- Easily maintained

- Easily modified to keep pace with change

- Able to accommodate a full-blown information system including, large database and terminal access across an IBM-SNA network

La-Z-Boy initially investigated KBMS and two other tools then on the market. In order to evaluate KBMS, an extensive "proof-of-concept" system was built. This system demonstrated that the tool could model the problem and that other functional requirements, such as database access and performance characteristics, were fully met. After careful consideration, the verdict was that KBMS was the one that could do the job.

La-Z-Boy's consultants confirmed the validity of a model drawn up by AICorp's developers showing decreased code size by orders of magnitude as follows: A system in Assembler language with 100,000 lines of code translates into 10,000 lines of code in COBOL; 1,000 lines of code in FOCUS and 100 lines of code in KBMS. Rules in KBMS are, in effect, multi-purpose, substantially reducing code size. For instance, a single KBMS rule understands forward-chaining and backward-chaining, and permits access to multiple databases. For La-Z-Boy's MIS people the comparison was dramatic, and their consultants concurred.

Prototype work for La-Z-Boy's KBMS production scheduling application began in late 1989. The initial goal was to develop embryonic approaches. La-Z-Boy was insistent that it be an application that could be changed without breaking down, and with enough flexibility so that it would function in every plant. With the corporate goal of lead-time reduction always in mind, the development team set out to build a scheduling system that would run close to when the actual production happens.

KBMS Clarifies Manufacturing/Shipping Connection

Early on, it became apparent that loading and scheduling of trucks was inextricably tied to manufacturing—from the dealer's order ticket, through the production lines to customer delivery. This and other time factors made automating the scheduling process more complicated than it seemed at first. It was a welcome challenge for AICorp because this manufacturing application seemed tailor-made for KBMS.

In the first phase, called very rapid prototype, or VRP, it was immediately clear that a small chunk of test data could not produce an insightful view of the solution. The nature of the problem, it turned out, is dependent on the large volume of tickets to be sorted. This complicated the test and evaluation, and required that KBMS operate equally well in both prototype and production development modes. In the test mode, it was learned that a virtually infinite number of configurations exists to schedule both production and shipping. There is no right or wrong answer, only better or worse.

Kirkwood and Boughey began to look at the problem in terms of "lays"—the furniture industry term for the sequence and number of layers of fabric to be cut—and "loads"—the sequence of finished pieces of furniture to be loaded onto several trucks headed for different destinations. The underlying question was, "If you had no constraints, what would be the ideal approach?" There was no ready answer.

For the prototype site, La-Z-Boy purposely chose one of its busiest and most efficient plants, one that also manufactures a full range of products. Traditionally, the plants made up the lays, or determined the order in which the various pieces of furniture would be produced. Only then, after production was nearly completed, did they start to plan a loading scheme for the trucks.

The implementation of KBMS called for pre-planning of the lays. To a degree, the test plant was doing that on an informal basis, so the KBMS prototype could prove to be a boon to linking manufacturing and shipping, and improving deliveries to dealers and customers.

The first step was to establish dummy production/shipping cycles, based on real, although not live, information, and test it against the day-to-day operations of the production schedulers. The results convinced Boughey and Kirkwood to focus on one aspect first: the loading and scheduling of trucks.

Taking the opposite perspective from the way in which most plants had operated before (scheduling a manufacturing sequence that seemed to make sense, and then thinking about the shipping afterwards), the developers started by focusing on the best way to send the product out. This meant first calculating the loads, or the groupings of products to ship.

Using both INTELLECT and KBMS, they developed sophisticated menus for the 3270 terminal. One graphic, for example, is a map of the United States that allows the scheduler to focus on a particular state and manipulate data easily, providing an excellent overview of the variety and volume of La-Z-Boy upholstered pieces slated for delivery around the country. A sub-menu provides more information about specific shipping zones within each state.

As much as it is an integrated problem, the initial phase needed to be attacked in chunks. The challenge for the knowledge engineers was to develop an algorithm showing how to plan the lays. The starting assumption was that there are good ideas on how to group the different colors, patterns and styles of fabrics, and that the system should be doing that for the user.

And, it should be flexible enough to accommodate changes.

A quick analysis of the situation showed that typically furniture scheduled to be shipped from a plant in a given week was being divided in half randomly. Half would go out by mid-week, and half would go out by the end of the week. In most cases, the first truck would not leave the dock until Wednesday.

The schedulers and dispatchers were making do, but knew their systems could be improved. System developers were uncertain about what would be an acceptable solution, but knew any reasonable improvement would bring tremendous benefits to La-Z-Boy. Just getting a few trucks to leave on Monday and Tuesday would be a terrific start.

Maximum Truckloads

The degree to which a truck is full was established as an important characteristic in scheduling. In order to get trucks out the door more quickly, developers needed to concentrate on certain trucks, specifically, those requiring the fewest distinct patterns or styles. Theoretically, these would be the easiest to fill and move off the dock. In the KBMS application, the amount of truckloads that could be manufactured at the same time became the variable "maximum trucks on dock," establishing the importance of freeing up precious dock space by concentrating on the "most full" trucks.

From a KBMS developer point-of-view, the approach to scheduling the lays was to focus on a key truck. For each distinct pattern on that truck, the system would request lays for patterns to fill it, and also match those for other trucks.

Ultimately, the KBMS application needed a "conflict resolution" by which it selects both the all-important "most full" truck and the patterns that fill it. Two packets, or subroutines, were developed—"Select Next Truck" and "Fill Lays"—that incorporated the process for resolving this conflict. These routines are smart enough to know when to apply, and with which data.

The "Select Next Truck" packet has an "agenda condition" that informs KBMS to inspect all unfilled trucks, and a "priority condition" wherein KBMS favors those that are most full (Figure 12-1). Once KBMS determines it is appropriate to invoke the "Select Next Truck" packet for a particular truck, the packet will focus on the different patterns needed for that truck, then execute the strategic agenda of the "Fill Lays" packet (Figure 12-2). The agenda condition in this packet in turn inspects all patterns that match the current pattern, and its priority function identifies trucks that are "most full," giving precedence to those on the dock.

Another factor, and a particularly vexing characteristic of the furniture industry, is the problem of less-than-truckloads, or LTLs. Because shipping is high bulk and low weight, it is terribly expensive to ship trucks which are not even close to being full.

La-Z-Boy appreciated the advantage of taking LTLs into account as early as possible in the manufacturing/shipping cycle. Now, with an overview of the relationship between scheduling lays and filling trucks, it is easier to plan efficient use of the Gerber cutter which has an average capacity of about 40 layers, depending on the fabric.

In the lay scheduling portion of the system, LTLs appear as simply another truck (or load) with many small groups of patterns. Initially, the system would only favor LTLs over other trucks if there was no room "on the dock" for a "regular truck."

In the prototype phase, KBMS developers made two key observations about the algorithm:

1) The system seemed to produce full lays at the beginning of the process, but towards the end of the manufacturing schedule, the lays were smaller. Thus, if the schedule could be more evenly distributed, manufacturing would be far more efficient.

2) There was no way to identify those trucks which, for whatever reason, were assigned a top priority. In some instances, it is imperative to get certain trucks out first, no matter how many distinct patterns they require.

The initial prototype resulted in fewer than 10 rules to accomplish the automatic lay scheduling process, due to the "declarative style of programming" inherent in KBMS. The developer need only inform the packets how to determine when they apply (agenda conditions) and how to resolve any conflict (priority function).

Consequently, employing these two new constraints was also relatively simple, because developers needed only inform the system how to take these factors into consideration, rather than code each case separately.

The resulting algorithm needed to utilize LTLs more often. In particular, patterns from LTLs would only be considered after the lay size reached a certain threshold. This approach solved the critical problem of efficiently planning fuller lays throughout the schedule. A second critical problem was solved by simply allowing the user to assign various priority levels to each truck (they would all default to the same priority). The "Select Next Truck" packet would then have to take this factor into account when calculating its priority function.

As the KBMS application is inevitably refined, changes to the algorithm are easily made by modifying the packet's strategic agenda. Because developers do not get bogged down with procedural code, prototypes can be built, and subsequently modified, quickly. The result is a production-quality application that can continue to be refined in a manageable way.

Additional Benefits

The implementation of the system has provided more status updating than anticipated, some on DB2 and some on IMS, with some important nuances. By sequencing and building lays, the new system minimizes the production time between the first item and the last item loaded onto a truck. Plants begin shipping orders sooner. There are fewer finished goods on the shop floor awaiting companion pieces. Product is arriving to dealers days faster, allowing La-Z-Boy to invoice

```
DETAILED RULE Screen          Edit Mode          Application: LA-Z-BOY

THIS RULE CAN BE EDITED, OR A NEW ONE CREATED VIA CREATE
MODE
Rule Name: FILL TRUCK          in packet:  SELECT NEXT TRUCK (t)

Rule:  IF T. TRUCK ID = PATTERNS. TRUCK ID &
          PATTERNS. IN LAY = 0
        THEN FILLING CURRENT PATTERN - YES;
          CURRENT PATTERN = PATTERNS.PATTERN ID;
          RUN




Author: OS2                          Date last modified: 19900609
1:Help/Expand    2:Synonyms    3:Return    4:Editor Menu    5:Lookup
6:Create Like                 8:First      9:Delete         10:Parent
```

Figure 12-1   This is an example of a rule in the "select next truck" packet.
It states that for each pattern in the selected truck (i.e. "T")
that has not been assigned a lay number, KBMS needs to
choose the pattern to be filled and asigned. It is then "RUN,"
which asks KBMS to execute its sttrategic agenda and return
to this rule. The developer need not code explicit loops or be
concerned as to where data is stored. For instance, "TRUCK"
could reside in DB2, while "PATTERN" is in VSAM and the
resulting assignments handled by KDB (KBMS's data storage
facility).

sooner and receive payment sooner—a significant gain for such a
large company.

For the first time, the system gives all La-Z-Boy users advance
knowledge of what goes into a lay. Furthermore, it significantly
reduces administrative labor. Early calculations by La-Z-Boy estimate that scheduling which previously required up to 50
"man-days" of administrative labor will be dramatically reduced — to
less than one "man-day." Also, because the screens now show real-
time data, people referring to them see a much more accurate
picture of reality.

In the field, the system is proving equally dramatic. For instance,
previously if an order did not arrive when promised, it was difficult
to find out why. Users could not determine why it was delayed, or
when it was likely to arrive. Now they can see both on an order/in-
quiry screen. This alleviates such major problems as out-of-stock
situations (La-Z-Boy's biggest concern, usually caused by unan-
ticipated demand for particular fabrics, which come from many
different mills).

DETAILED PACKET Screen   Edit Mode   Application: LA-Z-BOY

THIS PACKET CAN BE EDITED, OR A NEW ONE CREATED VIA CREATE MODE

Packet: FILL LAYS (p:PATTERNS, t:TRUCKS)   Agenda Type: STANDARD

Agenda Condition: FILLING CURRENT PATTERN = YES &
         P.PATTERN ID = CURRENT PATTERN & P.IN LAY = 0 &
         P.TRUCK ID = T.TRUCK ID

Goal Requirement:
Success Description:
Rule Prefix:
Exit Condition:
Entry Action: PRINT 'FINDING LAY FOR PATTERN', P.PATTERN ID, '
       & TRUCK #', P.TRUCK ID;
Exit Action:
1:Help/Expand   2:Synonyms   3:Return   4:Editor Menu   5:Lookup
6:Create Like   8:First   9:Delete
11:Rules

Figure 12-2   This is an example of the "fill lays" packet. KBMS will invoke the packet when filling the current pattern in order to find a match that has not been assigned a lay number, as well as identify the corresponding truck for that pattern. The priority function will choose the pattern designated for the truck that is most full.

As a result of the scheduling system now tied into KBMS, La-Z-Boy sales reps, whose biggest complaint was not being able to find out why an order was delayed and when it would arrive, can get answers and take appropriate action. Now, tickets get returned if they cannot get produced, because scheduling and production are so much more closely linked. The system, however, has significantly eliminated the probability of a ticket being returned because it cannot be manufactured.

Another way the system benefits the field is through its greater capacity to distinguish between different types of orders. The sales cycle in the residential furniture business is similar in many ways to that of the auto industry. Customers come into La-Z-Boy authorized furniture stores and either buy something off the floor (dealer stock) or order a customized variation of what they see (customer-sold).

As much as possible, La-Z-Boy has always given preference to customer-sold items. Now, KBMS has tie-breaking rules in the system to indicate priority. As competitive as the retail furniture business is, this is a crucial improvement.

Each customer who is buying a sofa or a chair has a different amount of time he or she is willing to wait for delivery. The combination of time saved and knowledge gained unquestionably means La-Z-Boy closes more deals.

After completing the prototype, decision, tooling and internal resource planning phases, La-Z-Boy was well-satisfied that KBMS would streamline the production scheduling process as anticipated. For AICorp, it was a tremendous challenge to take on such a complex problem, and rewarding to accomplish what was expected of KBMS and INTELLECT.

The most exciting thing about implementing KBMS for La-Z-Boy manufacturing and shipping applications is that it has such a dramatic impact on the way the company does business. As fine-tuning takes place, La-Z-Boy managers can determine optimal chunking of tickets according to geography, time, fabrics sequencing, and the mix of customer-sold and dealer stock orders. They are isolating why one schedule is better than another, and making the changes that will maintain La-Z-Boy's leadership in the furniture industry.

Author Biographical Data

Elizabeth M. Cholawsky, Ph.D., is Customer Support Manager at AICorp, Waltham, MA. Prior to joining the company as a knowledge engineer in 1988, she was an expert systems consultant with Applied Expert Systems, Cambridge, MA. Previously, Cholawsky had been a quantitative methodologist with the Central Intelligence Agency. She earned her Ph.D. at the University of Minnesota.

Don N. Harris is Principal Knowledge Engineer, AICorp, Waltham, MA. He joined the company in 1978 as a senior programmer. During his tenure he has had responsibility for a range of challenging assignments, including the role of knowledge engineer for the KBMS application at the La-Z-Boy Chair Company. He received a bachelor's of science degree from the University of Arizona, with a major in mathematics and a minor in computer science.

13

Forecasting Demand for Manufacturing and Deployment

An Expert System Enhanced Approach

Michael H. Manzano and Steven C. Rubinow
The Quaker Oats Company
Chicago, Illinois

Introduction

In recent years, the number of promotions, trade deals and other marketing events offered by manufacturers to their customers has increased dramatically, particularly in consumer goods companies. Because of the increase in promotions, companies are often plagued by a variety of operational problems that can result in large product costs and/or lost sales. Some of the manufacturing and distribution costs associated with large forecasting errors are as follows:

- high safety stock levels

- premium freight shipment costs

- abnormal intra-company movement

- inventory spoilage and obsolescence

- production overtime and idle times

- package and ingredient waste

- unplanned production line changeovers

- copacking penalties.

The Quaker Oats Company is a leading multi-billion dollar international food manufacturer of items in such categories as cereals, pancake mixes and syrups, snacks, frozen pizza, pet foods, sports beverages and others. Included in this array of products are thousands of individual UPCs (universal product codes) that can be manufactured in a number of facilities across the country, stored in a corresponding number of distribution centers and shipped to thousands of outlets where groceries are sold. The size and complexity of managing such a process effectively is readily apparent. Innovative companies are solving these problems by implementing logistics planning systems that integrate forecasting demand, with distribution (Distribution Requirements Planning-DRP systems) and production planning (Material Requirements Planning-MRP systems). Key to the success of an integrated logistics/manufacturing system is the ability to accurately forecast low level detailed demand (market, product, week). Such systems are essential to answer the questions of what to make, how much to make, where to make it and where to put it.   In this era of attempting to shorten production cycles toward a just-in-time perspective and attempting to be the low cost supplier in an industry without sacrificing quality, the ability to forecast a variety of differing scenarios for a spectrum of purposes is essential.

In addition, the function of such a system, regardless of the accuracy, provides a facility to furnish a common medium for communications among the appropriate parties in the marketing, sales, manufacturing, logistics and finance departments of the organization.   This allows a common format on which to construct weekly, quarterly and annual plans that are understood and accepted by all the groups that use these forecasts as the basis for subsequent efforts.

Like most companies, people in the departments referred to had already been forecasting for their needs even before the advent of a common and more refined forecasting system. This resulted in many flavors of systems being utilized; some sophisticated, some simple and many in between these two extremes. When the results of these methods were conveyed to other departments, in many cases, the forecasts were adjusted upwards or downwards based on the recipients' knowledge of the method used by the sender and its historical accuracy. Some of the forecasts tended to be overly optimistic by marketing and sales and were ratcheted downwards by manufacturing.

To assist in providing support for the very detailed requirements of manufacturing and logistics, several years ago, a model was developed to support these departments' needs alone. Although it increased the accuracy of the forecasts, the model had limitations in that it was hard to maintain, needed to be run by the technically adept maintainers of the system, and did not incorporate other necessary features in the system. In an attempt to improve on this model and provide a tool that was more accurate and could be used by anyone in the corporation who depended on forecasts to shape their actions, an effort was undertaken to build a comprehensive integrated forecasting system for use by the entire company, and one on which improvements using a variety of methods could be employed without retooling the entire system.

Model base development

There are several commercially available forecasting systems in the marketplace (both personal computer and mainframe based) that provide a library of forecasting techniques and facilitate this process with the use of expert systems to aid in tuning the forecasting method chosen and the forecast itself. After examining these products, it was decided that none of these would provide the foundation for the system that was needed and that would have the flexibility to meet both current and projected needs. It was therefore established that a system that was developed wholly internally and was adjusted to the company's particular set of criteria, was the only reasonable answer to meet the established requirements.

The first step in building this system is the construction of a model base that contains the numerous forecasting models that provide the analysis capabilities for a forecasting system. In this system, every one of the scores of brands has its own model in this model base which is then used to forecast the products that comprise the brand. The methodology used in the system employs multiple time series forecasting techniques, each of which projects a particular component of the forecast. The forecast components include what were judged to be the major variables affecting product demand, namely: seasonality, baseline, trend, trade deal and price change effects. In addition, it should be pointed out that these variables also were the only ones that could be supported by readily available databases that were timely and accurate. It was recognized that other factors are important to estimate demand. However, many of these factors were not available in a format that would provide routine input to the system and were thus relegated to a position of being used to manually fine tune forecasts based on the knowledge of these factors by the person using the system to achieve better forecasts. This decomposition methodology can be shown to significantly enhance low level detailed estimation. Presented below are the five distinct phases used in the decomposition methodology.

Deseasonalization of the time series

As a measure of seasonal variation in a time series, seasonal coefficients are derived using one of the following methods: modified census II decomposition, classical decomposition, ordinary least squares, and addressing additive, multiplicative and polynomial trend components. These coefficients are used to deflate the original time series and produce a deseasonalized series from which baseline and trend estimation can be performed.

Baseline and trend estimation

To measure deseasonalized baseline volume over the forecast horizon, the estimation procedures must express a representation of history as well as recent developments in the time series. Therefore, exponential smoothing techniques were selected and are used to create baseline and trend projections. Specific methods are selected as a function of baseline and trend time series volatility and were chosen from the following: Brown's one parameter linear exponential smoothing, Holt's two parameter linear exponential smoothing, Brown's one parameter quadratic exponential smoothing, Winter's three parameter linear and seasonal procedure, and Harrison's harmonic smoothing.

Optimal filtering procedures

In many time series forecasting methods, a tradeoff must be made between smoothing randomness and reacting quickly to changes in the basic pattern. Filtering procedures help exponential smoothing techniques identify when permanent deviations to the time series have occurred. The filtering techniques presently used in the decomposition forecasting system include: adaptive response mechanisms, optimum linear filters, including Kalman's, Wiener-Kalman, and Kalman-Bucy filtering. These filters are a class of linear, minimum error variance, sequential state, estimation algorithms.

Analysis of factor effects

Using stepwise regression, two exogenous factors were identified which significantly affect shipment volatility (price changes and the presence of trade deals). This module is used to identify period boundaries for the regressor variables of price change and trade deals. Assumed period boundaries include pre-deal, deal, post-deal, pre-price and post-price periods. The following analysis of variance procedures are used in the period boundary investigation: multiple comparison procedures, including Tukey, Bonferroni and Scheffe. The procedure selection is a function of factor level, sample size equality and estimated contrasts, pairwise or otherwise.

Multiple regression

Regression analysis is used to identify ordinary least squares parameter estimates for price changes, trade deals and trade deal discount rates. These regression parameters are used to inflate baseline volume estimates over the forecast horizon. The following regression techniques used in the decomposition system are available in the Statistical Analysis System (SAS) from SAS Institute, Cary, North Carolina: stepwise regression, square regression, REG procedure and GLM procedure.

The combination of all these methods has produced the basis for a system that, although it can be enhanced by methods described below, appears to be statistically complete for systems of this type. A major factor that can effect huge swings in the forecast is a sudden change in the buying behavior of customers because of timing or conditional changes, primarily related to promotional activity. These last minute changes can be input to the system through facile communications with the sales force to correct forecasted sales. In addition, a forecasting feedback system is also part of the system. In order to improve the forecasting performance of the individual providing the forecast, comparisons of actual versus predicted performance are readily available. This feedback system allows not only the assessment of overall performance but also attempts to present the various elements of the forecast to allow an appreciation of where the major sources of variance are. This knowledge should encourage the improvement of forecasts when the reasons for major variances are understood and these variances tempered the next time around.

The role of expert systems

Given a sound forecasting methodology from a statistical point of view, improvements to the results of the models are best driven through the use of expert system methodologies. Additionally, these same techniques prove invaluable in the maintenance of this system and in what has been referred to as "automated modelling". Several specific areas within the system have been targeted for enhancement by the use of these techniques and are indicated below. It must be noted that for the most part, these systems are intended to be used by expert or near expert users as opposed to novice users. The possible brittleness of some aspects of these models along with exogenous factors not incorporated in the system, currently requires that a somewhat knowledgeable individual be in control of the system.

Statistical procedure selection for forecast generation

In the section on model base development above, a variety of statistical methods are intentionally enumerated to indicate the choices that must be made in model construction. Anyone who has spent time in the area of mathematical model development quickly realizes that in addition to a good deal of quantitative knowledge, there is a certain artistic aspect to achieve the final desired outcome. This ability usually comes through experience, but opinions on the best course of action may differ significantly among individuals versed in the field. Based on the mathematical procedures accepted within the group developing the system, a knowledge base is constructed that assists the modeler in deciding the best path to take when selecting statistical routines. In the future, expert statistical systems may be constructed that will automate a significant portion of the model construction, but for the present, the modeler does the majority of the work.

Diagnostic assessments for model correction

After the model is in use, it needs to be re-examined periodically to see if it is still performing at optimum levels and if any recalibration or redevelopment needs to be done. A variety of statistically based diagnostic measures, as well as empirical observation, are used to determine this. Much like the process described in the section above, knowledge based assistance helps guide the assessment process.

Expert help for use of system

In a variety of cases involving the use of complex tools to generate improved business decisions, it becomes clear that some assistance is needed for those not sophisticated in the underlying techniques. The philosophy in the company has been, wherever possible, to build, and place in the hands of decision makers, robust, easy-to-use tools that facilitate the decision making process. This is in opposition to having a select group of individuals (forecasters) generate a final outcome to distribute to the appropriate individuals. Some basic, but often essential, guidance derived from a knowledge base is necessary for the optimal use and understanding of the system. Even if after a period of time, the end-user becomes sufficiently familiar with the workings of the system, employee turnover or frequent shifting of responsibilities cause the learning curve to begin anew.

Expert corrections to model for subjective and other factors

It is always recognized that no computer based model can easily incorporate all that is needed to make a sound business decision. On that basis, the output of the model is often adjusted by the end-users' knowledge of the current business environment and its potential impact on the generated forecast. The basic tenets for making these changes can be incorporated in the system in order to provide the experienced end-user with more rapid determinations of likely outcomes while presenting the less experienced user with some modicum of insight into the situation hitherto unrealized by this individual.

The formalism for representing knowledge in this system is best represented by simple IF-THEN rules that capture the basic judgments that are necessary to improve the system. It should be remembered that these systems are very data intensive and from a traditional information systems point of view, this management of the "data processing" elements represents one of the largest challenges for the on-going operation of this system. Because of this data intensity, a forward chaining paradigm is most useful to drive the inferential portions of the system. If the system were to develop more complex knowledge bases, a conscious decision to mix forward and backward chaining methods or even bi-directional search strategies would be warranted to increase computational efficiencies. In a system of this total size and complexity, but with a relatively simple knowledge base, the operational design goals have been met, to date, in most aspects of the system.

No commercial expert system shell or language is used to drive the knowledge based portions of the system, to again provide simplicity of maintenance and achieve operational effectiveness. The software product Express (available from Information Resources Inc., Chicago, Illinois) serves to provide much of the data base management, data processing, statistical routines and rudimentary expert system functions required by the system. The user interface of the system is graphical in nature and is personal computer based (developed in the personal computer version of Express) with seamless communications to the central mainframe computing facility where data, models and knowledge are stored in the Express environment. All maintenance to any component of the system is centrally administered and performed by experts in the particular area in question and is not routinely open to modification by end-users. Actual computational activities may take place either on the mainframe computer or the personal computer depending on the nature and requirements of the task at hand.

Conclusion

Currently the forecasting system is used by staff members in the marketing, manufacturing, logistics, finance and sales departments to varying degrees. Some staff use it as the primary development tool for their forecasts for the rest of the company, some use it to corroborate the results they have obtained from other methods that have served them well in the past. And like many systems that are designed to serve a large number of people in diverse areas, there are some others that ignore the model results and depend entirely on their own methodology. In any case, even those individuals in the latter category submit the results of their analyses into the system where it assumes the important role of forecasting repository, if no other purpose.

Results to date with the use of the system have been extremely positive, with forecasting error reduced an average of fifty percent using the results as they currently emerge from the system even before manual adjustments are made, if necessary, by the individual user. The impact of such reductions in forecasting error, if maintained or hopefully improved upon, can result in cost savings over current practices ranging in the millions of dollars per year.

Future developments for the system include extending, with reasonable accuracy, both the depth and breadth of the system's scope. While this system has specifically focussed on short-term volume forecasting, enhancements will address forecasting for longer time periods (at least 12 months) and also to be used in predicting financial revenues and allocating advertising and merchandising expenditures. Eventually, other aspects of production planning will be integrated with the system, automating as much as possible of the planning and subsequent plan execution process.

Additionally, the refinement of the methods by which the forecasts are generated is a principal objective. The inclusion of competitive effects, other promotional vehicles, retail movement, and retail inventory levels, among other factors, will allow the construction of a model that more broadly spans the entire manufacturer-to-market pipeline and is consequently more robust. Also, the size of the constituent knowledge bases will grow to further solidify the total methodology.

The true power of such systems is demonstrated when significant changes and improvements in business practices are facilitated and established as a result of their regular, prudent use. The system described herein falls into this enviable category.

Author Biographical Data

Michael H. Manzano is a Senior Decision Support Systems Analyst at the Quaker Oats Company. His principal responsibility is the development and implementation of robust models to support decision making in variety of functional areas in the company. He received a B.S. in Economics and Mathematics from California State University in 1982 and a Ph.D. in Economics from the University of Notre Dame in 1987. Michael has also authored several other publications in related areas of modeling.

Steven C. Rubinow is Director, Decision Support Services at the Quaker Oats Company where he is responsible for determining and meeting the decision support needs of most business functions throughout the company. This is accomplished through the creative use of data, models, knowledge and technology to address business issues.

Prior to joining Quaker Oats, he held positions in marketing research, portfolio management, and genetic and molecular engineering in a food service company, financial institution and food ingredient manufacturer, respectively.

He received the B.S., M.S. and Ph.D. degrees in chemistry and a M.B.A. from the University of Illinois as well as a M.S. in computer science from DePaul University where he is an adjunct faculty member in the computer science department.

14

CHRONOS:
Facilities Management

Ben Hitt, Ph. D., and Engene Mauro
Rawson Technologies, Inc.
Wellsburg, West Virginia

Introduction

The specific problem space is Martin Marietta, Aero & Naval Systems Facilities Department. Manufacturing, laboratory and design activities occur throughout a large campus. Vice Presidents, Directors and Project Leaders require modifications to the facility to accommodate new contracts. Master planners anticipate facility requirements and chart multiple contingencies. These issues need to be addressed:

How is the job done today? How can facilities management be simplified or done better? How it can be "computerized?" What design is best? How is the new system to be implemented? How are participants involved in the new system trained?

How is the job done today?

You can imagine the level of details involved. For this chapter let us take a high level perspective. My original bias, during the analysis phase, was: if effective planning was performed, then after a decision process, scheduled implementation would proceed. If during implementation, modifications were required to the plan, new plans would be released, and implementation would continue. We witnessed that implementation began before the "planners" released a plan, construction efforts occurred before a design had begun. Interestingly, some of the smoothest, on-time, on-budget activities occurred this way. It was time to dump our bias. How is the job done today? The one word answer -concurrently. While the "planners" were planning, other individuals had planning roles. These planning activities needed to be identified and included within the planning function. Essentially, some preparations for a large scale activity could be implemented on a "hand waive." As the formal plan was completed, design activities would near completion, material would be advance ordered and kited, then large scale construction activity was scheduled.

If the plan was gated to a contract "win," sometimes the contract would require start-up within weeks, requiring that facility modification be underway months earlier. If the contract was lost, contingency plans would optimally have provided for an alternate use for the already modified area of the facility, otherwise all the efforts would be a loss.

During times of plentiful large, long term contracts, planners could see years ahead clearly. As scrambling was required during budget cuts and shorter contracts, "fire drills" were the order of the day. Some processes were centralized, others were local. As the local need exceeded the centralized plan, furious requests were generated. Facilities Management was administered centrally, with local liaisons assigned to key projects. Frequently the local concerns competed for limited resources, requiring a decision. Simply stated, a request was made locally for a facilities modification (ie. Facilities Request). Followed by a Disposition process including a discussion of the request and its impact on other activities, leading to a accepted or rejected decision. If the request was accepted, it was prioritized, then placed on the design engineers schedule, followed by the construction schedule.

Four levels of engineering began, frequently concurrently:

- Layout

- Structural

- Mechanical

■ Electrical

Material lists were generated by each area, reviewed, then ordered in preparation for construction efforts following a related but separate scheduling mechanism.

Object Oriented Analysis

There are many intricacies to a particular analytical method. Within the problem space discussed we will relate two layers of five total layers.

Subjects:	Services:
Project Planning	Document
Facilities Requests	Request Summary
Disposition of Requests	Disposition Summary
Engineering Schedule	Engineering Schedule & Status
Material Requirement Planning	Material Schedule & Status
Construction Schedule	Construction Schedule & Status

Essentially the subjects aid our understanding by permitting compartmentalization. Services describes processes or outputs of the system.

All the subjects are interrelated, yet a system (electronic or paper & pencil) could be designed separately for each subject. If this analysis is correct, implementation and training issues will be easier to manage.

Services should inform upper and mid management of what they will get from this system (ie. What does the system do?). Details, aside, if a critical feature or requirement isn't included within these services, now is the time to know. A sanity check can be performed by explaining the subjects and services to management, then permitting management to make a wish list (list of requirements). The aim is to have each requirement contained within or among the services. Cost Analysis is apparently absent from the subjects and services. In this problem space the objects of cost is embedded in each subject and a cost analysis summary is a service contained within each service listed. To see these intricacies, management would need to look at the object, structure and attribute layers which coexist with the subject and service layers.

How can facilities management be simplified or done better?

Management may feel comfortable with a particular style for functional flow and system diagrams. We felt it was our job to convert our diagrams using an OOA (Object Oriented Analysis) methods to a form that each manager has comfort with. The mission was to ask management to identify bottlenecks, unnecessary approval steps, cumbersome processes and other impediments. Later We interviewed shop floor, construction, planning and engineering persons.

Within each subject, flow problems were identified and simplifications were proposed, leading to a proposed new system. Management was unwilling to simplify the existing process first, then implement a computerized system. Their preference was to design the new, improved ideas into the computerized system, then make one implementation for each subject.

How is the new system to be implemented?

Compartmentalizing the problem space into subjects permits concurrent engineering efforts in the system's design and implementation. It can also permit an evolutionary approach where basic structural elements are installed, then refinements are added later with involvement from users. Lets recognize that there are pros & cons to a complete implementation plan vs. an evolutionary approach. Either direction requires a complete answer to the implementation question.

Politically, a manager with territory in one subject area may prefer to wait and let another manager take the burden of being the first affected area (ie. to be infected with computers). While this politicizing doesn't lead to the best implementation strategy, cooperation during implementation is worth more than a better paper plan.

How can it be "computerized?"

Some companies have a global need to be paper-less. Such a mandate makes decisions regarding: should this be paper or computer?, moot. That was the case in this problem space. The goal was to computerize the entire process, from electronic time cards and purchase orders to automated scheduling. Regardless of the medium for information, some processes were beyond this authors AI talents. We identified our limitations and those of the available staff and proceeded to use AI in tried and true subjects such as scheduling with resource constraints and material control schemes. Planners were delivered an AI forecasting model that permitted detailed scenario descriptions or impact statements for each coarse of action, but the computers stayed out of the decision making business. Management believed that the systems' services should not be decisions but tracking and communication. A few years from now they may change their mind, and the system needed to be designed to adjust to these requirements, also.

What design is best?

Design issues are manifold. Objected Oriented Analysis doesn't have to lead to Object Oriented Design. In this case it didn't. Relational Database Management was chosen leading to various structured methods incorporating object oriented design elements in user interfaces and AI modules only. The bottom line is that the system works, its modular, its expandable, its fixable and many application design engineers can understand it. The intention was to avoid the cutting edge in design techniques so engineers could readily be found to continue working with it.

Rules of Thumb regarding requirements

If managers required one year of records maintained on-line, ten years became the requirement.

If 100 items were required per job, 10,000 items was the mark.

If 1000 items was to be the capacity, 1,000,000 was tested. Sure the system was over designed, but the costs involved were negligible when compared to reverse engineering efforts.

How are participants involved in the new system trained?

Some of the participants were involved in the analysis phase and had the opportunity to participate in the prototyping. These individuals became the hubs. They had ownership in the system and were eager to educate others.

Other participants wanted little involvement. As management increased the pressure to utilize the system and they ignored the request, transfers and layoffs resulted. Essentially, management's commitment to the system was strong enough to provide motivation to change the way business occurred.

Opposition was anticipated on the shop floor, yet some of the strongest support of the system was among these users. The system was easy to learn, not as cumbersome as the paperwork, and measured their performance fairly. Statistics from the systems were used by Performance Measurement Teams to validate their suggestions and decrease the implementation of innovations. In short, the miss-communications under the old system were significantly reduced with the new system, permitting employee productivity to be seen clearly.

Clear vision revealed management flaws and significant losses. Management's reaction was to own up to the failings and provide remedies. Employees appreciated management's honesty and efforts. Is the system to be credited with this wonderful picture, We think so.

Conclusion

The system became a tool that fit the users hand. As talk spread, other divisions became interested and migration plans were set. Now other companies have expressed interest. With easy customizing, deliveries are progressing.

CHRONOS: A Real Time Expert System Tool

The need for process control via expert systems is apparent even at the most basic levels of manufacturing. Small as well as large producers have recognized that intelligent computer guided manufacturing is necessary for them to compete on a global basis.

The problem is that many if not most manufacturing processes require human expertise for their successful completion. Various unpredictable and uncontrollable factors such as quality of raw material, weather and others, prevent utilization of strict algorithmic control. Human operators rely on their training and experience to adjust process parameters so as to produce a constant product. Even with experienced personal, the reactor times may not be fast enough to prevent major losses. Also while alarms and automatic shut down devices prevent catastrophic losses, the downtime to diagnosis and repair is the source of large losses in revenue due to lack of production. An additional an sometimes unretrievable loss is the loss of expertise when a long time operator either retires, moves or is otherwise lost to the manufacturer. The company must pay for training and low production efficiency while a new operator is being trained. Many other firms in areas other than manufacturing have turned to so called Expert Systems to save an retrieve employee knowledge and experience. These systems have proven valuable in areas such as insurance and mortgage underwriting, tax consulting and welfare management.

However, even today, only a few expert system products address real time processes in such a way to fill the need. Expert System shells are attractive tools for software and knowledge engineering. Applications developed with these tools have been used with great success in many venues. By providing the inference engine, they leave the developer free to concentrate on the constraints and relations in the problem, rather than being concerned with mechanistic concerns. The exclusion of expert system tools from the manufacturing and process control arenas is a result of the inability of most shells to include temporal reasoning and time-dated facts in their inference algorithms.

Also few shells are able to accept input directly from the environment. Rather, they rely on human input in response to questions presented on the terminal. Thus expert system process control has until now consisted largely of custom software developed in native programing languages such as C, Lisp, or Prolog.

Chronos is an expert system development tool that was designed to incorporate the characteristics of temporal reasoning and manage continuous acquisition of data. Each fact in the fact base has a set of temporal attributes:

1. Creation date or the point of entry into the base.

2. Starting date or the time at which a fact becomes valid.

3. Ending date or the time at which a fact becomes invalid.

4. Obsolescence Time or the amount of time a fact remains in the fact base before it is removed as garbage.

These attributes allow the expression of rules based on past, present or future facts using specific predicates such as " As Long As (fact) then", As Soon As (fact) Then" and "CLOCK". Chronos offers, by virtue of a procedural language, first order logic engine, the above predicates, and multi-tasking design both the ability to control real time processes and simulation of process over a range of virtual time modes. Chronos also has the advantage of being portable across a variety of operating platforms including Unix, MS-DOS and VAX. It is Microsoft Window 3.0 and X Windows compliant.

Chronos is new to the American Market but has found significant application in Europe. The remainder of this chapter is devoted to those applications.

The international engineering firm, Fives Cail Babcock (FCB), used Chronos in a process control function for a sugar refinery. FCB's client ran two refining ovens independently off a single steam generation plant. They faced problems with inconsistent starting material quality and quantity. The conditions of the charge were rarely the same across the two ovens. consequently, the efficiency of the steam generator facility was severely compromised to damp out the inconsistent and unpredictable requirements of the two refinement lines. Chronos monitors and controls independent steam heated refinery ovens. The application selects the cooking cycle start times and adjust pressures to assure sugar magma quality and reduce energy costs for steam generation. Chronos provided the advantage of being fully linked to the process and allowed for time dependant management of priorities.

Five Cail Babcock developed their application with only thirty first order rules. Depending on plant status, the fact base averaged one hundred facts. The FCB application examined thirty analog and digital acquisitions per minute and cycled through the process in ten seconds. FCB's client achieved substantially savings in energy costs by being able to move closely match the output of the steam generator to the oven requirements. Closer real time control result in a more constant quality of refined sugar and improved product.

The French Ministry of Defense Research Center (equivalent to DARPA in the USA) used Chronos in its satellite imaging and tracking process. Both applications are on SUN workstations. image processing was monitored to determine the exact position of a reference landmark in a satellite image. Images from prior passages of the satellite over the area must correlate to new pictures. However, a number of factors prevent the camera from ever photographing the same tract of land twice. The obvious solution is overlay photos so that landmarks in two or more pictures coincide. This is a time consuming and error prove process when done manually.

Satellite positions at a given time can be predicted accurately and expected image frames can likewise be predicted. An analysis of an image correlation surface must therefore conform reasonably to expected values based on predicted satellite positions. Thus definition algorithms are time dependant. Likewise image fusion is time related and used to confirm and update data.

The image processing monitor was developed in three months. The system examines sixty seven rules, including nine for management of temporal constraints against a fact base which averages 500 facts. The system has proven one hundred percent reliable within a thirty second time frame when the landmark is in the image.

Satellite position was predicted and modeled with Chronos via a Kalmar type filter. The use of the Chronos application improved accuracy and reliability in an environment where the filter operation is disturbed. A Chronos simulator runs on a SUN Workstation linked to an expert system on a PC.

The system virtually eliminated measurement deviations by diagnosing the different types of disturbances and taking appropriate action of filter parameters to connect for the disturbance. The system checks sixty four rules against an average of four hundred facts in a matter of seconds. The system required three months to develop and resulted in at least a three fold increase in overall performance.

In a further project the agency used Chronos to simulate the operation of a vehicle and its possible faults. The simulation was run on one PC linked Chronos running on yet another. Chronos was configured so that it took vehicle faults into account, alerted the driver and proposed alternate solutions. The system involved some two hundred commands and one hundred faults. The system has a response time of less than a second and required' four months to develop.

Vernolab is an international laboratory which carries out predictive analyses on motor oils. The analyses are automated and vary on the type of analysis required and results already obtained from upstream analytical stations. Chronos handles real time tracking of samples and sends routing commands to the sample conveyor and analytical stations. Chronos takes all constraints into account including station workload, sequencing, possible faults, etc. The system examines 30 first order rules against several thousand facts and cycles at one acquisition per second.

Telesystems is a large software engineering company. The firm is the principal telecommunication software firm in France. An expert system developed with Chronos supervises the computer rooms. It monitors the air conditioning, humidity, power supply and people access. By determining the cause of an alarm, Chronos effectively reduces environmental downtime.

Likewise Chronos monitors system performance and has achieved substantial reduction in system downtime by identifying the source of an alarm. The entire monitoring system runs on a PC connected via a local area networks. The response in both cases is less than one second.

Ciments Lafarge is the worlds second largest producer of cement. Mechanical and thermal constraints in a cement kiln can be very high and tight. Operation outside these constraints very quickly leads to expensive damage. The problem is that parameters can be measured only at strategic positions. No global model exists to compute the parameters where no sensors are located inside the kiln for example. However, experience with the process is extensive and a rule based system base on temporal facts has resulted in a more critical control of the kiln and the process. Chronos evaluates several models against the fact base. The models can be very simple or complex. The system performs 30 digital and analog data acquisitions every ten minutes and required six months to develop. The results

In all the cases examined successful completion of a process was dependant to a great extent on human operation with the expertise and experience to judge the status of the process. Prior to the Chronos application, the control of the processes was subject wide variation. Operators become sick and try to function with impaired health and subsequently impaired reasoning and response time. Furthermore, it is noted that different operators often control the same process in widely divergent ways and that experience among operators is at best inconsistent. As a consequence, variations in process productivity in terms of quality and quantity are the rule rather than the exception. A real time expert system offers a solution to the problem. Chronos eases application of such systems since it provides first order logic engine which tests rules against time dated facts.

Using this tool knowledge engineers can capture the expertise of skilled operators and construct systems to accurately control real time processes in reasonable periods of time.

Author Biographical Data

Presently, Eugene Mauro is Vice President of Rawson Technologies, Incorporated and project leader for D.O.D., N.A.S.A. and private industry contracts. Also, Mauro is President of the West Virginia High Technology Consortium. He authored the Facilities Management Information System and the Maintenance Management System applications, available on numerous platforms.

Previously, Mauro was a consultant to Marietta Corporation, PHH Corporation and Jiffy Lube International. He is a Former Chief Financial Officer and Applications Programming Leader for ten years at National Communications in Baltimore. He is a graduate of the Johns Hopkins University with BA degrees in Biophysics, physics and biology.

Ben Hitt, Ph.D. is presently Chief Operations Officer at Rawson Technologies, Incorporated. He is author of Troubleshooter and other natural language expert system and AI tools. He is former Executive Secretary for Merit Review in the Veterans Administrations Central office, Washington, D.C. Previously, Hitt was a research scientist and a member of staff faculty at Stanford University.

15

Expert Simulation for On-Line Scheduling

Sanjay Jain and Karon Barber
General Motors Corporation
Warren, Michigan

David Osterfeld
Electronic Data Systems Corporation
Warren, Michigan

Introduction

In recent years, the automotive industry has realized the importance of speed of new products to market and has mounted efforts for improving it. The Expert System Scheduler (ESS) facilitates these efforts by enabling manufacturing plants to generate viable schedules under increasing constraints and demands for flexibility. The scheduler takes advantage of the Computer Integrated Manufacturing (CIM) environment by utilizing the real-time information from the factory floor for responsive scheduling.

The Expert System Scheduler uses heuristics developed by an experienced factory scheduler. It uses simulation concepts and these heuristics to generate schedules. Forward and "backward" deterministic simulation are used at different stages of the schedule generation process.

The system is used to control parts flow on the factory floor at one automated facility. This highly automated facility is a testbed for

Reprinted with permission by the Association for Computing Machinery. "Communications of the ACM," October 1990: ©1990.

implementation of CIM concepts. The scheduler runs on a Texas Instruments (TI) Explorer II computer using software developed in-house utilizing IntelliCorp's Knowledge Engineering Environment (KEE) shell and the LISP language. The scheduling computer is networked to the factory control computer, which actually controls the plant floor. The TI Explorer II acquires current plant floor information from the factory control system, generates a new schedule and sends it back within a short time. The configuration allows fast response to changes in requirements and plant floor conditions.

Expert Simulation for On-Line Scheduling

The state-of-the-art in manufacturing has moved towards flexibility, automation and integration. The efforts spent on bringing computer integrated manufacturing to plant floors have been motivated by the overall thrust to increase the speed of new products to market. One of the links in CIM is plant floor scheduling, which is concerned with efficiently orchestrating the plant floor to meet the customer demand and responding quickly to changes on the plant floor and changes in customer demand. The Expert System Scheduler has been developed to address this link in CIM. The scheduler utilizes real time plant information to generate plant floor schedules which honor the factory resource constraints while taking advantage of the flexibility of its components.

The scheduler uses heuristics developed by an experienced human factory scheduler for most of the decisions involved in scheduling. The expertise of the human scheduler has been built into the computerized version using the expert system approach of the discipline of artificial intelligence (AI). Deterministic simulation concepts have been used to develop the schedule and determine the decision points. As such, simulation modeling and AI techniques share many concepts, and the two disciplines can be used synergistically. Examples of some common concepts are: the ability of entities to carry attributes and change dynamically (simulation + entities/attributes or transaction/parameters versus AI + frames/slots), the ability to control the flow of entities through a model of the system (simulation + conditional probabilities versus AI + production rules), and the ability to change the model based upon state variables (simulation + language constructs based on variables versus AI + pattern invoked programs). Shannon[1] highlights similarities and differences between conventional simulation and an AI approach. Kusiak and Chen[2] report increasing use of simulation in development of expert systems.

ESS uses the synergy between AI techniques and simulation modeling to generate schedules for plant floors. Advanced concepts from each of the two areas are used in this endeavor. The expert system has been developed using frames and object oriented coding which provides knowledge representation flexibility. The concept of "backward" simulation, similar to the AI concept of backward chaining, is used to construct the events in the schedule. Some portions of the schedule are constructed using forward or conventional simulation.

The implementation of expert systems and simulation concepts is intertwined in ESS. However, the application of the concepts from these two areas will be treated separately for ease of presentation. The following section discusses the expert system approach and provides a flavor of the heuristics. The concept of "backward" simulation and the motive behind it are discussed in the third section. The fourth section provides some details of the implementation and the plant floor where the scheduler is currently being used. The fifth section highlights some advantages and disadvantages of using the expert simulation approach for scheduling. Finally the last section highlights the synergetic relationship between expert systems and simulation.

Expert Systems Approach

Traditionally, plant floor scheduling has been a difficult problem to solve. Even after decades of research, management scientists have failed to find solution approaches which can be applied in practice for job-shop scheduling.[3] Most commercially available packages have not

1   Shannon, R.E. (1984). Artificial Intelligence and Simulation. In : Proceedings of the 1984 Winter Simulation Conference, (S. Sheppard, U. Pooch and D. Pegden, eds.). Institute of Electrical and Electronics Engineers, Dallas, TX, 3-9.

2   Kusiak, A. and Chen, M. (1988). Expert Systems for Planning and Scheduling Manufacturing Systems. European Journal of Operational Research 34, 113-130.

3   McKay, K.N., Safayeni, F.R., and Buzacott, J.A. (1988). Job-Shop Scheduling Theory: What is Relevant? Interfaces 18:4,84-90.

found generic application. The limitations on their applicability occur due to their lack of a good user interface preventing an "average" user from using the packages effectively or sometimes due to inability to customize the packages to requirements of a particular plant. Also, some of the math-based scheduling packages require large computation times in their search for a near optimum solution. The problem lends itself well to application of an expert systems approach. In recent years, several efforts have started utilizing this approach for solving scheduling problems.[4] The expertise of a human scheduler can be utilized to establish heuristic procedures which lead to schedules meeting the objectives of particular plants. The expertise is also utilized to customize the heuristics for meeting different objectives, or for providing a different set of heuristics for widely varying objectives.

The expert for this system was a plant floor scheduler with 20 years of experience in the field. He was very articulate in formulating the scheduling heuristics which took important plant floor issues into account. He was an excellent abstract thinker and capable of generalizing from specific situations to abstract application. His experience also guided the development to concentrate on prevalent situations on plant floors, rather than for all possible situations. His expertise was also used in evaluating the schedules generated using the heuristics. He guided the developers through iterations of evaluation and modification of heuristics until they met the requirements. The knowledge of the expert was put in the form of heuristics by knowledge engineers with AI and expert systems development background. Developers for this system had a strong background in manufacturing scheduling and simulation software development and were trained in use of AI tools and languages. The heuristics are used for the following decisions:

- Part dispatching, that is, which part order will be scheduled next from the candidate part orders. A part order is defined as an order for certain quantity of a particular part type. The quantity may be determined based on customer orders and batch sizing considerations.

- Machine selection, that is, which machine among the candidate machines will be used to process the selected part order.

- Interval selection, that is, which time interval in the window of time being considered on the selected machine is most suitable to process the selected part order. One of the trade-offs in this particular situation is setup versus Just-In-Time.

4   Kusiak, A. and Chen, M. (1988). Expert Systems for Planning and Scheduling Manufacturing Systems. European Journal of Operational Research 34, 113-130.

- Secondary resource constraints, that is, when to consider the constraints of labor, tooling, purchased parts, gages etc.

- Supporting events consideration, that is, when to schedule preventive maintenance, service parts and safety stock replenishment production.

The types of specific information which can be incorporated using knowledge-based technology is illustrated in choosing the machine for the next operation on a batch of parts. When multiple machines are available the following selection criteria are used:

1. The scheduler determines the ideal completion time for the batch of parts.

2. A window of time in which the operation on that batch can be scheduled is determined for each eligible machine. All machines which can perform the operation to be scheduled are eligible for consideration. The window is dependent on the process time for the operation and various user controlled parameters.

3. Each machine is checked for a block of idle time within the window which is large enough to schedule at least x% of the operation and any associated setups. "x" is machine and operation dependent.

Expert system technology facilitates the handling of complex reasoning. This is illustrated with the following heuristic which is used in conjunction with backward simulation[5]:

1. If there are intervals which will completely contain the run time (the time required to perform the operation and the setups), the interval which would yield the most synchronous schedule over all primary machines is chosen. Some consideration is also given to the secondary goal of setup minimization.

---

5   Barber, K., Burridge, K., and Osterfeld, D. (1988a). Expert System-Based Finite Scheduler. In : Proceedings of 2nd Annual Expert Systems Conference. Engineering Society of Detroit, Detroit, MI, 105-116.

2.   If there is no complete fit of the total run time (process + setup), but a partial fit of some minimum percentage exists, all preceding operations on the chosen machine are shifted earlier in time to make room for the one which needs to be scheduled. This results in having to shift all the affected operations across all the machines which precede those shifted on the machine being scheduled. The recursion techniques in LISP handle this quite well.

3.   If no fit is found on the primary machines, backup machines for the operation are examined. Backup machines are capable of running an operation but are not the machines of choice of plant floor personnel for the purpose. An operation is scheduled on a backup only when it can not be scheduled on primary machines in a desired time window and when allowed by the user. Both a complete fit and a partial fit are successively examined on backup machines, similar to the approach taken for scheduling on primary machines.

4.   If no fit of any kind is found, the window is shifted earlier in time and another iteration is done.

5.   In following the precepts of Just-In-Time, arrival of components is scheduled at the assembly cell on an as needed basis. For example, if an assembly requires three batches of a given component, the first arrives when assembly begins, the second arrives 1/3 through assembly, and the third arrives 2/3 through the process.

The primary objective of the heuristics is to meet customer demand. Secondary objectives may be setup minimization and work-in-process minimization. Setups are scheduled where possible during the transfer of parts from one machine to another to anticipate the arrival of corresponding parts from the previous operation. When more than one eligible machine is equally well suited to meet customer due date, or is within a minimal percentage of another, the machine which minimizes setups is selected. Tuning parameters are available to the user to generate schedules closer to objectives in his/her environment. For example, through these parameters the user can indicate a preference towards low work-in-process inventory at the cost of higher number of setups or towards a lower number of setups at the cost of higher work-in-process inventory.

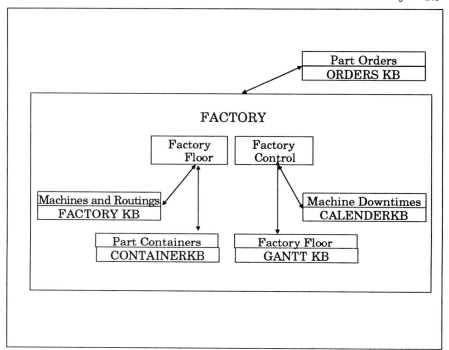

Figure 15-1   Correspondence between factory and its model used by ESS

The data and plant specific knowledge is organized into several knowledgebases (KBs) which are accessed by scheduling heuristics. The KBs and the corresponding information stored in them is represented in Figure 15-1. The knowledge used for factory control consists of knowledge regarding machine downtimes and the schedules under which the floor is currently operating. These are stored in CALENDAR and GANTT KBs respectively. Similarly, the knowledge about machines and routings is stored in FACTORY KB while information about part containers is stored in CONTAINER KB. The data regarding customer orders is stored in ORDERS KB. These knowledgebases together create a model of the factory. The model is used by expert heuristics and deterministic simulation concepts to generate the schedules.

Simulation Concepts

Simulation concepts have been implemented using object oriented programming. A model of the plant floor is developed using frames to represent parts, machines and operations (processes)[6]:

- Associated with each part is a process plan which identifies the sequence of operations or processes to be performed on that part. The part class is further subdivided into assemblies and component parts. The assembly sub-class can be used to define final assemblies as well as sub-assemblies. Assemblies identify how many of what components are required for assembly operation.

- Machines eligible to perform operations are identified by each operation.

- Associated with the operations are the setup and process times required for each part on which the operation can be performed.

The concept of "backward" simulation is used to construct the schedules. The concept has existed in a simple form in scheduling literature under the name "backward scheduling". In the library of AI techniques, a somewhat similar concept of backward chaining is used. Backward chaining works by starting from a goal state and working backwards to the initial state using production rules whose outcomes are goal state or sub-goal states. In AI literature, backward chaining usually does not include modeling passage of time. In backward simulation, the idea is to start with the goal state, and then simulate passage of time backwards to the initial state. In a plant floor scheduling context, the goal state is the end of horizon with all customer demands satisfied with production as close to due dates as possible. The events are simulated backwards, starting from the last operation of a part to its first operation.

The major motivation of using backward simulation comes from the thrust to implement the Just-In-Time philosophy. It is difficult to determine the release time for an order on the plant floor using forward scheduling or conventional forward simulation for complex scenarios. Several iterations will be required to determine correct release times for the hundreds of orders processed by a practical sized department on the plant floor. Queuing theory relations provide approximations to determine service times in simple multi-server networks, but few approximations are available for practical situa-

---

6   Barber, K., Burridge, K., and Osterfeld, D. (1988b). Expert System-Based Finite Scheduler. In : CAD/CAM, Robotics and Factories of the.Future, Vol. 2. , (B. Prasad, ed.) International Society for Productivity Enhancement, Southfield, MI, 291-295.

tions with multiple resource constraints and multiple routings. Also, the queuing theory approximations provide mean values which will not predict the release times of the orders as accurately as backward simulation.

The backward simulation considers known or deterministic machine unavailabilities similar to forward simulation. These known unavailabilities may be due to shift timings, current breakdowns or tool tryouts. Random machine breakdowns are not considered explicitly during scheduling, though some slack may be included to account for their occurrence. The ability to quickly generate schedules together with the ability to access real-time information allow generating new schedules in response to such machine breakdowns.

The use of backward simulation does require care in implementation of traditional dispatching rules. For example, while simulating backwards in time the job with the latest due date will be selected first to get the effect of the traditional dispatching rule "earliest due date first". Similarly, the job with the longest process time will be selected first to get the effect of shortest process time rule. As such, the traditional dispatching rules are not being used directly in ESS, though some of the heuristics are due dates based.

At times, a plant may receive orders which it cannot satisfy by the due dates desired by the customer due to capacity constraints. In conventional or forward simulation, this situation will be reflected by orders being completed later than their due dates. Backward simulation leads to order release times which are earlier than the beginning of the scheduling horizon in such a situation. The system will indicate that to meet the customer orders at the desired due dates the orders should have been released sometime in the past. In such a case, if the requirements or capacity are not adjusted, some of the orders will be made late. The new release times in such cases are calculated by intelligently shifting the schedule forward in time using idle time intervals on machines until it becomes feasible. The plant can advise its customers of the expected delay in completion of the orders. The system also allows easy interfaces to revise order due dates and quantities and to update machine availability for adding overtime. Either of these mechanisms can be used to explore available options.

Once customer orders have been scheduled, any remaining capacity is utilized by scheduling safety stock replenishments and low priority orders. Both the safety stock replenishments and low priority production orders are scheduled using forward simulation. These are placed within the idle time intervals left on the machines due to excess capacity. The schedule is also adjusted for scheduling preventive maintenance events and for honoring constraints of expected material receipts, labor, tooling etc.

The graphical representation of the schedules in ESS provides an easy way to validate the schedules. In addition to visual review, the schedules generated by ESS were validated and evaluated through a simulation model of the first application site. A very detailed simulation model developed using AutoMod and GPSS/H were used for this purpose. Schedules generated by several other commercial scheduling packages were also evaluated using the same simulation model.

The simulation model served to analyze the feasibility of ESS generated schedules. The impact of the schedule on plant resources which were not considered by the scheduler itself was evaluated. For example, the scheduler assumes that both the material handling system and the in-process storage capacities are unconstrained. The first application site was designed with adequate capacities in each of these areas. The simulation indicated that the schedules were feasible when these resources were constrained to actual capacities.

The quality of schedules was evaluated based on due date performance, machine utilization, and the synchronization of material flow. The results provided an evaluation of relative performance of the considered scheduling packages. It was demonstrated that ESS generated schedules met customer demand on time with a highly synchronous material flow. In fact, ESS was selected as the scheduling tool for this automated facility based on these evaluations.

APPLICATION

The Expert System Scheduler has been developed using the package KEE, which is a product of IntelliCorp. The simulation concepts and heuristics have been coded using LISP. The scheduler is currently resident on TI Explorer II hardware.

The system is being used at a highly automated facility involved in automotive component production. The factory consists of machining and heat treatment cells linked by an Automated Guided Vehicle System (AGVS). It also includes an Automatic Storage and Retrieval System (ASRS) for supplied parts and sub-assemblies, in-process work and manufacturing tools. More than 50 robots are used to load and unload parts in different processing cells. Tool banks are automatically transported to the cell by the AGVS where they are manually interchanged. All machine tools are numerically controlled. Each machining cell is fitted with in-process gauging equipment and incorporates statistical process control to ensure that all parts conform to specifications. Each cell is linked to the factory level computer via a broadband MAP network.

The plant floor is controlled through a Factory Control System (FCS) which resides on a Stratus 2000 computer. The factory floor status information is maintained by the FCS. In addition, the FCS contains current data on customer orders and on the expected material receipts. All this information is sent through a computer network to the TI Explorer II whenever the situation warrants a new schedule. The Expert System Scheduler is used to generate a new schedule within a short time, and the schedule is sent back to the FCS for execution. The FCS sends appropriate commands to the plant floor cell controllers for implementing the schedule. The information flows to and from ESS are shown in Figure 15-2.

A friendly user-interface is very important for acceptance of an expert system by the users.[7] ESS uses a very user-friendly mouse driven interface. The schedule is presented to the user in the form of a Gantt chart as shown in Figure 15-3. The figure shows the schedule for an 8-hour period for a small department on a plant floor. The user has an option to mouse-click on an operation and get more details as shown in the figure. The user can scroll the Gantt chart up and down if there are more than 24 machines, and right and left to look across the scheduling horizon. Successive operation of one particular batch of parts can be highlighted to follow its flow through the machines. In the figure the flow of a batch of parts of type '46gb' is highlighted.

The user can have the system generate the schedule step by step and follow its construction, or run all the steps together. The Gantt chart interface has been found very useful by the users, as it gives them an understanding as to how the schedule is built and the overall performance. It also gives the user a quick means of evaluating the quality of schedule.

Aside from the highly automated facility, where the scheduler has been controlling the production for several months, the scheduler has been evaluated by modeling and generating schedules for several factories which are not so highly automated. The system has also been transported to SUN workstations to allow more hardware configura-

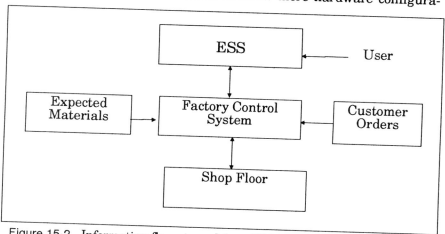

Figure 15-2   Information flows to and from ESS

---

7   O'Keefe, R.M., Belton, V., and Ball, T. (1986). Experience with Using Expert Systems in OR. Journal of Operational Research Society 37:7, 125-129.

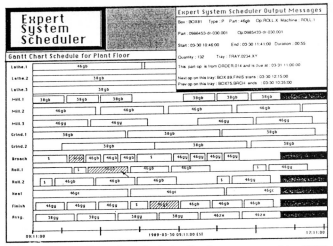

Figure 15-3   Sample schedule display generated by ESS

tion options at future sites. The evaluations and added portability have helped develop the heuristics for generic application.

Advantages/Disadvantages

There are some disadvantages associated with these systems. There are few people with expertise in expert system building tools and techniques, proficiency in LISP, experience in scheduling and simulation, and extensive manufacturing backgrounds. Knowledge acquisition for expert systems requires special skills. Developers of these early systems need to be selected carefully, and given extensive training. Although prices are coming down, hardware (LISP machines) and software are still expensive. Consulting by AI companies is very expensive. A technical challenge is to design and develop these systems in such a way as to eliminate the requirement of an AI expert to make most of the changes and enhancements. Additional issues are expected to surface as the system is implemented in multiple sites.

Interfacing AI hardware to general purpose hardware requires quite a bit of effort. Some of the speed advantage of AI hardware is lost due to time requirements for transferring data back and forth. At the first ESS application site, ongoing efforts have reduced the data transfer times substantially from their initial values. However, the total time for data transfer is still longer than typical schedule generation and review time.

The backward simulation approach can be used effectively for generating schedules. However, it cannot be used to examine the effect of random events. If robustness of a schedule is to be evaluated, it will have to be simulated in a traditional manner with random events incorporated. The fast response of the scheduler reduces the concern about the robustness of the schedule to a large extent.

Advantages of an expert system based scheduler include the incorporation of heuristics to tailor a scheduler to a particular business site. The scheduler can be easily customized to allow for operation peculiarities, business plans, operation goals, specific customers and order mix. Model based reasoning allows utilization of heuristics aimed at achieving the goals of synchronous scheduling, setup minimization, and machine dedication under changing factory conditions.

The ability to prototype rapidly is a benefit resulting from the use of an expert system shell. Some of the specific features which support rapid construction of systems are the developer interface tools, the inheritance features, the provisions for structuring knowledge, the modularity, and the ease of incremental development. Specifically, incremental development facilitates the understanding of a complex and ill-structured problem like factory scheduling. A limited module can be quickly developed for simulation testing to determine which additional factors need to be incorporated into the model.

Another advantage this technology offers is that the knowledge representation scheme is relatively generic. The frame based knowledge representation structure provided by the expert system tool eases the integration of the various knowledge bases developed for this phase of the project.

The interactive Gantt chart is a very valuable tool for both understanding new scheduling techniques and debugging the generated schedules. Understanding is gained of the scheduling algorithms and the performance of the schedule through graphical representation of relationships between batches on one machine and across machines. The graphical display makes it easier to spot inconsistencies or irregularities in a schedule. However, the graphics may be found more useful by people dealing with smaller plants.

This technology offers unusual flexibility for change. An expert system based scheduler can be easily customized to a particular application. Heuristics, or rules, can be incorporated to cover conditions at specific sites.

Perhaps one of the biggest advantages is the user acceptance of the schedule. The user can understand and believe in the system. The data on which the schedule is based is easily accessible and in an understandable form.

Conclusion

The application described here utilizes advanced concepts in AI and simulation modeling, together with the latest in computer hardware and graphics for effective real-time control of the plant floor. Though this application has been developed independently, development of such a system was hypothesized by Shannon in 1984. This application proves that the disciplines of AI and simulation modeling can be used synergistically for a practical purpose.

It is important to integrate the AI, Operations Research (OR) and computer systems software and hardware technologies to face the competitive challenges of today's market. Most of the problems can not be clearly classified as being from one discipline only. AI has the strength to use experts' knowledge in problems where experience can be effectively used to reduce solution time and efforts. OR has the strength to provide mathematical insights into a problem and using mathematical models to determine ways to arrive at optimal or close to optimal solutions. Advances in computer systems software and hardware technologies provide means to make both AI and OR methodologies faster and easier to use. Synergistic application of all these technologies to a given problem may lead to a much better solution with a much higher chance of use than with the application of any one of them by itself.

References

1. Shannon, R.E. (1984). Artificial Intelligence and Simulation. Proceedings of the 1984 Winter Simulation Conference, (S. Sheppard, U. Pooch and D. Pegden, eds.). Institute of Electrical and Electronics Engineers, Dallas, TX, 3-9.

2. Kusiak, A. and Chen, M. (1988). Expert Systems for Planning and Scheduling Manufacturing Systems. European Journal of Operational Research 34, 113-130.

3. McKay, K.N., Safayeni, F.R., and Buzacott, J.A. (1988). Job-Shop Scheduling Theory : What is Relevant ? Interfaces 18:4, 84-90.

4. Barber, K., Burridge, K., and Osterfeld, D. (1988a). Expert System-Based Finite Scheduler. In : Proceedings of 2nd Annual Expert Systems Conference. Engineering Society of Detroit, Detroit, MI, 105-116.

5. Barber, K., Burridge, K., and Osterfeld, D. (1988b). Expert System-Based Finite Scheduler. In : CAD/CAM, Robotics and Factories of the.Future, Vol. 2. , (B. Prasad, ed.) International Society for Productivity Enhancement, Southfield, MI, 291-295.

6. O'Keefe, R.M., Belton, V., and Ball, T. (1986). Experience with Using Expert Systems in OR. Journal of Operational Research Society 37:7, 125-129.

Author Biographical Data

Sanjay Jain is a Senior Project Engineer in Advanced Engineering Staff at General Motors Technical Center. He received a Bachelors of Engineering from University of Roorkee, Roorkee, India in 1982, a Post Graduate Diploma in Industrial Engineering from National Institute for Training in Industrial Engineering, Bombay, India in 1984, and a Ph.D. in Engineering Science from Rensselaer Polytechnic Institute, Troy, New York in 1988. His current interests are in the area of development and implementation of both math-based and artificial intelligence based scheduling packages. He is a member of IIE and CASA/SME.

Karon Barber is a Development Engineer in Advanced Engineering Staff at General Motors Technical Center. She received a B.S. degree in Physics and Chemistry from University of Alabama in 1962. She worked in the aerospace and petroleum industry before joining the Artificial Intelligence department at General Motors in 1985. She is currently the project manager for an expert system based generative designer, and is a member of the local arrangements committee for International Joint Conference on Artificial Intelligence and the program committee for the Engineering Society of Detroit Expert Systems Conference.

David Osterfeld is a Senior Engineer in the Manufacturing Consulting Division of Electronic Data Systems Corporation. He received a Bachelor of Business Administration degree in Quantitative Methods from Ohio University in 1975, an M.S. in Industrial and Systems Engineering from Ohio University in 1976 and an M.S. in Management Science from University of Dayton in 1980. His current interests are in the development and application of artificial intelligence based systems in manufacturing specifically plant floor scheduling.

Simulation, Process Modeling, and Resource Allocation

Intelligent Simulation
The New Generation of Expert Systems

Herman Bozenhardt
Artificial Intelligence Technologies, Inc.
Hawthorne, New York

Introduction

Due to the capital intensive nature of the process and manufacturing industries, the decision to build, modify, and automate is always a difficult one. Process simulation is a natural application for non-destructive analysis. Process simulation software has evolved to a high level of accuracy and plant fidelity. Along with this improvement, especially in dynamic simulation, comes the need for a high level of simulation expertise, computer skills, and resources to build unit level or plant models. As much as the traditional software tools improve, they have remained intensive efforts to use and maintain.

Independently, the artificial intelligence and expert system area has accelerated in its capability from the perspectives of 1) process modeling via object-oriented programming, and 2) skill capture.

In order to provide a simulation tool for process engineers, designers, and control engineers, ISIM was developed. ISIM provides the ability for any process engineer with no simulation or computer skills to "draw" a process flow on a screen, specify equipment parameters, and begin dynamic simulation. This paper covers the concept, design, and implementation of ISIM.

Simulation is the most basic of any engineering discipline especially in the process field. Chemical engineers traditionally have specialized in designing and developing mathematical models to represent plant environments. Focused primarily for process design and development, these models were built to replicate the simultaneous interaction of momentum, mass and energy transfer. As difficult as it is for a single unit operation simulation, plant level models with dynamics have always been the goal. Therefore, most large process companies and all engineering (A&E) firms have had dedicated simulation scientists. It is this required and dedicated expertise that has prevented smaller organizations and smaller subsidiaries of large organizations from acquiring the benefits of dynamic simulation. Simulation technology has primarily progressed in two major areas:

(1) Spreadsheet-like capability to make the design functions easier. This provides a static mass and energy balance. The macros or basic functions are programmed in Pascal or Fortran by the expert. These systems have been provided with computer-aided system engineering tools to ease the programming effort.

(2) Dynamic simulators built for testing control engineering systems. These simulators, most of which have grown out of a university environment, require the user to program the equipment models and the control schemes (Kalman filters, adaptive model based controllers, etc.). These systems provide a variety of trends and graphics for analysis.

Expert system technology, on the other hand, has rapidly accelerated from 1980 to the present;

■ In the early 1980's, the first generation of AI tools emerged from the university into the industrial environment. Tools such as OPS5, Prologue and LISP demonstrated data driven computation, knowledge acquisition, AI modeling, and intelligent code generation. They generally lacked the ability to integrate and ran on exotic hardware.

■ From 1984 through 1987, the second generation of AI tools replaced the first generation. These tools were characterized by KEE, ART, and Nexpert. These tools provided the ability to model and prototype industrial systems. These were the first deployable systems on standard industrial hardware. Their drawbacks were primarily in the area of performance (speed) and the development time.

■ From 1987 to present, a new "third" generation of expert system shells has been developed. These are large hybrid systems, such as Mercury KBE, which are performance oriented and because of the intelligent "CASE tools" reduce the development time.

It is from these large hybrid systems that process specialists have captured the knowledge of simulation experts, created generic equipment models with multiple mathematical techniques and tools, and "built-in" techniques for dynamic simulation and simultaneous solutions. These techniques are chosen, implemented and monitored automatically by the expert system. This new class of tool is called Intelligent Simulation "ISIM." This tool allows the user to simply describe the plant by specifying equipment and the system provides the expertise. This allows any user unfamiliar with simulation technology and modeling to generate and use rigorous dynamic process simulation.

This ISIM development has made an evolutionary leap in both the field of AI and Process Simulation. This technology collision may prove to be the single most significant development in engineering software in the 1980's or 1990's.

Object-Oriented Simulation Versus Traditional Simulation

A traditional simulation environment consists of simulation blocks linked together like a classical block diagram. The simulation blocks are Fortran-based subroutines with hard-coded, predetermined solution methods. There is a strict notion of input/output and the values that can be manipulated are usually single floating point numbers or Boolean.

A simulation of even modest size usually consists of hundreds of simulation blocks. With this complexity, it is very hard for a simulation to be extended beyond its original scope. For instance, if the original simulation modeled a variable as a constant and that value was used in many of the calculations of the simulation blocks, then it would be very hard to add any dynamics to that value. Much of the configuration would have to be reimplemented.

Furthermore, there is no notion of separate or independent "EQUIPMENT PARTS" or "MATERIALS" in a traditional simulation environment. The configuration represents the final equations after all of the interactions between the parts and the materials have been determined. Therefore, solutions to differential equations *must be developed and provided*. This predetermines the solution to the problem. The influence of any one part or material can often not be isolated and a substantial effort is usually required when a material (i.e., a feedstock, flow configuration, equipment failure or transitions of flow, energy or mass regime) is changed.

The ISIM system environment was created based on a CLOS object system. The object system provides modeling tools, techniques and utilities for implementing a simulation using an object-oriented methodology. In this representation, each component of the simulation is implemented separately. The implementor specifies how each component interacts with the rest of the simulation environment. When these interactions are well defined, a flexible simulation environment exists that allows for unseen combinations of components (i.e., process equipment) and the addition of new components with little or no impact on the original simulation.

In ISIM, the simulation code associated with the parts and materials is isolated. Simulation blocks still exist, but instead of representing mathematics they represent physical devices. These blocks are modules which describe how a piece of equipment operates on materials. Thus, instead of manipulating numbers, they manipulate objects that represent the materials. These instances are called simulation variables. The main advantage of manipulating simulation variables instead of numbers is that there is code associated with the simulation variables that determines how that instance will react to actions of the simulation block, thus isolating the effect of the material. Therefore, a heat exchanger impacts Btu's and pressure drop, and the receiving materials determine the impact on temperature and flow. This code is called simulation methods and, depending on the object methods, contains a variety of functions, such as:

- integration methods of differential equations

- equations of mass transfer (PDE)

- equations of heat transfer (PDE)

- equations momentum transfer (PDE)

- heuristics or rules

- database calls/searches

- probabilistic functions

- simultaneous equation solvers

- lead/lag transfer functions

- equipment performance characteristics

■   instrumentation characteristics

These methods provide the ability to represent performance and function that can't be described by logic driven programming.

For example, a reboiler adds thermal energy to liquid in a distillation tower causing a change in temperature and state for those materials. In ISIM, the transfer function of the reboiler will have no reference to the heat capacity of the material, even though that value is crucial in determining the effect it has on the material's temperature. Instead, the reboiler will add thermal energy to the material and then the simulation variable modeling the material will change its temperature accordingly. This allows for the reboiler to manipulate unforeseen types of materials at the same time.

Furthermore, if a high shear mixer in a reactor adds the same amount of heat energy to the material that the cooling jacket removes, then the material simulation variable instance will not change its temperature. (NOTE: If the material's temperature was raised and then lowered, it would be mathematically incorrect and that error would propagate, since the invalid temperature would be used to compute the heat flux between the material and the cooler.) In order for this coupling to be modeled correctly in a traditional simulation (without explicit foresight), one piece of code would be written that takes all the factors into account (i.e., a heat balance). Therefore, the separation of parts would be lost and the ability to add or remove these mixers, coolers, and auxiliary equipment would not exist.

This implies that the simulation variable provides the coupling between simulation blocks. The presence of the mixer in the configuration is isolated from the jacket since each one can easily be removed or added. Yet, the effective coupling between both parts is still simulated. The simulation variables are the material objects which are being processed. They have the characteristics of flowing fluid. To provide separation, identification, resolution and dynamic accuracy, simulation variables are bundled into groups or "buckets" of objects. These material objects are the equivalent of dv's or dm's. The volume of each object and the number of objects per bucket provides the numerical resolution needed in the simulation.

The object-oriented approach provides the most modular approach to software development. The development of a simulator of this capability could not be done with traditional computing without several years of effort. The architecture allows easy modification, additions and deletions from a standpoint of 1) a new process equipment object, 2) new simulation techniques, and 3) incorporation of traditional computing or external data into the existing methods.

The methods used will minimize the amount of code by several orders of magnitude. The AI tools and debugging routines, plus the low volume of code, allows documentation, understanding and modifications to be easy.

The effort to "construct a plant" from a process library will be trivial given that the engineer is familiar with the process and has an existing flow sheet. A typical process unit of a plant could be constructed graphically in 2-3 hours after a short ISIM tutorial.

ISIM in a Chemical Engineering Environment

Although rarely thought of in exactly these terms, the ultimate design tool for practically any engineering endeavor would be a computer program that is capable of simulating, in a simple, straightforward way, the responses of an "as built" engineered facility. In the immediate past, a prime reason for not generating many such programs is that even a small facility is usually very complex. The effort required to write the programs needed for all the subsystems can be significant; it requires the expertise of specialists over an extended period of time. Nonetheless, many have had the experience of dealing with a single subsystem in a plant wherein a correct answer was critical for one reason or another, and to assure correctness, a detailed computer model was written so as to check detailed design parameters and consequent performance.

In addition to the design issues, the ultimate tool for determining the responses of a facility to operator control, and/or to computer control, would, as before, be a computer program capable of simulating the responses of a real plant to control signals issued to the field devices. As before, the cost of customized programming for every such plant would be high. Nonetheless, the complexity of modern control systems and the potential consequences of errors in those signals provide powerful incentives to provide such simulators. Again, many such simulators have been written for special situations.

The issues surrounding control system checkout, including interlock management, prompted an inquiry into whether, using current programming techniques, it would be possible to put together a generic computer program which was capable of creating simulators for many different plants. Quite literally, we wanted a program that could generate programs, and which could be used by engineers with little programming experience.

One critical aspect of the design of such a program was that the users primary task would only be to tell the program the engineering parameters associated with each plant component, and how the components are related; i.e., to create a database. No further programming, that is, writing code in some computer language, was to be required of the user. Our objective was to provide a basic set of plant components such as pipe, valves, etc. so that it would be possible to create a simulator for most any one of our plants without the need for additional specialized components.

The effort required to create such a database is not trivial, and the person doing it needs to be free of the burdens of detailed programming. A complete job for a real plant involves, among other things, three dimensional piping layout, purchase specs for components such as valve coefficients, etc., plus engineering judgement about how best to simulate special situations that don't quite conform to the simulated objects. Additionally, one must also provide such values as estimated heat transfer coefficients, heats of reaction, fluid densities, viscosities, etc.

Once such a base is created, the program should respond to commands such as, "start with tank x containing y feet of fluid," "the fluid is (say) water at 70F," "open a valve," "start a pump," "accept the following set-point," etc. Such commands were to be issued either manually from a graphics screen associated with the simulating computer, or alternatively, from the controlling computer system via a communications link by (say) a plant operator at the console, or under program control.

The program should in turn compute all fluid flow rates, compute all temperatures, maintain all material and heat balance relations, and handle any special calculations associated with a particular kind of device of defined chemical reaction. Among the devices (Objects) were heat exchangers, vessels with and without jackets, pumps, agitators, pipes, elbows, tees, meters, controllers, control valves, certain types of reactors, etc. The results of all such computations were to be accessible to the control screen, or should be reported back to the controlling computer system via a communications link. the net result might be expressed quite simply as follows: the controlling computer system should "think it was dealing with a real plant," and would thereby perform its entire set of control functions.

Investigation into available technology for creating such a program led to the possible use of "object-oriented programming," and in particular to the LISP-based product of Artificial Intelligence, Inc., named "Mercury KBE." This particular product makes it possible to create the specialized objects needed to simulate the corresponding real-hardware components of a chemical plant and to achieve exactly the objectives indicated above. The original version of the product "ISIM" made use of components which were relatively simple in the engineering sense. The original objective of the project was to test the concept and determine whether the specification for such a program could realistically be met. It turns out that they can.

One of the virtues of object-oriented programming is that the objects can range from elementary to sophisticated. The immediate objective for "ISIM" was to make them only as sophisticated as need be to prove the concepts as applied to a real chemical plant. Attention has since turned to development of more sophisticated "objects." For example, the original model of a linear control valve could only simulate flow of an ordinary liquid. The current version can also simulate the flow of a compressible gas up to and including sonic velocity, and can, in fact, recognize the difference in fluids that are passed to it. The original heat exchanger object was a simple 1-1 counter-current exchanger; a more recent member of the class acts like a condenser.

Ultimately, it is conceptually possible, and perhaps realistically possible to make the objects so sophisticated that the difference between computed model function and reality as reported by real sensors, is practically nil. When such a simulator is constructed, then it in fact can be regarded as the ideal "expert system."

It is considered as unlikely that "expert systems" as popularly understood can possibly anticipate all the possible situations that can arise in a complex chemical plant, whereas, an ideal simulator could. This concept is not new, and, in fact, it is reported to be in use with some electronic systems where device performance is substantially linear. In a certain sense, this approach has been used already in the chemical industry, for example, in the HTRI programs commonly used for heat exchanger design and evaluation. It is believed, however, that with the capabilities of the "Mercury" product, particularly its ability to work with large database systems such as "Rdb" and "ORACLE," combined with its ability to run "foreign" programs, that it will be possible to put such sophistication into the context of a complete chemical production unit.

Such as "expert system" could be synchronized with, and run in parallel with, the production unit that is simulates. With such an arrangement, many plant operating problems could be dealt with quickly and efficiently. Currently, commonly available computer systems would probably be strained to meet the computing loads that would be imposed by "real time" applications, but it is believed that such limitations will be removed as machines get bigger and faster.

Overview of Applications of ISIM

Because of the flexibility afforded by an "expert controlled" simulation environment, several traditional applications present themselves, plus new, but previously unachievable applications can exist.

- Control Engineering functions which have traditionally required dynamic simulation environments for easy overlay of model-based, expert system based or other specialized control. Although traditional PID control is considered a natural and common application, a nondestructive test is worthwhile. All too often, control engineers have sacrificed the numerical rigorousness of the system for correctness of the response curve to a pulse or step input. This type of system allows the control engineers to have a "test bench" for control schemes which could be readily extracted and placed in actual plant control. Absolute accuracy can be realized by including hydraulic line elements which are usually ignored.

The control engineering applications spectrum has the opportunity to widen with the use of ISIM. Batch control applications with the specific overlay of discrete on/off upon the continuous functions can be tested. The testing of all control related functions can be done within the ISIM environment by having the ISIM system function solely as the "plant" and being hooked up to a DCS or PLC system. In this way, the actual software techniques of the DCS can be implemented and validated.

In addition, the DCS and/or PLC system software revisions (operating, database, configuration, root level control) can be tested against a plant simulator before final installation. This is important because software revisions, system "fixes," and corrections for known bugs, when loaded on older applications, often cause incompatibility and software induced faults if not corrected.

- Process design and development are the traditional users of static mass and energy balance. The rigid models or more recent spreadsheet applications are good for nameplate design parameters. However, designers treat operation fluctuations with sensitivity runs on their static models. This assumes transients have no impact and processes respond stepwise. It is these assumptions that cause dramatic plant upsets. With ease of use, dynamic analysis of the process will take on its true meaning. This provides improved designs, better process developments, plant technology "step outs," plant troubleshooting environments, and process scoping studies.

- Plant level maintenance can use this tool to quickly assess problems and remedies and minimize wasted effort and inadequate "patchwork" on a problem.

- Training of operations personnel can use ISIM to familiarize them with the plant from a macro level, impact analysis, sensitivity, and down to the "feel" of a valve and the impact of stroke % on flow rate.

Using communications tools, a trainee could run the plant from his/her DCS/PLC console. This prepares the trainee to run the plant from the automation console they will be using. In addition, operator qualifying exams and tests could be run by the operator at the DCS/PLC console with the "trainer" inducing upsets, "breaking" actuators and "breaking" instrument at the ISIM interface. This type of trial under adverse conditions will quickly identify skill level.

■ Finally, most organizations have their plant "as built" knowledge base distributed in the expertise of the designers. Most drawings are not updated, VLE data is separated from the plant level user due to complexity, and various systems retain analysis tools. ISIM provides a unique concept to unite "as built's" with other design data and to be kept evergreen for control studies, debottlenecking exercises and for daily analysis. ISIM can serve as a rigorous collection point because it implements the simulation technology for all the users.

This widespread level of use was heretofore unheard of because, as attractive as dynamic simulation is, it was too difficult to use. The ISIM system manages the expertise, technology and complexity of simulation in order for any engineer to use it.

ISIM Support Features for Process Manufacturing

In order to provide a realistic process environment when an assembly of a plant is made, a number of supporting tools supply features to modify simulation objects and variables.

■ Line Network Solver—This feature analyzes the piping network constructed with all the components (reactors, heat exchangers, valves, etc.) and, when material is flowing, it calculates the individual pressure drops (caused by each part of the plant) and the flow rates. This technique uses an on-line network parser, an iterative solver for simultaneous nonlinear equations and has rigorous chemical engineering models for viscous fluids.

■ Object Control—This provides the ability to generate and dispose of material objects so they behave like materials and fluids in a process. This permits and controls circulation loops.

■ Configuration Toolset—This graphical editor allows placement of process equipment on a process flow sheet. Then, based on equipment type, the user "bolts" up equipment together and the correct alignment of methods is automatically made. As the equipment is linked, the direction of flow is accounted for. On "creation" of any process equipment, a form is displayed for the user to fill in the parameters required to fulfill the model.

Benefits

There are several desirable characteristics associated with building an intelligent simulation for any industrial process. The underlying simulation methods and modeling techniques can be data driven from the declarative component of the knowledge base. This means that by changing any of the attributes or objects which are undergoing simulation causes the optimal simulation procedures to be automatically configured and utilized. This removes the need for either modification of the procedural code or running a sub-optimal simulation.

An important characteristic of an intelligent simulation is its intimate knowledge about the domain. Each module or unit operation can be modeled independently by a domain expert and inserted into the configuration without any modifications to any other code. For the user to be able to install it, to use it, or to modify it, it would not be necessary to know anything more than the process flow, as opposed to intimate knowledge of computer or simulation technique.

The knowledge base can utilize heuristic, as well as procedural simulation methods. This is an important characteristic of the ISIM because not all of the underlying physics can usually be captured, modeled or computed using purely closed-form analytic representation.

Conclusion

The fusing of object-oriented programming as a basis of expert system technology to capture a scope of modeling technology and chemical engineering has generated a new generation of expert systems and a new generation of simulation technology. The technology is a carefully architected set of tools for the professional chemical engineer. As the industrial world realizes the benefits of expert systems, and the ease of use of the paradigm specific tools, such as ISIM, a broad range of applications will follow. The fundamental science of simulation in the hands of many users may signal a new era in manufacturing excellence and higher quality.

Traditional simulations and simulation environments are characterized by their inflexibility and lack of domain knowledge with respect to a particular problem. Even the traditional simulation environments which are easier to use require a high degree of customizing and procedural programming to fulfill the requirement of any real world problem. The ability to easily modify and enhance simulation methods is impaired because the systems are inherently logic driven.

The area of Intelligent Simulation (ISIM) provides a new and unique technology based on data driven methods. The concept of ISIM allows the user to build a manufacturing unit from an inventory of "plant hardware" developed in software.

The range of applications is varied; however, the most substantial benefit is to computer applications and design/development disciplines.

Author Biographical Data

Over the last 14 years, Mr. Bozenhardt has held various engineering and management positions in major corporations implementing integrated control engineering, artificial intelligence and operations research based solutions to operation plant environments. Mr. Bozenhardt's main contribution to the technical community is the demonstrated infusion of highly profitable technical solutions into operating plants. The result of these efforts are the subjects of his numerous papers which touch many industries. Mr. bozenhardt is a recognized expert in the areas of control engineering, distributed control systems, artificial intelligence, and plant computing environments. He is a graduate of Polytechnic in New York and a member of AICHE, ISA and ISPE.

17

Object-Oriented Knowledge-Based Approach to Process Modeling

Herman F. Bozenhardt
Artificial Intelligence Technologies, Inc.
Hawthorne, New York

Introduction

The field of Artificial Intelligence, especially object-oriented programming (i.e., relational data-driven computer science techniques), has opened up a new world to those who are concerned with process modeling. The confines of logic driven computing and traditional algorithms provide severe limits to representation of knowledge and decision methods. Those organizations who must deal with control problems and quality problems often settle for inadequate solutions because of the enormity of the problem or the overwhelming need for nontraditional or expert methods.

This paper concerns itself with modeling processes with AI techniques and its application within process model development, statistical analysis, statistical control, and dynamic process simulation. The orientation is chemical operations, specifically batch oriented.

Due to the increased desire for competitiveness of the American industries, a serious commitment has been made to quality control and subsequent quality improvement. Quality improvement can be implemented in all aspects of manufacturing, processing, and management. This is realizable because specifications, composition, profitability, time to complete, or other measures of "goodness" can be assessed and evaluated. Quality itself is an entity which is attributable to all aspects of doing business.

Statistical Quality Control and Statistical Process Control are different industrial applications which contain similar technological approaches:

- Observe quality via sampling.

- Assess the measure of quality.

- Predict the position and direction (velocity and acceleration) of the quality parameters.

- Apply a control law on the process or procedure to position the quality within the acceptable and controllable bounds.

The science of statistics is used to sample the quality to provide representative values. Statistics provide the clarity on the quality by suppressing measurement "noise" or biasing. This provides a truer assessment for the next step—the time variant nature of quality, or the current quality position and its apparent changes in predictable intervals of time.

The real issue and controversy in the industrial and commercial world is what to do as a next step. Most companies simply provide charting of quality and are at a loss as to what control laws to apply and at what process independent variable or combinations thereof to apply it to.

Today, America's industrial "mania" is in SPC; however, it is in a superficial, low impact, low ROI approach. Too many companies are satisfied with charting and an "open loop" approach. Engineering analysis, statistical review and investment in control systems are required for successful quality control. The essence of SPC is control;

- The direct manipulation of independent variables which impacts quality.

- The progressive modification of operating procedures and equipment as "assignable causes" are identified, to eliminate them.

- The direct manipulation is a calculated increment to direct and maintain quality directly on target.

The control calculation required in a SPC/SQC methodology, or any modern control concept, requires a process (manufacturing) model or a control model (e.g., PID algorithm). The PID algorithm is appropriate for simple, fast, linear-like systems. It is in the manufacturing process modeling where the effort is required. Modeling requires three main ingredients:

1) sufficient and representative data to cover the quality and control space to be examined;

2) statistical and mathematical sciences to generate the models. (More recently, AI methods to represent non-algorithm models);

3) process domain knowledge to direct the selection of the variables from known causalities. In addition, process knowledge is necessary to validate the model's structure and feasibility with known theory.

These areas are the foundations of statistical projects which require statistical experts, process domain experts and substantial manpower.

The technology to prepackage statistical expertise (and experience), along with inquisitive tools to draw domain knowledge of the process and fuse them into validated empirical models, is the subject of this paper. AI techniques provide a coherent method from which to build a single paradigm specific (statistical model building) tool.

Traditional Modeling Methods

Statistical projects in manufacturing and commercial industries have been in use for nearly one hundred years. Statistical techniques of regression analysis, goodness of fit, and noise characterization are some of the nearly one thousand techniques one can use. These techniques are often iterative and are dependent on previous results and decisions determined from experience or via research. Also, they often lead to statistically low confidence values of regression coefficients.

During the commercialization of computers, the techniques of statistical sciences have been the ones of most commonly coded routines. These have generally been specific and dedicated routines, specifically designed to handle a specific process or procedural problem under investigation. Today, many of the manufacturing, engineering, and commercial software contain a large statistical component. Over the last twenty years, several software companies have emerged who have specialized and succeeded in providing a library of coded statistical software routines and goodness of fit tables.

These software provide libraries of routines indexed by menus with copious "help" screens. To supply market appeal, "window" user interfaces and database interfaces are provided routinely.

These software are only of value to the engineer who is familiar with and skilled in statistical sciences. The availability of multiple methods and quick calculations can take a modeling project from several months down to three to four weeks. The critical effort in this case is the mastering of all the statistical methods that are required as one explores a given problem.

The statistical methods uncover correlations and model frameworks and must be built upon and cross-checked with goodness of fit tests. This requires a learned technique of searching, evaluation, and renewed searching based on an acceptance or rejection of a hypothesis or an evaluation against a criteria. A necessary skill for the search is experience and expertise in statistics.

Traditional statistical projects also contain a high level of iteration between the statistics (i.e., "pure numbers") and process expertise. Statistical regressions can generate equations with a 90%+ expected variance that correlate to inappropriate or orthogonal variables strictly due to happenstance. A tremendous amount of "sanity" evaluation and model validation is needed. The statistical expert and the process expertise must be in close proximity to ensure the model is a useful representation which fits a theoretical form. If not, a "model" can become just a collection of equations which happen to fit a particular problem at a particular time, regardless of the true interactions of process variables. When on-line in a real process, these systems deteriorate rather quickly.

Expert Systems in Statistical Modeling

The concepts of statistical modeling are like any other expertise in that modern AI technology can capture and represent it. If one can automate the statistical expertise, then the opportunities to provide a tool for quick, definitive statistical model development can be at hand.

Two noteworthy attempts to do this in the past have been made and succeeded. First, in the 1970's, Procter and Gamble commissioned a research effort to capture statistical science technologies and build a plant optimizer for generic use. Although written in FORTRAN, the Procter and Gamble internal tool and its commercial venture (released in 1984) provided a major step in statistical skill capture.

In the late 1970's, Exxon Refining engineers experimented and applied a statistical learning tool to the optimization of large-scale furnaces. In the early 1980's, the technique was refined by Exxon Chemical into what today is called the Hyperplane search. Hyperplane explored process objectives with a limited set of statistical tools. These tools assessed the best change in operations to increase profitability, and then searched for the variable (or variables) to change and then determined the magnitudes of each change. The entire search mechanism was driven by adaptive statistical methods. This method was done on-line in real time in DCS host computers (using a language called BICEPS) and was applied successfully to chemical reactor optimizations.

Today's object-oriented, integrated AI shells provide an ideal data driven environment for capturing expertise. The following aspects support an efficient way to build an expert statistician in a contained system:

- rule-driven/statistical rule-based decisions

- searching methods to find the best variables

- sorting criteria/sorting applicable tables

- number crunching

- objects and methods that acquire dynamic changes and that "learn" and adapt to a problem

Ultimately, a considerable library of statistical techniques that can be "related" to one another are created. These techniques use objects to pass data, criteria, thresholds, and variables dynamically between themselves. Each object can be inspected and supervised by other objects or rules to direct problem solving skills in the library at the problem. This allows for automated statistical reasoning.

Anatomy of Intelligent Statistical Process Analyzer

Intelligent Statistical Process Analysis (ISPA) contains three major cooperating expert subsystems. The first major subsystem of the ISPA is the Statistical Master. This is an object-oriented model of the expert procedure to develop statistical models. The following are examples of some of the techniques which are represented and modeled:

- Variable Screener screening on independent variables for cross correlation/interrelation, analyzing for redundancy, and inspecting associated data for missing values.

- Degeneracy Check generates a multidimensional vector which allows only feasible model forms to be built.

- Variable Selector this method screens all independent variables by correlation coefficient, noise level and a vector of process facts passed to it by the "process expertise" (covered later) to the dependent variables.

- Linear Regression System to apply a first pass regression model.

- Nonlinear Regression System applies specific model forms.

- Univariate Modeler examines singular variable causality.

- Multivariate Modeler includes an adaptive simultaneous equation solver to expand to the size the model requires.

- Goodness of Fit Test applies R, F test, T test, etc., as needed by each module.

- Final Model Compiler parses a large matrix of variable combinations and permutations based on all preliminary findings. Then, after generation, a model PEV (percent of variance) maximizer and "the process expert" sort out the best fit (maximum % explained variance) for a validated model.

The second major subsystem is the Process Expert. This cooperating member is an intelligent knowledge acquisition system and a control on the model design. This asks the user about his process and then self-generates an expert system for use in model validations and variable selection. The object system utilizes an easy to use form interface and database system to inquire and gather information such as:

1) correlations expected which dependent variable is affected by which independent variable and the type of response expected; direct or inverse.

2) reliability indicates the conviction the human process expert has in his information on correlation. The highest rating tells the system to be exacting on the model form and the relation of variables to one another. If the statistics show opposite results, they and the data that generated them should be considered erroneous and will be eliminated from future consideration. This will also affect any conclusion reasoned about currently by the "tainted" data. Lower reliabilities are provided at several levels which causes similar rejection procedures if they do not pass respective "goodness of fit" criteria. Therefore, a model generated from raw data can be allowed to be counterintuitive to the process expert, if it passes validating statistical tests and the process expert's confidence in his own understanding is

low. The lowest reliability allows a process correlation to be set by the statistics alone.

3) importance provides a multiple level of rating a variable or variable pair (dependent-independent). Depending on their rating, some variables will be given top priority in all searching sorts and attempts to attach these variables or pairs to the final model. This forces important variables, regardless of value, into models at the exclusion of less important ones. This allows a user to build a model exclusively on their requirements.

All three major criteria form the basis of how variables are treated and parsed for model development. In all cases, information is entered in by forms prompted by the system. Actual creation of the "process expert" is done by the system using the knowledge and constraints input by the user.

The last major subsystem is the Object-Oriented Database. The object-oriented database allows specific data to be passed from module to module for use. It minimizes database access and the use of associated computing resources. It allows direct and simultaneous inspection and manipulation of data via the statistical or process master subsystems.

The Model Development Algorithm

In order to drive the model development and access the knowledge base of either the process master or the statistical master, an algorithm was created to provide the underlying mechanism.

Initially, all the independent variables are considered as potential starting points. The degeneracy checking, correlation checking, and process information checking provides a feasibility and priority for analysis.

The algorithm forces the exploration of variables and permutations and combinations of variables which look promising (i.e., high correlation, passes F test, R* test, etc.). Each time it finds a variable set it measures it against its goal of increasing and, ultimately, maximizing percent of explained variance.

The potential variables or combinations thereof are subtracted by nonconformance to statistical tests and negation by the process expert.

The system builds a tree of potential model variables and combinations via search and variable explosion. The tree is pruned by either cooperating member. The remaining variable sets are incrementally added to an equation and at each step tested for goodness of fit and value to the model.

The result is a model which:

1) is a best statistical fit that has been validated many times,

2) has been validated by the process expertise throughout development, and

3) is a model containing the maximum percent of explained variance possible in this dataset.

System Implementation

The system was implemented using a workstation hardware platform. The software required a VMS operating system, but was transported to a Unix environment for other users. Due to the size and complexity of the system, a large AI environment Mercury KBE was selected as the development platform, along with Digital's Rdb relational database.

The concept of the paradigm specific tool was fully developed. The user is completely shielded from statistics and need only manipulate four external parameters which control the depth and efficiency of the search space, such as:

- maximum power (exponent in model) of the model (default = 5)

- linear or nonlinear final form of the model (default = nonlinear)

- normal distribution and derivatives confidence limits; 90%, 95%, 99% (default = 99%)

- minimal threshold for increment in percent of explained variance for new model components (default = 1%)

The system is deployed with all menu, form and scroll box interfaces and requires no computer skills. The user is guided through:

1) loading and editing data

2) choosing the variables to be considered

3) selecting the four parameters previously mentioned

4) starting the "run" and reading the "conversation" between the process expert and the statistical subsystems

5) reviewing the relationships via the plotting functions provided

The output is a detailed point by point analysis of:

1) data, variables, relations and causality

2) cross correlation and potential degeneracy from independent variables; analysis for variable exclusion

3) relations of dependent variables to independent variables, their conformance to significance, normal distribution, and process expertise

4) model development

5) model validation; actual versus predicted values are printed out

6) percent of explained variance measurement and point by point deviation analysis

System Application

The ISPA was exposed to two difficult problems that had been previously abandoned as too difficult. The first problem was a major chemical company which analyzed a complex reactor operation for final product quality control. The results showed the generation of major models which are directly attributable and usable to improve their product quality. The models will be used directly in computer based recipe control. An additional result was the high validity of the automated (DCS controlled) plant models. This can directly help justify plant automation prospects. The effort saved several man-months of dedicated research.

The second problem was the analysis of an extensive, unedited database on the quality of a steel product produced from a continuous caster of a major steel company. With a single run of several minutes, the system found several significant relations in heat balance around the caster's mold which could predict the occurrence of a condition which is a severe quality problem. The problem was previously assessed as too large and complex for a statistical project. One part of the problem developed a set of relations and models identical to what a metallurgical engineer did. ISPA solved the problem in two minutes, while the engineer took three months with a modern traditional statistical package.

Both applications saved substantial manpower and provided a solution to a problem which could have taken years of statistical consultation to solve.

The nature of a paradigm specific tool, especially a complex and domain rich tool, allows for a substantial and dramatic benefit when focused at its domain of expertise. The intelligent statistical model builder can be focused at many manufacturing and commercial quality problems, as long as the quality is measured and some causality exists in the independent variable data.

This tool can help "close the loop" on many SPC/SQC/control problems with immediate and significant results. As an example, a supervisory control system can use the model to predict quality and advise/adjust control parameters. In addition, many manufacturing, design, development, pilot and research problems could be solved with such an identification tool.

Control Extensions

The Intelligent Statistical Process Modeler can be coupled with the Hyperplane concept for an intelligent control system. The ISPA tool can develop a model for the Hyperplane's search (or other optimizers) as it attempts to push manufacturing to an optimum. Extensive statistical methods, greater speed, and more models at each decision node could present an optimizing controller with more depth and options to improve (cost, quality, etc.). Hyperplane drives the process by inspection of the coefficient sign and magnitude in each model. More significantly, the ISPA can present nonlinear models to a Hyperplane method which would be more rigorous, oscillate around the optimum less, and converge quicker than the current linear regression technique which it, as well as other optimizers, depend on.

Conclusion

The paradigm specific tool ISPA has taken a major step beyond generic AI shells and traditionally developed intelligent statistical software systems. The major reason why this technology became more successful is through the use of a process expert/cooperating expert system ("built-in") to validate relations and intelligently "prune" the statistical search space. This directs model form, variables to be included, and correctives of the final model. Previous systems contained no process expertise and, therefore, became burdensome for validation or generated models which lacked robustness.

Two major reasons why the technology was created now are:

- Advances in computing technology, such as relational databases, standardized platforms and cluster technology, provide easy access to data and the ability to supply results to many end users.

- The advances in AI shells to support a generally complex, multiparadigm, integrated and cooperating expert system. Previously (prior to 1986), the environment to create such a tool did not exist.

The technology within the system emulates a team of master statisticians working on a project with an expert. This is an example of a paradigm specific tool (PST) which is a member of a new generation of expert systems. This tool and other PST's focuses the power of expert systems to solve a specific class of problems while insulating the user from AI, computer skills and the problem science domain.

Modern AI techniques have evolved to the point where large knowledge bases can be implemented and extensive expertise can be captured. The newest generic improvements are in paradigm specific tools.

The need for model building from actual business data is universal for those industries looking to cut costs or improve quality. A paradigm specific tool was built around statistical model building. Extensive statistical tools were coupled with a self-defining process tool and an object-oriented database.

The tool was tested against some difficult problems with exceptionally good results. This validates the benefits of a paradigm specific tool which can be focused generically at a class of problems.

Author Biographical Data

Over the last 14 years, Mr. Bozenhardt has held various engineering and management positions in major corporations implementing integrated control engineering, artificial intelligence and operations research based solutions to operation plant environments. Mr. Bozenhardt's main contribution to the technical community is the demonstrated infusion of highly profitable technical solutions into operating plants. The result of these efforts are the subjects of his numerous papers which touch many industries. Mr. Bozenhardt is a recognized expert in the areas of control engineering, distributed control systems, artificial intelligence, and plant computing environments. He is a graduate of Polytechnic in New York and a member of AICHE, ISA and ISPE.

Randomized Heuristic Search Approach to Cutting Stock Problems

Yuval Lirov and Moshe Segal
AT&T Bell Laboratories
Holmdel, NJ 07733

Introduction

This paper relates to resource allocation and, more particularly, to allocating a common resource among a number of demands for the resource. The methodology described in the paper is widely applicable, but for expository reasons we restrict our attention here to "cutting stock" problems, which usually arise in a manufacturing environment, where the common resource and the demands for the resource have geometric meaning or interpretation. For example, (R.W. Haessler, 1988) describes a cutting stock problem in a paper mill, which produces large rolls of paper, and specifically a heuristic procedure for solving a one-dimensional roll trimming where large width rolls of paper need to be cut into a number of smaller width rolls of paper to meet customer demands.

In the process of cutting a large roll of paper, there is usually a large number of alternative settings for the knives which are used to cut the large roll into the number of smaller rolls. Ideally, it is desired that the knives be set and, perhaps, reset several times in such a way that each large roll would be cut into a number of smaller rolls without leaving any waste and at the same time the demand for small rolls must be satisfied.

Unfortunately, the ideal settings of the knives are seldom achieved for at least two reasons. One, the sum of the widths of the smaller rolls may not equal the width of the large roll. Second, even if there exist patterns for which the sum of the widths of the smaller rolls does equal the width of the large roll, the number of alternative settings for the knives to be examined may be large and the time to find the ideal selection of patterns to meet customer demands may be prohibitively long.

In light of the above, it is common to settle on a solution which may have some waste and satisfy customer demands within a tolerance, but which can be found expeditiously.

Summary of the Method

The trim problem, if to be solved optimally, is known to be NP-hard and requiring prohibitive computational costs even for small numbers of constraints. The basic conventional method of solving similar problems is via an integer linear-programming formulation (Papadimitriou and Steiglitz, 1982) which is resolved either by a branch and bound method or by a relaxation and a subsequent heuristic. This approach assumes linear objective function and linear constraints, and thus restricts the class of technological constraints that can be dealt with. Therefore, a method dealing with an arbitrary set of constraints is sought.

We propose an improved method of dealing with such problems. Our method consists of searching a specially structured state space which allows us to solve efficiently a wider class of stock-cutting problems. This special structure is achieved by using an expert system approach (Winston, 1984) which consists of representing the problem-solving knowledge as a set of rules. The rules are chained using a separate inference engine to obtain new results. These rules can be manipulated separately and independently from the inference engine. We also combine a probabilistic algorithm (Metropolis and Ulam, 1949) for generating the cutting patterns and a search algorithm for the optimal set of such patterns. Additionally, such a special structure allows the implementation of a learning aspect of the solution-seeking process whereby the problem-solving experience is utilized for the next similar problem thus speeding up the entire solution process. We will discuss this issue in a separate report (Lirov, 1992).

Our method includes generating a set of cutting patterns as candidates for a recommended solution, setting goals or constraints for a pattern to meet before the pattern becomes a candidate for the recommended solution, searching the set of patterns and determining those patterns that meet the goals, and appending those patterns that meet the goals to the recommended partial solution. We generate feasible cutting patterns using a random procedure. We then calculate the number of times a pattern should be repeated in the solution: a multiplicity ratio. The demand for small rolls are then reduced by the specific number of rolls appearing in the cutting pattern. The generating and reducing are being repeated until the demands are met whereupon the resultant multiplicity ratios and feasible patterns can be provided as the recommended solution. The cutting pattern generation procedure and the calculation of the multiplicity ratios are done in integers and thus alleviates difficulties commonly encountered when using linear programming based approaches.

Application of the Method to Paper Mill

To aid in understanding the principles of our method, we choose to use a paper mill production process. Consider a large roll of paper (log-roll) having a width W. For purposes of our description, we assume that the diameter of the large rolls are identical and therefore play no role in the problem. Consider also that log-rolls are to be cut into a number of smaller width rolls, each smaller roll having its own width W_i, $1 \leq i \leq I$. The total demand for rolls of width W_i is D_i. In many cases, industry practices allow for satisfying the demands within a tolerance. Thus a solution which produces d_i^* rolls, $d_i \leq d_i^* \leq D_i$ is considered to be acceptable. Clearly customers demands for small rolls can be viewed as a demand vector D with I demand elements D_i where each demand element D_i could be the sum of the demands of several customers, while the smaller widths can be viewed as a width vector W with I smaller width elements W_i. Hence, the total demand (measured say in inches of log-rolls) is $\sum_i D_i W_i$ and it is worth noting that the lower bound on the number of log-rolls needed to satisfy the demand is equal to $[\sum_i d_i W_i W]$.

One should also note that any cut of the log-roll can include one or more cuts of any specific width W_i, e.g. a single cut of the log-roll could include a number of cuts of the same width W_i.

Goals and Constraints

We now define the term "cut pattern". A cut pattern identifies a specific setting of the knives so as to cut a specific pattern of smaller rolls from the log-roll. A cut pattern may be expressed as a cut pattern vector Y of integer numbers and of dimension I and where each respective element Y_i of cut pattern vector Y identifies the number of smaller rolls of width W_i which are to be cut from the log-roll of width W during one roll trimming operation, i.e. during one cut of one log-roll. A "feasible cut pattern" is defined to be a cut pattern that satisfies certain constraints. One obvious constraint that must be satisfied by a feasible cut pattern is that the inner product of Y and W must not exceed the width W of the log-roll, i.e.:

$$\sum_i Y_i W_i \leq W \qquad\qquad (4.1)$$

Another example of a constraint that must be satisfied by a cut pattern which is due to the finite number of cutting knives is that the total number of small rolls in the pattern must not exceed R, i.e.:

$$\sum_i Y_i \leq R \qquad\qquad (4.2)$$

A third example of a constraint (goal) is that the amount of waste (trim loss) should not exceed $W-\underline{W}$. This constraint taking into account (4.1) translates to:

$$\sum_i Y_i W_i \geq \underline{W} \qquad\qquad (4.3)$$

Still other constraints can be imposed on a cut pattern before it may be treated as a feasible cut pattern. The other constraints (possibly non-linear) typically arise in response to process or technological limitations and to manufacturing and transportation costs. However, for purposes of our example we choose to use the constraints of equation (4) in this description. If a cut pattern vector Y meets all of the constraints of a particular application, it is accepted as a feasible cut pattern.

We now define the term "multiplicity". Note that a feasible cut pattern Y may not by itself satisfy all of the demand D. In such an event, it will be necessary to select again the same or perhaps different feasible cut patterns to assure that the demand for D_i rolls of width W_i is satisfied. If two or more cut patterns are identical, then it is said that there is a multiplicity of that cut pattern. In particular, if there are M occurrences of a specific feasible cut pattern, the multiplicity value of that cut pattern is said to be of value M.

We also associate with each feasible cutting pattern its "trim loss" which is defined to be $T = W - \sum_i Y_i W_i$. Solutions composed of cut patterns with small T (perhaps zero) to reduce waste and large M to reduce the set ups of the cutting knives are preferred.

Method Description

The problem to be solved can be restated to be the timely generating of feasible cut patterns together with their respective multiplicities so as to satisfy the demands of customers on the one hand, consistent with reducing the total trim loss of all patterns in the solution on the other hand.

The solution method described here consists of a recursive procedure and includes four sets of rules, and five memories, which can be arranged as illustrated in Figure 18-1, according to the principles of Semantic Control Theory (Lirov, 1987 and Rodin, 1988).

The first memory contains the customer demand vectors D and d. Taking into account the demand vector D, Goal Identifier generates a multi-dimension random distribution function of a cut pattern as the multi-dimensional random vector and stores the multi-dimension random distribution function in second memory. In the next step, taking into account the random distribution function, Goal Selector randomly generates a number of feasible cut patterns, which are stored in third memory. With the feasible cut pattern vectors Y and the customers demand D at hand, Adapter generates multiplicity values M for the respective feasible cut patterns and stores the multiplicity values in fourth memory. Using the multiplicity values M and the feasible cut patterns Y and the customer's demand D, Reducer generates residual demands RD and Rd, which are defined latter, and selects a recommended feasible cut pattern Y, and its corresponding multiplicity value M and residual demands RD and Rd. The recommended feasible cut pattern and its multiplicity are stored in Data Base for later use i.e., they become part of a potential final solution. The respective residual demands RD and Rd from Reducer are substituted for the demand D and d in first memory and the arrangement iterates until Goal Identifier detects the customer's residual demand RD to be equal to or less than a design parameter, $D-d$, which reflects an allowable and acceptable tolerance for the demand, which in an ideal case would be zero. A candidate for a final solution is found when $RD \geq 0$ and simultaneously for the first time $Rd \leq 0$.

Goal Identifier

Goal Identifier generates a random distribution function over the integer lattice of a multi-dimensional interval, which would take the shape of a hyper-cube with vertices specified by the demand elements D_i, when viewed as coordinates in an I dimensional vector space. Further details on generating random distribution functions over a multi-dimensional interval may be found in many standard textbooks on computer simulation such as (Rubinstein, 1981).

Goal Selector

Goal Selector then randomly generates a number of feasible cut patterns in the following manner. Firstly, Goal Selector generates a cut pattern vector Y with elements Y_i for each I from 1 through I according to the random distribution function stored in second memory. Consider a cut pattern vector Y to be a random vector distributed according to the random distribution function in second memory.

Secondly, Goal Selector analyzes the generated cut pattern to determine whether or not the generated cut pattern satisfies the imposed constraints or goals such as the constraints represented by equations (4) as well as any other constraints or goals, i.e. Goal Selector determines whether or not the generated cut pattern is a feasible cut pattern.

Thirdly, Goal Selector repeats the above two stages until a predetermined number of feasible cut patterns is obtained and the generated number of feasible cut patterns is stored in third memory.

To exemplify the above three stages, it ought to be noted that the first and second stages could be intertwined or could be done concurrently as a parallel process to more timely generate a predetermined number of feasible cut patterns. For example, starting with an empty cut pattern vector Y, i.e. all elements of the cut pattern vector are zero, one could generate a discrete probability distribution vector P having probability elements P_i defined as:

$$p_i = \frac{D_i}{\sum_{j=1}^{I} D_j} \tag{5.1}$$

Next, we randomly select an integer i according to probability P_i. Thereupon, element Y_i corresponding to the selected integer i of the empty cut pattern vector Y is incremented by the integer one and the updated cut pattern vector Y is analyzed to determine whether or not the imposed constraints (2) are satisfied. In the event that they are, the pattern is saved as a feasible cut pattern, the initial customer demand vector D is reloaded, and we re-start the iterative process with a new empty cut pattern. In the event that (2.1) and (2.2) still have slack and (2.3) is still not satisfied, the discrete probability distribution vector P is updated by decrementing demand element D_i by one and recalculating the distribution vector P and thereafter randomly selecting another integer i accordingly to update probability element P_i. The process of generating a feasible cut pattern iterates for at most a predetermined number of random draws. It terminates with either a feasible pattern in hand or with a decision to abandon the trial cut pattern. The process is then restarted from the beginning until a predetermined number of feasible cut patterns is generated.

Adapter

Adapter generates a multiplicity value M for each of the number of feasible cut patterns stored in third memory by the following methodology. Firstly, a multiplicity ratio X_i is generated for each value of i from 1 through I for which there is a feasible cut pattern element Y_i that is greater than zero:

$$X_i = \frac{D_i}{Y_i}, \text{ for } i \ s.t. \ Y_i > 0 \tag{5.2}$$

Secondly, having generated a number of multiplicity ratios X_i, Adapter selects the smallest multiplicity ratio X_i corresponding to a feasible cut pattern Y, truncates any fractional remainder from the smallest multiplicity ratio X_i from equation (5.2), and assigns the integer part of the smallest X_i as an upper limit $\overline{M}$ of the multiplicity value M for that feasible cut pattern Y, i.e. $\overline{M} = [\min_i X_i]$. Accordingly, the multiplicity value for a feasible cut pattern Y is symbolized as M where M can have any value between zero and the upper limit $\overline{M}$, i.e., $0 \leq M \leq \overline{M}$ (In most cases one will set $M = \overline{M}$).

Thirdly, Adapter repeats the above two stages until a multiplicity value M is generated for every feasible cut pattern Y and the generated number of multiplicity values is stored in fourth memory.

Reducer

Reducer now generates residual demands RD and Rd by using customer demand D and d from first memory and the number of multiplicity values M from fourth memory and the number of feasible cut patterns Y from third memory to update the customer demand in first memory and to select one multiplicity value M as well as its corresponding feasible cut pattern Y to be stored in Data Base as a part of the recommended solution.

Firstly, Reducer generates residual demand vectors RD and Rd for each feasible cut pattern Y by subtracting the vector product of the feasible cut pattern Y and its scalar multiplicity value M from the customer demands D and d, or

$$RD = D - YM \tag{5.3}$$
$$Rd = d - YM \tag{5.4}$$

where D is the demand vector, $D-d$ the permissible demand tolerance vector, Y is the feasible cut pattern vector, and M is the multiplicity scalar value.

Secondly, Reducer evaluates each of the residual demands in the following manner. Each residual demand vector RD also has I residual demand elements RD_i. Reducer selects the largest residual demand element in the residual demand vector, called $MaxRD_i$, and divides the value of the largest element $MaxRD_i$ by the multiplicity value M of the feasible cut pattern Y that was used in generating the residual demand RD, according to equation (5.3). The resultant quotient is called the evaluation ratio ER for that feasible cut pattern Y, or:

$$ER = \frac{MaxRD_i}{M} \tag{5.5}$$

and the evaluation ratio is temporarily stored. Of course, other forms of evaluation ratio ER are possible. For example, any function of $MaxRD_i$ and M would be a satisfactory evaluation ratio, consistent with the intent of (5.5). That is, to be able to highlight good patterns to include in the final solution with small $MaxRD_i$ and at the same time large M.

Reducer repeats the evaluation process for each of the number of residual demands RD (there being a residual demand vector RD for each feasible cut pattern Y), which results in a corresponding number of corresponding evaluation ratios.

Thirdly, Reducer selects (a) the residual demand vector RD, (b) the feasible cut pattern vector Y used to generate that residual demand vector, and (c) the multiplicity value M of that feasible cut pattern vector, all of which are in one-to-one correspondence with the smallest quotient generated in the evaluation process, i.e. the smallest of the ERs, the evaluation ratios.

Fourthly, Reducer (a) replaces the demands D and d stored in first memory with the residual demands RD and Rd selected in the third stage above, i.e. with that residual demand corresponding to the smallest evaluation ratio, and (b) stores, in Data Base, the feasible cut pattern Y and its multiplicity value M selected in the third stage above as a part of the recommended solution. The purpose of the reducer is to select cut patterns which will result simultaneously in maximal reduction of yet unfulfilled demand and having solutions with large multiplicity ratios.

Numerical Example

We now refer to Figure 18-2 for a numerical example that illustrates the principles described above. Consider log-rolls of width W equal to

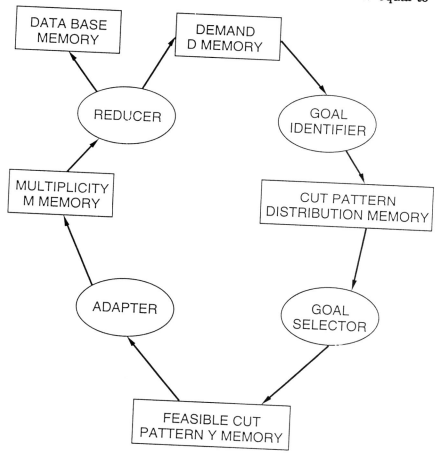

Figure 18-1   Semantic Control Architecture

91 units. $\underline{W}$ equal to 90 units and R equal to 5 rolls. Next, consider a requirement to cut the log-rolls into D_i=13 smaller rolls, each of width W_i=30 units, and into D_2=23 smaller rolls, each of width W_2=20 units, and finally into D_3=17 smaller rolls, each of width W_2=10 units. Let us also assume that all the demands have a tolerance of one unit each, (D_i-d_i)=1. Using the above symbology:

$$W=91, \underline{W}=90 \qquad\qquad\qquad\qquad (6.1)$$

$$R=5 \qquad\qquad\qquad\qquad (6.2)$$

$$W=(31,20,10)=(W_1,W_2,W_3) \qquad\qquad\qquad (6.3)$$

$$D=(13,23,17)=(D_1,D_2,D_3) \qquad\qquad\qquad (6.4)$$

$$d=(12,22,16)=(d_1,d_2,d_3) \qquad\qquad\qquad (6.5)$$

Note the symbology of equations (6) has been transferred to, and is shown in, node A Figure 18-2 for use in this numerical example.

Goal Selector generates a number of feasible cut patterns. In so doing, Goal Selector randomly generates a number of cut patterns Y, two respective ones of which are shown in nodes B and C of Figure 18-2. Goal selector analyzes the cut patterns Y to determine whether or not the cut patterns Y satisfy certain constraints or goals, e.g. the constraint represented by equations (4). In this case, note from the W vector in node A and from the Y vector in node B that:

$$\underline{W} \le YW \le W, \quad \sum_i Y_i \le R$$

$$1\times30+2\times20+2\times10=90\le91 \qquad\qquad (10.1)$$
$$1+2+2=5\le5 \qquad\qquad (10.2)$$

Hence, the cut pattern Y in node B is a feasible cut pattern. But, as mentioned earlier, there are likely to be a number of cut patterns generated and each generated cut pattern is to be examined for the purpose of determining which one or ones of the generated cut patterns is or are also a feasible cut pattern or patterns. In our example, node C illustrates still another feasible cut pattern.

Adapter generates multiplicity value M for each feasible cut pattern Y. Note that, by use of the multiplicity ratios X_i from equation (5.2), eight repetitions of the feasible cut pattern in node B is the maximum possible number of cuts using that feasible cut pattern without exceeding the customer's demand D. In this example, we will assume $M=\overline{M}$.

Reducer generates residual demand vectors RD and Rd by using the initial customer demand vectors D and d from first memory and the number of multiplicity values M from fourth memory and the number of feasible cut patterns Y from third memory to update the initial demand D and d in first memory and to select one multiplicity

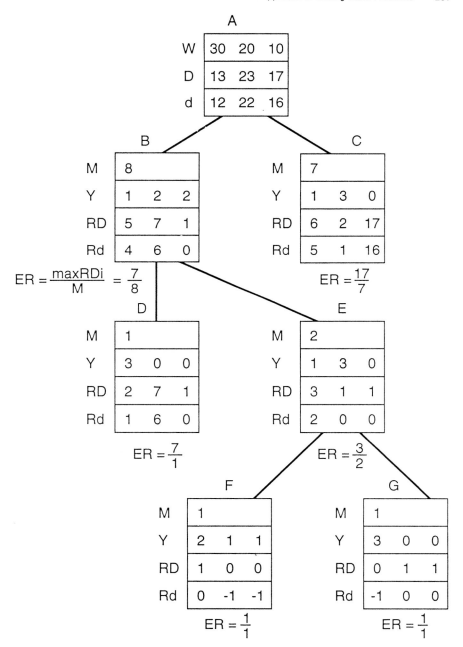

Figure 18-2   Randomized Search Trace

value M as well as its corresponding feasible cut pattern Y to be stored in Data Base as a part of the recommended solution. In this numerical example, Reducer generates the residual demand vectors RD illustrated in nodes B and C by determining the difference between the demand D in node A and the amount of that demand D that is satisfied by the multiplicity value M of the feasible cut pattern Y when that multiplicity value is applied to the feasible cut pattern. As to node B, the residual demand RD that remains to be satisfied after eight, i.e. $M=8$, cuts of feasible cut pattern$Y$ (remember that cut pattern $Y=(1,2,2)$ was randomly generated and was then determined to be a feasible cut pattern) is a residual demand vector RD of $(5,7,1)$. Now as to whether node B or whether node C will be used as a part of the recommended solution is conditioned on the evaluation ratio ER, here in our example, on the smallest ratio of the maximum residual demand element $MaxRD_i$ to the multiplicity value M, i.e. upon the smallest evaluation ratio determined pursuant to equation (6). Here, note the maximum residual demand element of node B is seven, i.e. its $MaxRD_i$ is $RD_2=7$, while the maximum residual demand element of node C is 17, i.e. its $MaxRD_i$ is $RD_3=17$. In accordance with our methodology, the procedure recommends a solution with the smallest evaluation ratio, in this case the evaluation ratio of 7/8 which corresponds to the feasible cut pattern in node B.

The foregoing process then identifies node B as being a part of the recommended solution. Thereafter, the arrangement repeats the process to obtain nodes D and E. Since node E has a feasible cut pattern and has the smaller evaluation ratio, it, i.e. node E, is added or appended to the recommended solution and again the process repeats to nodes F and G.

As earlier mentioned, the arrangement repeats the methodology until Goal Identifier detects some predetermined tolerance that is deemed acceptable. In this example, we assume that the acceptable tolerance is set in such a manner that no residual demand element shall exceed a value of one. That means that either of nodes F or G can be added to the recommended solution since they both include feasible cut patterns and since no element in their respective residual demands exceeds one i.e. $(RD-Rd)\leq 1$.

We use Figure 18-3 to illustrate how the iterative methodology combines the results to obtain a completed recommended solution. First, node B of Figure 18-2 recommended eight multiplicities, i.e., $M=8$, of feasible cut pattern $Y=(1,2,2)$ be cut from the log-roll in smaller widths $W=(30,20,10)$. Translating Figure 18-2 into Figure 18-3, note that the recommended solution includes the knives being set to cut the log-roll for one width of 30 units, two widths of 20 units, and two widths of 10 units. That cut is repeated eight times, each cut being of the aforesaid normalized length.

Second, node E of Figure 18-2 recommended two multiplicities, i.e. $M=2$, of feasible cut pattern $Y=(1,3,0)$ be cut from the log-roll in

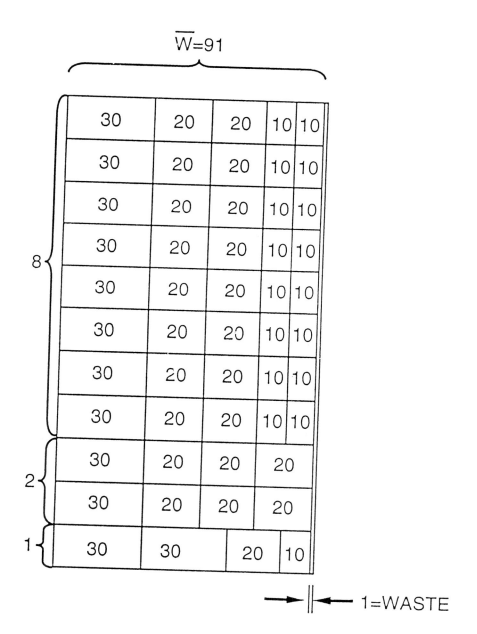

Figure 18-3   Optimal Stock Cutting Example

smaller width $W=(30,20,10)$. Translated into Figure 18-3, note that the recommended solution includes the knives being set to cut the log-roll for one width of 30 units, three widths of 20 units, and zero widths of 10 units. That cut is repeated two times. Third, node F of Figure 18-2 recommended one multiplicity, i.e. $M=1$, of feasible cut pattern $Y=(2,1,1)$ be cut from the log-roll in smaller width $W=(30,20,10)$. Translated into Figure 18-3, note that the recommended solution includes the knives being set to cut the log-roll for two widths of 30 units, one width of 20 units, and one width of 10 units. That cut is performed once. Note here that we could have, alternatively, used the solution shown in node G as a part of the recommended solution.

Note also that the trim loss using the recommended solution in our numerical example is only one unit for each of the eleven cuts of the log-rolls. It is also worth noting that the procedure might end with several potential final solutions having associated trim losses and multiplicity ratios for the corresponding cut-patterns. A final recommendation could then be made using an objective function which takes into account the relative importance of minimizing waste and setup expenses.

Note that the total trim loss is 11 (every log-roll has a trim loss of 1). Also note that the number 11 of the log-rolls is optimal since it is impossible to obtain any cut schedule containing only 10 log-rolls. The optimality of 11 log-rolls can be shown from the following consideration:

$$\sum_i W_i d_i = 30*12 + 20*22 + 10*16 = 960$$

and

$$\sum_i W_i D_i = 30*13 + 20*23 + 10*17 = 1020.$$

Since $W=91$, we obtain the following bounds for the number of log rolls: $\frac{[960}{91}, \frac{1020}{91]} = [10.55, 11.21]$. But this interval contains only integer 11 which proves the claim for this problem.

Conclusions

We describe a randomized solution method to cutting stock problems. Randomization is applied at the feasible cutting pattern generation stage to alleviate the shortcomings of the traditional Linear Programming or deterministic heuristic search approaches. The method has been implemented in PROLOG and tried out with encouraging results on several problems of practical size.

Acknowledgement

The authors are grateful to E. Bahary for numerous discussions on the subject.

References

R. W. Haessler,"Selection and Design of Heuristic Procedures for Solving Roll Trim Problems," Journal of the Institute of Management Sciences, Vol. 34, No. 12 (December 1988), pp 1460 - 1471.

Y. Lirov, Artificial Intelligence Methods in Decision and Control Systems, Doctoral Dissertation, Washington University in St. Louis, MO, 1987.

Y. Lirov, "Knowledge Based Approach to Cutting Stock Problems," forthcoming in Computers and Mathematics with Applications, special issue on Cutting Stock Systems, editor Y. Lirov, Pergamon Press, 1992.

I. Metropolis and S. Ulam, "The Monte Carlo Method," Journal of the American Statistical Association, 44(247), 335-341, 1949.

C. Papadimitriou and K. Steiglitz, Combinatorial Optimization, Prentice Hall, 1982.

E. Rodin,"Semantic Control Theory," Applied Mathematics Letters, vol 1, no. 1, 1988., pp. 73-78.

R.Y. Rubinstein, Simulation and the Monte Carlo Method, New York: John Wiley. 1981.

P. Winston, Artificial Intelligence., Addison-Wesley, 1984.

Author Biographical Data

Yuval Lirov is a Member of Technical Staff in Bell Laboratories in Holmdel, NJ. He studied Mathematics at the University of Vilnius, Lithuania, Operations Research at the Technion, Israel, and Artificial Intelligence and Control at Washington University in St. Louis. A winner of the AIAA-1987 Outstanding Achievement Award, Dr. Lirov has authored over 50 technical publications and patents; he has recently edited a book on Applications of Logic Programming in Decision and Control. In AT&T he develops expert systems for manufacturing and telecommunications applications.

Moshe Segal is Supervisor, Operations Research Methods Group AT&T Bell Laboratories. He and his group developed several generic optimization algorithms and programs; queuing analysis tools; workforce scheduling packages; models and codes to optimize communications systems and networks; and more recently a variety of operations support tools for manufacturing.

He received a BSc and Ingenieur degrees from the Technion, Israel Institute of Technology, and he holds a Dr. of Engineering degree in Operations Research from the Johns Hopkins University. He is an author of several papers and patents. Mr. Segal is currently a member of the editorial boards of Operations Research Letters, Discrete Applied Mathematics and Annals of Operations Research.

6

Diagnostics

A Fault-Tolerant Neural Network Applied to Nondestructive Inspection

S. H. Simon
LTV Missiles and Electronics Group,
Dallas, Texas

Introduction

This overview presents research and development of a personal computer (PC) based neural network system which can be applied to the problem of non-destructive inspection and testing. A general outline of neural network basics is presented. The neural network approach uses learning from examples. The goal of this article is to present a general system which may be applied to damage detection of mechanical structures using two different methods.

Mechanical structures, such as aircraft, manufacturing tools, vehicles, and even household appliances, have maintenance schedules to extend their performance lifetimes. Routine maintenance allows the operator to regularly notice and repair any problems which come up during normal wear and tear on the machinery. However, inspection and maintenance require that the machinery be taken out of operation, resulting in unproductive downtime. Also, problems may arise before scheduled maintenance, resulting in unscheduled, unproductive downtime.

This raises a common and accepted dilemma. If routine maintenance is scheduled too frequently, the result is expensive, scheduled downtime. If maintenance is scheduled too infrequently, the result can be disastrous and very expensive in terms of downtime or loss of the machinery. Usually, experienced operators can provide some intuition about optimized schedules. But their experience is limited. And unforeseen events can result in unexpected, unscheduled problems. Also, during detailed inspection, trained experts are needed to interpret the results of the inspection instruments. Not only are these experts costly, but they may also be in high demand, and therefore not always available, resulting in further delays.

It would be beneficial if a way to schedule maintenance only when needed could be found. When damage or wear is detected, the machinery could be taken out of production, inspected, and repaired. It would also be helpful if an automated system to interpret the results of inspection were available.

This work describes how neural networks might be applied in these ways. Because of the flexibility of neural networks, they may be applied in different ways. First, neural networks can be connected directly to the machinery and provide a continuous update on the health and maintenance status during operation. Second, they can be used to augment the availability of trained experts by interpreting the routine output from inspection instruments, such as X-ray, N-ray, ultrasound, etc. We will develop these ideas further in this paper.

Neural Networks

A neural network (NN) is a software or hardware implementation of a parallel, adaptive matrix which takes multiple, simultaneous inputs, processes the inputs through a mathematical threshold function, and returns an output. In simpler terms, a NN is an adaptive filter that is not programmed but is trained by showing a range of examples. The NN uses these examples to develop an internal representation by regression. The internal representation is a "mental model" generalization of the presented examples. The NN uses this mental model as a pattern to match other patterns against. It can then recognize other patterns within a range of tolerance. For example, we have trained with images of military vehicles at 0 and 90 degree orientations. This range of two example images per vehicle was sufficient for the NN to learn the vehicle and be able to recognize images at other angles such as 35, 75, and 110 degrees. Therefore, our NN pattern recognition system has a some tolerance within its match capability.

A NN has a degree of tolerance because it is composed of many, redundant nodes. Each node in a NN sees a different part of the situation. These parts may overlap. The different nodes make a decision based on their part of the problem, then a consensus of these decisions is used to make a final, global decision or recognition. For example, in an inspection situation, one node might see a section of material which has no defects, while three other nodes might see a defect. Therefore, the consensus is that the material contains defects. Also, weighting of node observations can be used. For example, if three nodes see no defects and only one node sees a defect, but greater weight has been assigned to defect patterns, then the single node may win the consensus of decision and the material will be classified as defective.

The weighting is set up during training. The NN is presented with examples of sensor output from defective and nondefective material. The NN nodes record the inputs and sum them. When the sum of the inputs exceeds a given threshold value, the nodes generate an output signal which is propagated to other nodes. As the various examples are presented, the paths between nodes are strengthened and weakened based on a training algorithm. The weakening and strengthening is done by assigning weights as defined by a training algorithm. The weights represent the NN memory of learned patterns much the same as coefficients represent a curve in a regression equation.

In Figure 19-1, we present the mathematics for the behavior of nodes in a NN. Inputs X1 through Xn are multiplied by Weights W1

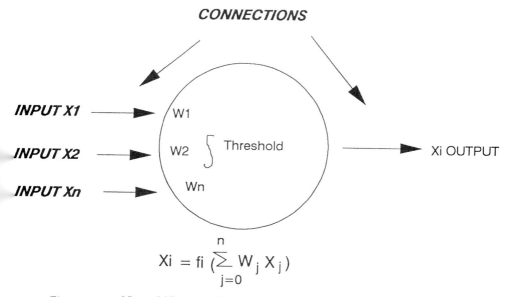

$$Xi = fi \left(\sum_{j=0}^{n} W_j X_j \right)$$

Figure 19-1   Neural Network Node - When the result of the summed inputs exceeds the threshold, the output Xi results.

through Wn for this specific node and then summed. The output Xi results when the weighted sum of the inputs exceeds a given threshold within the non-linear transfer function fi.

NN nodes are connected in layers. Although other network configurations are discussed in the literature, for our discussion, there are three layers as shown in Figure 19-2. The input layer is connected to the external world or to the sensors. The middle (or hidden) layer is used to consolidate the results of the input layer, and the output layer provides the results of the consensus propagated from the previous layers. A rough analogy is provided by comparison with a legal trial. The two lawyers gather information from the external world much the same as two nodes in the input layer. The twelve person jury consolidates the information from the layers, similarly to the middle layer. And the judge makes the final verdict, somewhat like the output layer. In practice, there are usually many more nodes in the input layer than in the middle layer, and the number of nodes in the output layer usually corresponds to the number of possible outcomes.

Weight assignment and signal propagation among nodes and between layers is defined by a training algorithm. The training algorithm is created based on theories which describe learning mechanisms. Various psychologically-based learning mechanisms have been proposed and evaluated in the literature. Some of these mechanisms were used to recreate different aspects of biological systems in the NN models. Some of the goals in the literature included optimizing learning, learning quality, and learning time. The different mechanisms configure a specific NN model to perform a specific function. The mechanisms may be broadly classified in two groups: supervised and unsupervised techniques.

In supervised learning, the weights are adjusted to minimize the difference between expected and real outputs. Learning is guided or

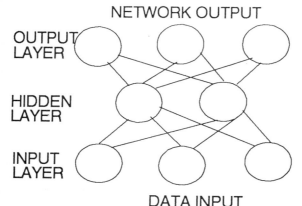

Figure 19-2   Neural Network - The input layer is connected to the middle (hidden) layer, which is connected to the output layer.

directed by identified examples. Weights are increased as the error between expected and real events is decreased. If the NN sees a defect and misidentifies the defect during training, then the weights are decreased. However, if the defect is identified, then the weights are increased. The amount of increase varies with how accurately the NN recognizes the degree of defect.

In unsupervised learning, the weights are adjusted by examining the inputs only, with no regard for expected outputs. Either the difference between real and recycled input is minimized or the entire NN is adjusted for optimized execution with the given inputs. For each training example, the weights are modified so that all nodes result in a similar, consensus behavior.

In both types of learning, when training is completed the network configuration and weights are fixed. The NN may now be exercised in recognition mode. The NN will yield a learned response when the input matches or is close to the examples presented at training. Although deceptively simple, this is a revolutionary concept.

A NN is an adaptive system which can modify its internal structure to optimize its performance based on past experience. It is not programmed, it is taught. The NN's unique ability to learn by example reduces the characterization of a set of sensors to a feasible and finite processing task for the computer. Therefore, using a NN as the nerve center of a sensor system is a natural consequence.

In addition, NNs provide functionalities not easily obtainable with other technologies. For example, a NN is not programmed and it does not need to have a precompiled knowledge base. All that is necessary is that the NN be provided with a set of training examples to learn and adapt from. The NN will automatically modify its internal structure to optimize its performance based on these training examples.

Because of their flexibility and adaptable training, NNs can be used for a variety of tasks. One such task is non-destructive inspection. We will discuss two types of inspection: inspection through continuous monitoring of the machinery and inspection through various imaging techniques.

Smart Structures: How to Monitor the Maintenance Health of Machinery

The term "machinery" can apply to a wide variety of mechanical structures, including aircraft, manufacturing tools, vehicles, and even household appliances. In the aerospace industry, the study of the use of an array of sensors to monitor the structural health and maintenance schedule of an aircraft is called "Smart Structures." In this discussion, we will use the term smart structures as an umbrella to refer to a full suite of smart machinery. Our smart structures work will be presented in general terms applicable to most other mechanical structures.

Introduction to Continuous Monitoring

Smart Structures are mechanical devices or vehicles which contain an array of embedded sensors. The signals from the sensors are used provide information about the safety, structural health, maintenance schedule, or performance lifetime of the structure.

Because of the number of parameters to be monitored and the size of the structure, a large number of sensors are needed. An area of investigation has been how to monitor all sensors close to real time using cost and weight effective methods. The conventional approach of running a finite element model (FEM) can be costly and bulky, requiring a supercomputer to process the sensor signals. It is not clear that this approach, if viable otherwise, would even be time efficient.

Another approach is the use of a model-based expert system. This approach reduces the need for high-powered number processing typical of a large supercomputer. However, the a priori knowledge needed for this type of system is immense. For example, the equivalent of a FEM simulation might be run to develop the model for the expert system. However, once it is run, the model does not require a supercomputer, but can be handled as a set of heuristics in the expert system running on a smaller computer. Other possible limitations include difficulties with noisy, faulty, or missing sensor signals. Also, once the system is developed for one structure, it may not be easy to transfer the system to a different structure. It would be more efficient to have a system which could learn and adapt to the specific implementation structure without a great deal of processing or reprogramming. This flexibility exists in the form of NN. We chose to base our Smart Structures approach on NN, specifically because of this flexibility.

General Schematic of a Smart Structure Application

A smart structure application consists of three subsystems: 1) Structure; 2) Sensors; 3) Computer. Although we are using aircraft components, the structure can be virtually anything. We have envisioned starting with sailing vessels, tooling machinery, automobiles, and even the human body. The sensors used, depend on the structure and desired application. For example, to measure stress and strain, we use strain gauges. To measure heat, we might use thermocouples. We might also use accelerometers, EEGs, or ECGs, etc. depending on the desired application and measured variable. The results of the sensors are interpreted by computer.

The computer forms the basis for sensor collection and interpretation. We have chosen to use a NN running on a PC, because of portability, ease-of-use, and adaptability. With a NN, we do not need an expert in sensors to build a model. We do not need to reconfigure the software if we add sensors. And we do not need to reprogram the entire system if we change sensors or structures. For all changes, we only need to retrain the NN to handle the new data or configuration. This means that the NN used to interpret strain gauge information, can also be used to interpret information from a microphone simply by retraining. If the sensors signals are collected through an analog-to-digital converter and placed in a buffer for the NN, then no other changes are needed. In fact, we are currently using a NN which interprets video images. The same system, trained for different goals, can interpret X-ray images. We will discuss this idea in a following section, on image processing. The major point is that the NN-based smart structure system is flexible and applicable to a variety monitor and maintenance situations.

Fault Tolerance

In a multi-sensor system, we can have a variety of faults and misreadings. The faults can arise from faulty sensors, failed sensors, noise, and spurious events. With a small number of sensors, some of these errors can be anticipated in a model of the system. As the number of sensors grows, the complexity of the anticipated model would grow combinatorially. This explosion in complexity, coupled with the unpredictable, spurious events can quickly make a model-based system inadequate. The NN approach uses real examples, not predicted examples. So with sufficient examples, noise and spurious events become part of the internalized NN model. The NN can handle failed and faulty sensors by taking a global consensus of all sensors. At some point, even the NN system will no longer function, but performance will degrade gracefully as a function of sensor failure. The system will not fail abruptly simply because one or two sensors do not work property.

Although the NN provides a level of fault tolerance, a useful capability would be the ability of the NN system, not only to function, but to identify cause and location of faults. We have investigated this idea, somewhat. A NN usually does not register a fault. It simply works correctly or does not work correctly. However, that is not quite accurate. What, in fact, happens is that the NN provides an interpretation of a global pattern from the collective inputs. If it cannot recognize that pattern, it fails. So the way to identify faults is to take advantage of the global nature of NN.

We have considered two methods. The first method looks at two levels of networks. A global NN connects to an array of local NNs. Each local NN is sampled for validity. The second method is more elegant. In many situations, there are symmetric configurations of sensors. Symmetry is a local effect which may be definable during sensor placement. While the global effects are identified by the NN, a separate routine can sample the results for these symmetries. A failure to find symmetry indicates a problem with a set of sensors. By correlating with other symmetries, the routine could locate the local problems. In our own work, we located a set of switched sensors by using this method.

Potential Development Application of the Smart Structures Concept

Using the smart structure concept, a fault tolerant, manufacturing machine monitor could be developed rather inexpensively, then generalized for other machinery. Experienced machinists can sometimes diagnose the state of repair of machinery by the sound or feel of the equipment. Vibration sensors can record changes in sound and vibration ("feel"). A set of sensors can be placed both strategically and redundantly to record vibrations during normal operation. The resulting signals, along the state of wear and repair may be recorded on a PC. This data can be used to train a NN to monitor the machinery. The trained NN can then be connected directly to the machinery and provide a continual output of the progressive wear and tear on the system by identifying sampled vibration signals. Once this system was developed, integrated, and debugged, it could be modified and generalized to work with different machinery and different sensors. Based on our own experience, the initial prototype would require about 6 months, using the appropriate hardware and software expertise. Further development would require another 6 months to integrate the hardware and build a software sampling system. Final debugging, unforeseen additions, and subsequent deployment will require about 6 to 9 more months for a range of roughly 1.5 to 2 years from project initiation to completion, assuming management support and dedicated staff.

Image Recognition: How to Locate the Problem

If a structure is thought to be faulty, the fault must be located. There are a number of mechanisms for non-destructive inspection. Parts may be inspected visually, by ultrasound, by X-ray, by N-ray (neutron), etc. All methods require an expert for location, identification, and interpretation of a fault. Usually, each method requires a different specialist. We have drawn the analogy between these techniques as all being types of image recognition, which we can do with a NN system. The same NN system that works for smart structures can be adapted to work with image recognition. Rather than interpreting the results of a set of sensors, the NN will interpret collective results of a set of bit patterns. Whether these bit patterns come from video, X-ray, ultrasound, etc., makes no difference.

As in the smart structure discussion above, a set of examples or bit patterns with identification is collected on computer. The NN is trained to recognize and identify the patterns. Then, the trained NN system is connected with the inspection system to locate faults and interpret the results. Other technologies attempt to recognize patterns by collecting all the bits or pixels, then looking for specific patterns within the collection. Usually these other methods also work sequentially, one pixel at a time, so processing time can grow quickly as the number of pixels grows. The NN looks at the global collection of pixels and then processes many of the pixels simultaneously, in parallel. Therefore, processing time is much less than by the other technologies.

We have conceptualized image recognition as being similar to the multi-sensor analysis in smart structures. Rather than interpreting a set of sensors, the NN is now interpreting a larger set of pixels. With that conceptual change, most procedures transfer between smart structures and image recognition.

General Schematic of an Image Recognition Application

An image recognition system consists of three subsystems: 1) Image; 2) Sensor; 3) Computer. Although we are using video images, the image sensor can be virtually any type of sensor. We might also use accelerometers, X-rays, video cameras, or ultrasound, etc. depending on the desired application and measured variable. The results of the sensors are interpreted by computer.

The same NN that we use for the damage detection concept in smart structures, we can also use for damage detection in image recognition. Although a PC may be used, we have chosen to use an Apple MacIntosh because of the availability of powerful, yet inexpensive software. As in the previous concept, we do not need an expert to describe the theory involved in damage detection. All that is needed is for the expert to provide a set of example pictures with representative damage situations. The NN can use these example to generalize within each category of damage and provide the necessary tolerance used to flag potential areas of damage on a test image. Although a prototype NN system may not be infallible, it has the potential to relieve the expert from the more routine inspection jobs.

This leaves the expert free to address the more difficult, and frequently, more interesting inspection jobs which are too complex to train the NN to perform. However, an additional advantage of the NN system is that it does not need to be as specialized as the expert. We intend to investigate the synergy involved in using a NN which has knowledge about a variety of sensors to see if it can generalize its knowledge and understand more about the more complicated types of faults and damage possible. It is clear that interpreting X-rays is not very different from interpreting N-rays. The NN system has the potential to generalize across the fields, if we are clever enough to know how to present the training examples.

Potential Development Application of the Image Recognition Concept

Inspection of potentially damaged equipment using X-ray, N-ray, and ultrasound requires the expertise of a trained technician. Experienced technicians are an expensive commodity. NNs could be used for image recognition to handle routine inspection tasks, leaving the experts free to handle the more difficult situations, perhaps as identified by the NN. The NN can be trained to recognize various damaged conditions after being trained by examples provided by expert inspectors. The trained NN system could then identify when it recognized damage, when it saw no damage, and when it was not clear. The technician could screen the results in the beginning to determine the limitations of the system. As with the smart structure system described previously, when this system has been developed, integrated, and debugged, it could be modified and generalized to work with different image processing systems. Based on our experience, the initial prototype would require 2 to 3 months, with the appropriate graphics expertise. Further development would require about another 4 months to modify the training sets. Final debugging, additions, and production would require an additional 9 months to build credibility, for a total of about 1.5 years from project initiation to completion, assuming management support and dedicated staff.

Conclusion

We have presented an overview of two potential applications of NN techniques to non-destructive inspection. Both techniques may be based on the same central NN, and both may be generalized. In our own work, we have built the first level prototypes of each system, but used the ideas with slightly different applications, not appropriate to this discussion.

We hope that this brief overview will interest other workers in pursuing the fault tolerance and adaptability afforded by NNs.

References

A. K. Caglayan, S. M. Allen, and S. J. Edwards, Hierarchical Damage Tolerant Controllers for Smart Structures, 1989, Wright Patterson AFB, Report AFWAL-TR-89-3009.

M.L. Minsky, S. S. Seymour, Perceptrons, The MIT Press, 1969.

J. C. Royer, A. Merle, C. de Sainte Marie, An Application of Machine Learning to the Problem of Parameter Setting in Non-Destructive Testing, Proceedings of the Third International Conference on Industrial & Engineering Applications of Artificial Intelligence & Expert Systems, Charleston, S. C. , 1990, pp. 972 980.

D. E. Rumelhart, G. E. Hinton, J.J. McClelland, Parallel Distributed Processing , Vols. 1, 2, 3, The MIT Press, 1986.

L.G. Valiant, Functionality in Neural Nets, Proceedings of AAAI-88, Saint Paul, Minn., 1988, pp. 629 634.

Yoh-Han Pao, Adaptive Pattern recognition and Neural Networks, 1988 Addison-Wesley, Boston, Mass.

Author Biographical Data

Hank Simon is the manager of Artificial Intelligence (AI) in the Research Department at LTV, Missiles Division, leading a team of talented professionals in various applications involving AI, neural networks, expert systems, and robotics. He has been working in software design for more than 15 years, and AI for 10 years.

He has a Ph.D. in Artificial Intelligence, along with degrees in chemistry and physics. He has published more than 25 papers in AI and more than 10 papers in Nuclear Physics.

Amethyst:
Vibration-Based
Condition Monitoring

Dr. Robert Milne
Managing Director
Intelligent Applications Ltd
Kirkton Business Centre
Livingston Village
West Lothian
Scotland, UK

Introduction

Amethyst automates the interpretation of rotating machinery vibration data in order to facilitate condition based maintenance. This results in considerable time savings, greatly reduces the needed skill level and provides considerable benefits to the end user organization. Amethyst has been delivered to a large number of companies in the UK and the USA, resulting in considerable cost savings for industry. Because it is an easy to use product, it has the potential to address a large number of companies throughout industry world-wide.

Every item of rotating machinery such as a pump, compressor, fan, motor and shaft all generate vibrations as they rotate. Although these items are carefully constructed and balanced, very small vibrations occur due to imperfections in the bearings, the mounting of the equipment, the shape of the shaft, build up on the blades of fans, or within the pump, or other problems due to the materials.

In traditional industries a critical item of equipment such as a pump is overhauled and rebuilt at regular intervals of perhaps once a year. This approach, known as calendar based maintenance, helps to prevent failures and is far superior to maintaining pumps only when an actual failure occurs.Such a failure could lead to significant cost implications of the company if they lost a critical item of equipment at the wrong time. Although calendar based maintenance is superior to failure based maintenance, it still can be very costly. It is common that a pump is overhauled when there is nothing wrong with it. Sometimes this introduces further failures. This also generates tremendous cost wastage. It is also common to have a pump fail before its scheduled maintenance period, resulting in a catastrophic failure in spite of the preventive maintenance program. To improve on these problems most modern corporations use condition based maintenance activities[1].

Given that every item of equipment vibrates as it rotates, one must also recognize that as faults develop, in almost all cases, they increase the vibration at certain frequencies. This is commonly understood by noticing that a pump is shaking much more than it should be. A healthy, properly working pump will exhibit very little vibration. The goal of condition based maintenance is to monitor the condition of the machine through vibration characteristics, and only maintain it when the vibration pattern indicates that there is a problem. When this is working properly,

machines are only maintained when their condition warrants. As a result, unnecessary strip down and repair is eliminated and failures that would happen before the calendar maintenance period are also detected. This leads to considerable efficiency in the cost savings of maintaining equipment, it also helps to lead to a tremendous increase in factory availability by reducing unplanned downtime[2].

In large installations where the rotating machinery is critical to the safety and operation of the plant, hand held data collectors are used to collect vibration signatures. The hand held data collector is a small computer and spectrum analyzer combination. It is controlled and managed by a database on an IBM PC compatible computer. The engineering maintenance personnel visit each pump and collect a vibration signature from all the bearings, the gear box and the main shaft of the item of equipment. This is normally done by placing the transducer with a magnetic mount onto the bearing and then triggering the spectrum analysis. The spectral data is then stored in the hand held computer until the engineer returns to the main database. The vibration signatures are then downloaded into the database. The database software provides two fundamental capabilities. The first is

1   Neale, M. Condition Based Maintenance -A Review of Methods and Economics. The Factory Efficiency and Maintenance Conference, PEMEC 88, Birmingham, 27-30 September, 1988.

2   Hill, J.W., and Baines, N.C. Applications of an Expert System to Rotating Machinery Health Monitoring. Institution of Mechanical Engineer Conference, Heriot-Watt University, September 1988.

to organize all rotating machinery for a particular plant into a series of routes. It maintains a calendar based scheduling system so that each machine is visited at a regular interval. At the start of a typical day a route of 20 or more machines would be downloaded to the hand held data collector. This would then tell the engineer which machines to visit and which measurement points to take. The engineer will then spend the remainder of the day visiting the machines and collecting the data. Depending on the scale of the plant it can require a full 8 hour shift in order to collect all the vibration signatures. In large scale operations collecting the data is a full time job.

The other main capability of the PC database is to provide a means of alarming and displaying the vibration signatures. There are a number of methods to indicate that the vibration pattern is above a threshold. This indicates that the vibration has increased enough that there may be a problem. It is important to realize that there is no indication of what the precise problem might be, as will be described later.

Traditional Diagnostic Methods

The current traditional software provides extensive capabilities to display the spectral graph of any of the spectral measurement points. It also provides the capability to compute a trend and see over time how the overall vibration is changing. In general, for every measurement point an overall vibration and a spectrum are collected. The overall represents a single number such as .314, which represents the total vibration among all the frequencies. The spectral vibration is a plot usually at 400 line resolution showing the vibration and individual frequencies. The software generally alarms and manipulates on the overall and yet the spectral vibration is what indicates the problem.

The key part of condition based monitoring is to determine two items. The first is that the vibration indicates that there is a problem with the machine and so some action should be taken. The current data collection packages and software do this very well. The second, and more critical activity, is to determine the exact problem. This impacts the repair that is necessary. Some problems are very simple to resolve, others require a shut-down of the plant and major overhauls. It is very important to understand what repair is needed.

The problems addressed by Amethyst include bearing problems, possibility requiring a replacement of a bearing; gear problems, reflecting gear teeth damage and gear misalignment, possibly requiring an overhaul of the gearbox; blade problems to do with pumps and fans, possibly requiring a simple cleaning of the blades or a more expensive replacement of the blades; problems of balance of main rotor shaft requiring a particular balancing action to be conducted; problems of misalignment of the machine, generally requiring the machine to be bolted down more properly; problems of misalignment between the driving motor and the prime unit such as a fan, requiring a basic alignment sequence to be performed; problems of cavitation in pumps; problems from hydraulic or aerodynamic difficulties; problems from resonances; and finally, and perhaps most importantly, problems from the bad use of the data collector and bad data collection. These problems indicate that there might not be a problem with the machine at all, but that the operator did not do the data collection properly and accurately enough.

Amethyst diagnoses problems on the following fixed or variable-speed machines.

Fans & blowers   Spindles   Pumps (centrifugal)

Motors-AC & DC   Gear drives   Compressors (centrifugal)

Generators   Belt drives   Steam turbines

Paper machines   Chain drives   General machines

The primary value of the condition monitoring system is derived when the end user knows whether to tighten a bolt, re-balance the main shaft, replace a bearing or overhaul the gearbox.

In order to determine which fault it is, it is necessary to examine the vibration spectrum. Each physical part of the machine will vibrate at a different frequency. The main shaft rotates at what is called the fundamental frequency, if there are 6 blades the vibration will occur at 6 times the fundamental frequency. Bearings are generally around 20 times the fundamental frequency and gear mesh problems at about 50 to 75 times the fundamental frequency.

Traditionally, the operator must look at the FFT (fundamental frequency) graph to determine for each peak what physical part of the machine it represents. He must also have knowledge of the overall vibration level and the amplitude of each frequency. There are standard guide-lines for when the amplitude is considered to be severe and he must make the calculations to determine whether a high level of vibration represents a problem or not. In diagnosing many problems, it is necessary for the operator to also look for side band vibrations. These are vibrations at a frequency such as the gear mesh frequency plus or minus the RPM frequency. This again requires some careful calculations.

The only tool provided by existing software products is an XY cursor capability utilizing the graphs. Once the end user has identified the peaks and their frequencies he must then apply fundamental knowledge about rotating machinery and how faults manifest themselves.

One of the major limitations for most companies attempting to use condition monitoring is having adequate knowledge of how to interpret the data in order to use the system properly [3]. For large corporations, experienced and trained mechanical engineers perform, this diagnosis. The basic training course will cover simple problems, but is not adequate to cover common, but more complex deviations. As a result, there is a classic problem of not having enough skilled personnel with enough experience to actually interpret the data. Yet this is the most critical and valuable part of the entire condition monitoring system.

In most major manufacturing plants they have a number of people trained in the interpretation of the rotating machinery vibration patterns. However, these people are senior in the corporation and although they are able to perform this task they have many other responsibilities. As a result, their time is too critical to be spent diagnosing mundane problems. In many other corporation, there are not adequate personnel with the experience to interpret condition monitoring data.

The purpose of Amethyst is to automate this diagnosis and interpretation of the spectral data. This results in allowing lower skilled people to use vibration based condition monitoring, providing cost savings to existing customers and eventually widening the base of companies which can use condition monitoring. As can be imagined it is a very slow process to actually diagnose a fault. On a typical 8 hour route collecting data, a dozen measurement points will often be in alarm and require diagnoses. It can easily take an experienced person 4-8 hours to manually go through the steps needed to develop the diagnosis. Amethyst does the same diagnosis fully automatically, the end user must spend approximately 30 seconds starting the analysis and then a report is produced with the results as shown in Figure 20-1. This leads to a direct times saving of 4 hours potentially for every route collected. This also represents a reduction factor of 500 in the time needed for the task. For large corporations in active use, this represents a half man years savings immediately.

---

3   Hill, J.W., and Baines, N.C. Applications of an Expert System to Rotating Machinery Health Monitoring. Institution of Mechanical Engineer Conference, Heriot-Watt University, September 1988.

```
┌─────────────────────────────────────────────────────────────────────┐
│ Route STEEL     Total number of faults: 3              Alarm:IN/S     │
│                                                        Alarm: g/SE    │
│ Machine SCREW COMPRESSR                                               │
│                                                                       │
│                                                                       │
│ Pos 5    Dir H    IN/S                                                │
│ Mach Type pump-horz cent        Bad bearings with MAJOR faults       │
│                                 BSF and/or BTF Sidebands Exist        │
│ Rotating Speed RPM        1525  Hydraulic or Aerodynamic             │
│ Other Shaft RPM           1775  Misalignment                         │
│ Overall Ampl              0.179    IN/S                               │
│ Alarm Limit               0.314    IN/S                               │
│                                                                       │
│ bpf0                     8237.93                                      │
│ bpf1                    11587.07                                      │
│                                                                       │
│ GMF1                    74725.00                                      │
│ GMF2                        0.00                                      │
│                                                                       │
│ Blade pass               6100.00                                      │
│ Belt pass                   0.00                                      │
└─────────────────────────────────────────────────────────────────────┘
```

Figure 20-1   An Example Diagnostic Report

The Knowledge Based Approach

Amethyst involved two software packages integrated together to an interface of a third software package. The IRD Mechanalysis condition monitoring software, known as 7090 is used as the main database. The expert system rules are implemented in the Crystal expert system shell from Intelligent Environments. The Violet product from Intelligent Applications is used to provide the interface between Crystal and the IRD database[4]. It is also used to provide the extra functionality needed such as the manipulation and extraction of the vibration spectrum. For those of you who are not geologists, an Amethyst is a Violet colored Crystal.

Intelligent Applications worked with IRD Mechanalysis to perform two major software tasks. An interface was developed between the Violet software package and the IRD Mechanalysis database. Violet is able to access virtually all information in the database. It had to know how to integrate with the database, to scan all points on a particular route, to identify whether the point was in alarm or not, and to be able to pick up different types of measurements. This

---

4   Milne, R.W. Artificial Intelligence for Vibration Based Health Monitoring. Journal of Condition Monitoring, vol.2, no.3, 1989, pp. 213-218. Published by BHRA, The Fluid Engineering Centre. Editor, Ruth Higginbotham BLib.

portion of Violet is written in C and is standard software for accessing a standard database. Violet then provides a number of access functions which are callable from the expert systems shell to access the information which has been extracted from the database.

In order to ensure that the end-users did not become disenchanted with the system, they were not exposed to it until we were happy that the systems test and the final test efforts had been completed. Because the problem that the system is solving is critical to their own functionality, end-user acceptance was relatively easy to obtain. This is one area where the end-user was not involved to the extent that normally would have been expected. This was not a major problem in the system because it was dominated by many of the other issues, such as the integration and development of the knowledge base.

Knowledge Engineering Methodology

There is considerable interest in the appropriate methodology for the development of expert systems. Over the years a variety of techniques have been discussed. Two of the most common were various types of interview/protocol analysis, perhaps involving a recording of an expert solving problems and the analysis of a transcript. Many times these are directed through a variety of "what if" scenarios asking the expert what he would do in this situation. In other cases, the expert is observed through the natural course of his duties and then an analysis is made to discover the rules.

For the development of Amethyst, none of these approaches were followed, rather a very structured system analysis of the problem and how best to develop the diagnosis was conducted. Mr. Pete Bernhard, the expert from IRD Mechanalysis, and Dr. Robert Milne, the knowledge engineer from Intelligent Applications, sat down together to begin the development of the rulebase. As a first task a list was made of the machines of interest. This included basic motors, pumps, compressors and fans.

A second list was drawn up of the common vibration problems of faults which were desired to be identified by this system, this included many of the faults currently being covered by Amethyst. A third list was then drawn up which detailed the measurements that were available. This included a list of measurement points for a typical configuration, i.e., the radial and axial measurements at both ends of the shaft. It also included the overall vibration level and the vibration spectrum at each measurement point. Finally, it included specific features such as the amplitude of a vibration at a given frequency, the existence of harmonics, sidebands and broadband vibration. It should be noted that the first two lists were drawn up primarily from the marketing viewpoint. It was felt that to make

Amethyst more customer targeted, the coverage of machines and faults identified was necessary.

The third list was drawn up from the knowledge of what was available within the database and of the basic characteristics of a vibration spectra. The next stage of the rulebase development was to provide a matching for each type of machine with regard to which faults were possible. For example, aerodynamic problems do not occur on an electric motor, just as DC source problems do not occur on a typical fan as shown in Figure 20-2. The development of this matching between machines and faults provided a good way to organize the data and plenty of opportunity to explore different possible ways of organizing the rulebase. After this analysis it was decided to organize the rulebase by potential fault areas, each fault then being sub-divided as appropriate for the different types of machines it was relevant or not relevant to.

The next stage of the process was starting to develop the actual diagnostic rules relevant to each machine. The expert, Pete Bernhard, was very eloquent and able to clearly state many of the standard vibration diagnostics and how they apply. The knowledge engineer, Rob Milne, had a reasonable background in mechanical engineering and signal analysis, and was able to rapidly learn and understand each of the potential faults -both how they would appear in a vibration spectrum and also physically what was wrong with the machine and how this resulted in the given vibration signature as shown in the rule in Figure 20-3.

In fact, considerable time was spent tutoring each other in their respective fields as the knowledge engineer learned more and more about vibration analysis, so the expert was taught about basic knowledge engineering and expert system concepts. At the end of this process, the expert was able to rapidly articulate a high level of abstraction -the many diagnostic rules needed and how they would apply. This process has continued so that two years after the initial session the expert is able to work with Crystal and write his own diagnosis without the help of the specially trained knowledge engineer.

The initial knowledge base was developed as a "yes/no" question style in order to provide a demonstration system and be able to test some of the ideas. The team at Intelligent Applications then supplemented the existing Violet rules with new abstraction functions that were necessary to provide the coupling directly between the diagnostic rules and the database. This overall methodology worked far more efficiently in this situation than any number of possible example problems or interviews or protocol analysis.

Machine Database - Modify Measurement

Place the block over the
selection you wish to edit
and press the enter key.

Route: RT061
Machine: CHEMI WASHER
Position: 1
Direction: H

Current record no: 1

Machine type: ac motor
RPM: 1184
Speed type:c   constant speed
Drive type: direct

Axial measure point ref: A IN/S
Other brg reference: H IN/S

Transducer mount: magnet
Transducer type: 970
Constants record number:

Bearing type: Plain

Next record
Previous record
Specify a record

Copy from prev point
Overhung rotor: No

Quit

Figure 20-2   Typical Machine Data Entry Screen

IF   the overall vibration is too high
AND A MULTIPLE OF ixrpm is too high
AND the overall axial vibration is more than 50% of the horizontal
THEN misalignment is likely

Figure 20-3   A Typical Simple Rule

The Expert System Shell

The expert system shell used for the development of Amethyst is Crystal from Intelligent Environments in London. Crystal is currently the leading expert system shell in the United Kingdom. Crystal is a backward chaining rulebase shell with extensive user interface and screen painting facilities.

Intelligent Applications has used Crystal for a large number of applications, many of which are on-line, high speed applications in the manufacturing area. Crystal is extremely fast in its inference mechanism. As a result, this makes it suitable for large diagnostic rulebases running at relatively high speeds directly connected to data acquisition systems.

The rule entry in Crystal is done through a screen and function key driven rule editor. The rules are similar in concept to a traditional "if/then" rule although there is only one item on the right hand side of the rule, that is the rule name itself. As a result, it is necessary to combine the test from the left hand side of the production rule with the actions and side effects desired for the right hand. Although this may seem a little awkward to the production system purist it is isomorphic in its behavior, we have never considered this to be a limitation or suffered as a result.

The rule editor is used to edit one rule at a time. During the editing, the rule name is at the top of the screen and its expansion or contents is presented down the screen. It is possible to move very quickly from rule to rule. For example, if the first condition of a rule is the second rule "to determine whether the overall vibration is high", by simply pressing the F10 key the rule editor allows you to edit the overall vibration test rule. By pressing the F8 key, one moves back up through the rule tree in order to see other rules in which this rule is used.

A fully implemented Crystal rulebase is similar to a very large tree, which is exactly what one would see during the execution time of a backward chaining production system. Part of the efficiency of Crystal, however, comes from the fact that the rules are linked together through this rule editor in a very natural way. As a result, the system does not need to search through a large number of productions to find what the next rule might be. Rather the rules are effectively ordered through this graphics editor. This is what results in the extremely high execution speed.

The basic data structures in Crystal are numeric or string variables and arrays. These data types are completely adequate for expressing all the mathematical, numeric and string relationships desired in the program. Although there are no objects or other complex data structures in Crystal, extensive use is made of arrays to simulate objects. This requires a small amount of set-up, but is similar in syntax to accessing the individual slots of a typical object.

Crystal has an extensive screen painter utility which was used to rapidly develop the various user displays for input and output of the system. It is a classic cursor and function key driven system for painting regions, drawing boxes and providing shadowed areas. As a result, with minimal effort, a robust user interface and very elegant screen presentation was possible, The same screen paint mechanism can be used to format data and reports for report files. This was very important for the easy development of the typed reports.

A very important feature of Crystal is its advanced debugging system. At any point during the execution of the rulebase, it is possible to interrupt the execution and enter the trace and debugger area. One is able to then move through the entire execution of the rulebase and see which rules have been tested and whether they have succeeded or failed. In this way, it is possible at any point to move back through the entire execution process and evaluate whether the rulebase is working as desired. At the same time, it is also possible to examine the values of all variables and arrays. This provides one of the most effective on-line debugging systems that we know of for an expert system shell. This has proved exceedingly useful during the debugging stage in order to allow us to rapidly trace down any possible bugs in the system.

Although Crystal is considered primarily a rulebased expert system shell, it is actually a complete programming environment. It has a complete set of commands to provide assignment, incrementing variables and testing conditionals. In actual fact, a small percentage of the total Amethyst software are the diagnostic expert system rules. The majority of the lines of code are used in gaining user input, controlling the looping and the flow of the software, providing editing facilities and the report generation facilities. The most fundamental of these is the ability to restart a rule. This command provides for the basic looping mechanism. The tree structure nature of the rulebase and the ability to restart a rule allows the user to develop "do while, do until and for next style looping". This provides a very well structured programming environment as there is no equivalent to the "go to" statement. If one desires at a particular point to execute a specific rule, that rule is just added in as part of the current rule. In order to be highly efficient, Crystal uses a concept similar to the Lisp programming language by using pointers to rules, rather than explicitly replacing the body of the rule in line. As a result, inserting a rule at many places in the overall rule tree, does not increase the size of the code. This feature combined with the other programming language features, provides for a very efficient fourth generation like programming language environment.

It was necessary to implement various mechanisms for causing rules to fail explicitly in order that it had the correct behavior for dealing with exclusive sets of problems or combinations of problems. In general, we would always make the system find all possible rules that match. To accomplish this, the behavior of the backward chaining system was altered, such that it found all rules rather than stop on the first rule.

The tasks which were better performed by the C code were implemented as standard software. The resulting expert system only manipulates the expert knowledge level and not any of the fancier database accesses. The resulting system therefore is a good integration of expert systems technology with standard programming and database technology[5].

The total expert system rulebase is composed of 780 rules according to the Crystal expert system shell. Of these 780 rules, 275 rules are for diagnostic interpretation. The difference illustrates the number of rules needed to control and organize such an expert system.

---

5   Thomas, G.B. and Thomas, R.C. An Expert System Interface to a Suite of Rotordynamics Programs, Institution of Mechanical Engineers Conferences, Heriot-Watt University, September, 1988.

There are also many rules needed for initialization and editing of the user options.

Violet

The Violet software from Intelligent Applications provides the key added functionality for developing expert system applications based on vibration spectral data. As powerful as Crystal or any expert system shell is, it does not provide inherently, the capability to access vibration spectral data. Although Crystal has interfaces into standard dBase, these interfaces were not designed for the large scale database access and manipulation necessary in the Amethyst product. More importantly, it is not efficient to represent the large vibration spectral arrays as numeric arrays within the expert system shell. The primary place for this efficiency would not be adequate, as when it is necessary to examine many sections of the array. Although this could be done using the programming language capabilities of Crystal, it is clearly not a typical application for a production rulebase system. In order to provide this extra functionality efficiently, Violet is used. Violet provides two primary capabilities; the first is access and control of the database for the vibration spectral data storage; the second is the manipulation and extraction of the loaded vibration spectrum.

Violet controls the access to the database through an extensive library of dBase access routines implemented in C. For example, when it is necessary to cycle over all machines on a data collection route to identify those in alarm, Violet is able to access the indexes to the route definition files to locate the appropriate route. It then accesses the indexes for all the machines on that route to identify each machine. It then uses the indexes to go into the very large databases and load up the spectral data and other machinery point descriptions. As part of this process, it is necessary for Violet to retain the last machine analyzed in the current route so it is able to provide the next machine and control the looping sequence. The version of Violet in use with the Entek data collector database accesses over 20 different database files during its execution.

The other primary purpose of Violet is to provide for high level data abstraction. One expresses a rule such as unbalance as unbalance of the shaft by stating that if the vibration at the fundamental speed has almost all the vibration from the machine, then there is unbalance. In other words, if the vibration at the RPM dominates the spectral signature, then unbalance is present. Violet provides the functionality to determine whether vibration at a particular frequency dominates the spectrum. Another important diagnosis is to determine whether a particular frequency such as the frequency of blade rotation has a large number of harmonics. Violet directly provides a function to determine whether a peak has a large number of harmonics, this allows for a very efficient stating of the logic of the rules. At the same time, Violet is very efficient at processing the spectrum and determining whether particular conditions are met. Violet also provides a number of facilities for searching through the spectrum looking for specific symptoms. For example, it is sometimes necessary to know whether there is a large peak at a high frequency. Violet is able to look for a large peak from anywhere above 6 times the running speed to the maximum frequency in that spectrum. This requires a careful searching of the majority of the spectral data. Other faults require looking for a broad band of vibration anywhere at high frequency, again Violet provides the functionality to achieve this as shown in Figure 20-4.

Violet, in conjunction with Crystal, actually represents an applications specific toolkit. Rather than being just a general purpose expert system shell. Violet provides the extra functionality and data abstracting needed to develop vibration diagnostic oriented expert systems. The only part of the system that would change would be the access into the database or vibration data storage area.

This concept of an applications specific shell is becoming increasingly important. It should be recognized that a large part of any

---

Replace Bearing Immediately:
IF   The overall vibration is too high
              Violet: read overall
              Violet: read alarm limit and test
AND   There is a spike energy alarm
              Violet: read overall
              Violet: read alarm limit and test
AND   Bearing inner race frequencies are preset
              Violet: read overall
              Violet: read alarm limit and test
              Violet: compute threshold and test
AND   Bearing outer race frequencies are present
              Violet: read overall
              Violet: read alarm limit and test
              Violet: compute threshold and test
AND   inner race has sideband
              Violet: test for sidebands
OR   Outer race has sidebands
              Violet: test for sidebands
AND   There is a haystack of vibration
              Violet: test for haystack
AND   Report fault: Replace bearing immediately

Figure 20-4

application is the control and looping mechanism needed to process the collection of data. Although, strictly speaking, this is not part of the Violet software, it is part of the Violet/Crystal combination. For any application, there are a large number of standard rules used in the Crystal expert system shell to provide the framework and control around Violet. Once the basic diagnostic rules have been removed, the resulting combination of Crystal, Violet and a basic number of control rules, provide the framework for the rapid development of other applications. For a new application, perhaps sixty or seventy per cent of the software will already be developed and the remaining forty per cent, typically the diagnostic rules, are quickly implemented.

The machinery monitory system was tested on a large number of machinery databases. Data for every fault was identified and the system was tested to be certain that every fault did occur. For several large databases, the system was also checked to make sure that the diagnosis it developed was the same as a human expert would have developed. As many combinations of user options and data examples were used as possible to make sure that the system was robust.

The primary goal of the system was to have an accurate and rapid diagnosis. The accuracy was checked by considering a large number of case examples. The speed was such a dramatic improvement over previous systems that no formal detailed analysis was required.

Structure of the Knowledge Base

The knowledge base is divided into four main sections: load, utility, diagnosis and explanations. The load knowledge base is primarily a control knowledge base and contains no diagnostic rules. It is used for the initial top level control of the system. It also provides for the editing of simple, but important areas such as contents of the report header and the diagnostic contents. The load knowledge base is also responsible for collecting up the basic information from the user such as whether he would like to analyze a point or a complete route.

The utility knowledge base is used for the development of the machinery database. As such, it is not a diagnostic rulebase. It is an extensive user screen editing capability developed using the screen painter and fourth generation language environment of Crystal. Utility provides for the automatic creation of machinery databases from the existing users database. It provides facilities to edit, examine and alter any machine point. It also provides facilities to rapidly duplicate machine descriptions. Because of the large size of the databases and the number of options available for each machine description, the utility knowledge base is of considerable size.

The diagnostic rulebase has two main sections: the first section is the overall control, the second section is the diagnostic rules themselves. The overall control is responsible for getting the current measurement point from the route, determining whether it is in alarm or not and then loading up the appropriate data. Control will then determine whether to analyze the next point on a route and continue with the looping process. The diagnostic section contains the diagnostic rules themselves; each rule will be a combination of high level logic and access through Violet to spectral data. The explanation knowledge base provides for the output to the screen, file or printer of the explanations of each possible fault. Once again, it does not contain any expert knowledge; rather, it is a program that is given a list of which faults to explain and provides pre-canned explanations. It would not be appropriate for the end user to attempt to read the actual diagnostic rules from the expert system. Instead a specially prepared paragraph of text is provided for each fault which gives a high level, easy to understand explanation of the fault. It also provides suggestions with regard to further actions and tests as shown in Figure 20-5.

Benefits of Using Amethyst

Currently, Amethyst reduces a task which should take from 4-8 hours to only a few minutes. Most large companies, particularly those using Amethyst, currently conduct this type of analysis every day. If we take the lower end of that estimate, 4 hours a day, this results in one half

---

REPLACE BEARING IMMEDIATELLY
Most of the symptoms of bearing deterioration are present. Ball pass frequencies for inner and outer race are present. There were RPM sidebands of ball pass frequencies. There may have been a haystack too. Exact sypmtoms found are listed in the diagnostic message. To avoid possible bearing failure, replaced as soon as possible.

LOOSENESS
Axial vibration is NOT high (more than 50% of the amplitude at 2 x RPM or 3 x RPM is greater than 50% of the amplitude at 1 x RPM.
Then looseness is the problem.
Check at the machine for-
    Visual signs of looseness. Tighten loose parts and recheck vibration.
    looseness is directional. Vertical vibration is largest for looseness of
    the base or at the bearing split.
Study the machine with pickup locations are described in your IRD textbooks.

Figure 20-5   Example of the Explanations

man year of savings directly. In actual fact, the savings can be much higher by allowing better use of personnel and preventing other problems as discussed below.

Because condition monitoring is fundamentally oriented towards preventive maintenance, it is extremely difficult to accurately assess the cost savings. This is because one is preventing a catastrophic failure. Because condition monitoring is generally successful, these failures are almost always prevented. As a result the costs are not incurred. In reality, the cost savings of condition monitoring represent the savings of a number of catastrophic failures, resulting in considerable down time, machinery damage and replacement costs. Because Amethyst increases the effectiveness of condition monitoring, it helps to magnify already established condition monitoring savings.

Early beta test copies were available for Amethyst in January 1989. There were approximately 5 beta test sites who used the software between February and April 1989. As the versions of Amethyst evolved, some of these beta test sites would cease operation for a short while and then continue with an updated release. Officially, Amethyst started shipping as a product on the 1st of May, 1989. In the first 8 months, over 100 copies were shipped. In addition, a large number of copies had been distributed to IRD for internal corporate use. These copies are used primarily for demonstrations and sales promotion, although some are being used to assist in IRD's consulting activities. After only one and a half years, over 200 copies have been delivered.

The Broader Impact

A major impact of Amethyst is that it has the potential to affect the many thousands of companies conducting condition monitoring. It is not a large application at a high price requiring extensive development time. It is a simple, but effective, application at a low cost that is very easy to use. Although the initial customers have tended to be very large corporations, the price and usability of Amethyst is well within reach of every manufacturing facility in the United Kingdom that uses condition monitoring currently. The product clearly has the potential to impact a very large number of small manufacturing operations.

There is another secondary impact of Amethyst that must be assessed over the long term.

Currently one of the major reasons preventing companies from using condition monitoring is that they have to train the expertise in collecting and analyzing the data[6]. It is estimated that only a small percentage of the potential condition monitoring market has been penetrated to date. There have been estimates that suggest that only 10% of the companies that could benefit from condition monitoring currently do so.

One of the major restrictions on those companies is having the skilled expertise available. Amethyst does not help in providing the expertise to collect the data, but does provide a simple means of interpreting the data. As a result, Amethyst has the potential to help the condition monitoring area mature another step. This could result in a considerable increase in the number of companies able to benefit from condition monitoring with the resulting consequent savings to UK industry.

If all companies were using condition monitoring properly, not only would there be tremendous direct savings in engineering maintenance cost, but there could be considerable improvement in manufacturing efficiency, throughput, energy cost and reduction in product cost.

General AI Strategy

An important question to ask of any application is "why was an expert system used instead of just traditional mechanisms?". For the answer to this it is useful to look briefly at the history of vibration analysis of condition monitoring. As recent as only ten years ago, most vibration measurements for condition monitoring were done with a swept filter or similar analysis. Most of these simply provided an output on a piece of paper, the expert then examined this plot or graph to diagnose the faults of the machines.

The advent of microelectronics provided two major changes and a great opening up of the condition monitoring world. The first change was the availability of low cost PC computers that could store the large databases cost effectively for the end user. The other change was the ability to provide microprocessor based data collection and analysis units. This provided for a new market of hand-held vibration data collectors.

By 1987, when the initial development of Amethyst was beginning, the market had just become accustomed to the microprocessor control data collector and the PC based database. Many people were now becoming comfortable with collecting and examining graphically the large quantities of data. The market, however, was just starting

---

6   Milne, R.W., Artificial Intelligence for Vibration Based Health Monitoring. 4th European Conference on Non-Destructive Testing, September 1987, London, England.

to realize that the process of examining the data they had collected for faults was becoming unmanageable. For large plants, with many thousands of machines, it was impossible to provide a detailed analysis needed on each point. As a result, a considerable need was developing for automatic analysis. At the same time, many companies were now able to afford condition monitoring equipment, but did not have the trained expertise in vibration diagnostics. Very rapidly, there were not enough experts to go around.

At roughly the same time, PC based expert system shells were evolving rapidly. Again, as recently as 1986, high speed expert system shells able to run on a standard IBM PC were not generally available. The few products that were around were implemented in other AI languages such as Lisp and normally required extended memory and exhibited slow execution speeds. The advent of products such as Crystal opened up the low cost way to develop these expert systems on standard PCs. At the same time, most expert system applications that had been developed were consultative in that they asked the user to provide input at diagnosis time.

Intelligent Applications was very active in pioneering on-line applications of these systems where all the data needed for the diagnosis came directly from the data acquisition system and not from the end user. The Violet software was one of the first packages available to provide this on-line connection to vibration databases. The final important factor that brought things together was the demands of two companies to increase their sales. Intelligent Applications was looking for a partner and an outlet in its capabilities to develop vibration based expert systems. IRD Mechanalysis was looking for ways to expand its product range and provide more value to the customer. The first time both parties got together and discussed the combined capabilities, the possibilities were recognized and enthusiastically development work began.

Although a system such as Amethyst could have been developed with traditional software, it would have required considerably more development time and resources. In this particular situation, however the market was just maturing so that the time was right for this type of application. The main reason that the expert systems were used was that people involved with expert systems were the first to realize that this next stage in the development of condition monitoring was possible. If the market was much older, or a more traditionally oriented company had thought of it, it is possible that the expert system would never had resulted. In retrospect, however, the use of expert system technology to develop and expand this portion of condition monitoring has proven to be very effective and a good choice.

Conclusion

In this chapter we have looked at Amethyst, an expert system for the diagnosis of faults in rotating machinery. Amethyst was constructed with straight forward rulebase technology and classic expert system development procedures. By being carefully focused on a well defined application, it is now having a widespread impact. Not only are many engineers able to use sophisticated expert system technology, but a major revolution is taking place with regard to how they perform their day to day tasks. The technology is already prompting further development and the lessons learned from this application will certainly lead to many new and better capabilities.

References.

1) Neale, M., Condition Based Maintenance -A Review of Methods and Economics. The Factory Efficiency and Maintenance Conference, PEMEC 88, Birmingham 27-30th September 1988.

2) Hills, P.W., Vibration-based Predictive Maintenance Systems for Rotating Machinery. The Factory Efficiency and Maintenance Conference, PEMEC 88, Birmingham 27-30th September 1988.

3) Hill, J.W., and Baines, N.C. Applications of an Expert System to Rotating Machinery Health Monitoring. Institution of Mechanical Engineer Conference, Heriot-Watt University, September 1988.

4) Milne, R.W., Artificial Intelligence for Vibration Based Health Monitoring. Journal of Condition Monitoring, vol 2, no3, 1989 p213-p218. Published by BHRA, The Fluid Engineering Center. Editor, Ruth Higginbotham. Blib.

5) Thomas G.B., and Thomas, R.C., An Expert System Interface to a Suite of Rotordynamics Programs. Institution of Mechanical Engineers Conference, Heriot-Watt University, September 1988.

6) Milne, R.W., Artificial Intelligence for Vibration Based Health Monitoring. 4th European Conference on Non-Destructive Testing, September 1987, London, England.

7) Milne, R.W. On-line Artificial Intelligence. The 7th International Workshop on Expert Systems & Their Applications, Avignon, France, 13-15 May, 1987.

8) Milne, R.W. Artificial Intelligence for Vibration Based Health Monitoring. Journal of Condition Monitoring, vol.2, no.3, 1989, pp. 213-218. Published by BHRA, The Fluid Engineering Centre. Editor, Ruth Higginbotham BLib.

9) Milne, R.W. Artificial Intelligence for Vibration Based Health Monitoring. 4th European Conference on Non-Destructive Testing, September 1987, London, England.

10) Moore, R.W. Expert Systems in Process Control: Applications Experience. 1st International Conference on Applications of Artificial

Intelligence to Engineering Problems, Southampton, England, 15-18 April, 1986.

11) Neale, M. Condition Based Maintenance A Review of Methods and Economics. The Factory Efficiency and Maintenance Conference, PEMEC 88, Birmingham, 27-30 September, 1988.

12) Thomas, G.B. and Thomas, R.C. An Expert System Interface to a Suite of Rotordynamics Programs, Institution of Mechanical Engineers Conferences, Heriot-Watt University, September, 1988.

13) Winston, P. Artificial Intelligence, Addison Wesley Press 1984.

Author Biographical Data

Dr. Robert Milne is the founder and Managing Director of Intelligent Applications. He is considered one of the most knowledgeable persons in the UK with regard to expert systems applications in manufacturing.

He has a B.Sc in Electrical Engineering and Computer Science with special emphasis on Artificial Intelligence from the Massachusetts Institute of Technology (1978) and a PhD in Artificial Intelligence from Edinburgh University (1981). Dr. Milne has over 75 technical publications.

He is a member of the American Association of Artificial Intelligence, the ACM, AISB and the British Computer Society.

He is currently on the national committee for the BCS Specialist Group on Expert Systems. He is also the co-organizer of the BCS DTI Manufacturing Intelligence Award. He is a Chartered Engineer.

21

KLUE: A Diagnostic Expert System Tool for Manufacturing

Gerald Karel and Martin Kenner
3M Software and Electronics Resource Center
St. Paul, Minnesota

Introduction

The Minnesota Mining and Manufacturing Company (3M) manufactures literally thousands of unique products. Common to all these production processes is the need to transfer critical process knowledge from the expert product developers and engineers to those people actually operating and controlling the manufacturing process. Therefore, the potential for diagnostic expert systems in 3M is enormous. In early expert system projects at 3M, rules were used as the primary knowledge representation. It was soon discovered that program control within the rule system and the connectivity of the rulebase became increasingly difficult to observe as the number of rules in the system increased. Importantly, the maintenance of the rulebased expert system required the constant attention of specially trained knowledge engineers, creating an extra layer of intervention between the expert knowledge and the users of that knowledge. KLUE is a

Reprinted from the Proceedings of the 1988 ASME International Computers in Engineering Conference and Exhibition, July 31-August 4, 1988. San Francisco, California.

diagnostic expert system tool that addresses these problems by presenting the knowledge in the form of a decision graph. In KLUE, both the program control and the diagnostic strategy are explicitly represented. Domain information is added or modified by direct operation on the decision network. In applications within 3M, the explicit nature of this knowledge representation has enabled the transfer of the maintenance of the diagnostic systems to the process engineers themselves, providing strong encouragement for the transfer of AI technology to the manufacturing divisions in 3M.

The large number of complex manufacturing processes at Minnesota Mining and Manufacturing Company (3M) has provided a potential for the widespread use of diagnostic expert systems. These expert systems could provide a mechanism for the transfer of critical knowledge from the process engineers to the process operators, as well as provide appropriate advice at those times when process problems, if not handled quickly and correctly, could lead to expensive delays in production. However, while the opportunity for expert system use is great, acceptance of the technology at 3M requires that the systems adopted be mature product implementations; the emphasis is on the use of the expert system as a tool to improve production.

As with any new technology, maturity of expert systems can only be realized when the implementations move from controlled use of the software by specialists in the field of expert system development into the hands of end users, whose specialties lie elsewhere. This move requires a level of robustness and maintainability not available in most current systems. KLUE, the "Knowledge Legacy of the Unavailable Expert," is an expert system tool developed by the authors to address the requirements of 3M manufacturing areas. It provides an integrated system for development, maintenance, and operation of a diagnostic expert system in a manufacturing environment. It has a proven record of success in field implementation, and has been extended into other diagnostic fields, and even into non-diagnostic problem domains.

Issues of Knowledge Representation and Control

The development of an expert system in any domain requires the study of that domain, as well as the expert's knowledge, on a level independent of implementation details, a level referred to as the "knowledge level"[1]. Using the knowledge level analysis, the system designer selects a mechanism for the representation of the expert's knowledge so that the capture of that knowledge can be accomplished

---

1   Newell,A., 1981,"The Knowledge Level," AI Magazine, Summer '81, 1-20-33.

as readily as possible, with the understanding that the more direct the mapping from the formulation used internally by the expert to the representation selected, the greater the validity of the system when implemented. The idea, then, is to select some means for representing the expert's knowledge that is very close to the manner in which the expert seems to internally represent and process information.

Rule-based representations have demonstrated success in several domains, with MYCIN the classic example[2]. In these systems, the knowledge of the expert is captured in the form of IF ... THEN ... ELSE sequences. Writing rules is a deceptively simple process, and the transfer of expert knowledge to rules generally seems to be a straightforward task. The success of rule based systems has lead to a general acceptance of rules as the representation of choice, especially for diagnostic systems. However, though the formulation of expert knowledge into rules is apparently straightforward, the control within which a rulebased system processes the knowledge is extremely complex. The sequence of rule selection and application, often called an agenda, is crucial to the successful solution of the problems to which the system is applied. The construction of this agenda and the control of dynamically changing information generated as the system performs are problems that become increasingly difficult as the number of rules in the system increases. Maintenance, the additions to and deletions from the knowledge base, alters the agenda formulation, and since the inference control is implicit in the rule organization, the effect of the maintenance actions may not be readily apparent. To overcome this difficulty, many large, rule-based systems require special "browsing" tools to facilitate tracing through rule sequences in an effort to describe the control mechanisms being used.

Several expert systems have attempted to address this specific issue through the use of representations that separate the knowledge from the inference control. NEOMYCIN[3] was a rulebased system descendant from MYCIN that was actually composed of two rule sets, one used to represent the expert knowledge and the other used to control the inference process. ESE/VM[4] allows rules to be grouped into structures called Focus Control Blocks (FCB).

2   Buchanan, B.G., and Shortliffe,E.H., 1984, Rule-Based Expert Systems, Addison-Wesley Publishing Company, Reading, Massachusetts.

3   Clancey,W.M., 1986, "From GUIDON to NEOMYCIN and HERACLES in Twenty Short Lessons: ORN final report 1979-1985," AI Magazine, Vol. 11(3), 40-60.

4   Hirsch,P., Katke, W., Meier, M., Snyder, S., and Stillman, R., 1986, "Interfaces for Knowledge-base Builders' Control Knowledge and Application-specific Procedures," IBM Journal of Research and Development, Vol 30(1), 29-38.

Within each FCB are control language statements which explicitly prescribe the desired control mechanisms to be used with the rules in the FCB. CENTAUR[5] separated control and knowledge by using two different representation mechanisms, frames for the knowledge and rules (as applied to a task agenda) for inference control. PROSPECTOR (as described in Waterman, 1986, as well as Buchanan and Shortliffe, 1984) utilized yet another representation scheme, this time using a semantic network as the representation of the inference process, but with this network actually constructed from rules that contained the domain knowledge and other rules governing Bayesian probability application as a basis for inference. In all of these systems, control was represented more explicitly than in MYCIN. And in the cases of NEOMYCIN, ESE/VM, and CENTAUR, the control strategies used were flexible and determined by the developer. Still, we felt that it would be difficult to deliver dynamic applications based on the rule representation found in these schemes.

It is important to note that this in no way indicates failure of any of these expert systems, only the inappropriateness of the representation selected. Appropriateness becomes an issue when the system is evaluated not strictly for its ability to solve a particular problem, but when it is considered as a tool that is to function in the problem environment. It becomes apparent that the selection of an appropriate representation involves more than simply examining representational capability. All of the representational schemes used in the systems mentioned possess equivalent representational power[6]. It would seem reasonable that a system should utilize a representation that can provide a means for the capture of expert knowledge and also a mechanism that appropriately represents the inference control decisions necessary to utilize that knowledge. The more appropriate the representation of both knowledge and control, the more readily the acquired knowledge of the expert can be captured by the system. MYCIN and its descendants have succeeded because the maintenance tasks that are necessary for the continued growth and function of the systems have been performed by specially-trained knowledge engineers who facilitate the process by bridging the relatively wide gap between the expert knowledge and the representation. As Clancey (1986) has shown, control in these rulebased systems does not map directly from the control used by the domain experts. The mapping of the expert's knowledge into the representation has been a process in which the knowledge engineer performs a translation of the knowledge as represented by the expert into the representation as used by the system. Without the knowledge engineer, the mapping becomes very difficult.

5   Aikins, J.S., 1983, "Prototypical Knowledge for Expert Systems,: Artificial Intelligence, Vol, 20, 163-210.

6   Nilsson,N.J.,1980, Principles of Artificial Intelligence, Morgan Kaufman Publishers, Los Altos, California, 361-412.

To assure the acceptance of expert systems as productive tools, this requirement for specially trained maintenance personnel must be eliminated. To accomplish this, the representation selected must appropriately model the knowledge structure and the control used by the human expert, in order to avoid that intermediate translational step. The responsibility of the knowledge engineer is to select a representation that is capable of successfully capturing the knowledge of the expert and that also makes the processing of that knowledge readily apparent.

We have consistently observed that in manufacturing, human diagnosticians generally seem to compose solution strategies as a traversal of a decision graph. Selecting a representation scheme that models this decision graph more directly than a rule-system does provides a more simple mapping from the expert's knowledge to the representation, with control issues made explicit in the capture since control should also map directly from the expert to the representation. For this, the semantic network seems the best possible representational scheme. Control in the processing of the knowledge is explicitly represented by the paths through the graph, with the semantic links providing these connections. Given the explicit nature of the control and the appropriate mapping of the expert's knowledge to the semantic network, verification and maintenance of the system become much simpler tasks. The selection of an appropriate representation has eliminated the need for constant intervention by the knowledge engineer in the transfer of knowledge from the expert to the system.

Requirements for Manufacturing Diagnostic Expert Systems

Before describing the common requirements often encountered in the development of diagnostic systems for manufacturing, we should first provide definitions for a few of the key job classifications. Operators are those people running the process for the purpose of generating final products or the components that make up those final products. Operators are trained to operate the equipment on the factory floor and to inspect the manufactured pieces. Process engineers have several responsibilities. They must define specifications that will ensure products of acceptable quality, and then install measuring devices to monitor the key process parameters and performance characteristics. They will have a thorough understanding of the functionality of the equipment and will establish the proper setpoints of the machinery. In short, process engineers strive to increase the percentage of products successfully manufactured. When an operator notices a significant drop in yield, or an abnormal process measurement, the process engineer is often called to diagnose the problem.

Often, the problem will be the result of some machine failure. If so, a maintenance person will be called to continue the diagnosis and perform the necessary repairs, since the maintenance people hold a detailed understanding of the workings of the various machines.

Much has been written about the benefits of expert systems in manufacturing[7]. Most importantly the expertise of process engineers and maintenance people is preserved and disseminated to each other and to the operators. Operators will then be able to handle many of the common process problems,freeing the process engineers to develop improvements to the process. It is important to note that although the operators,the process engineers, and the maintenance people are the ones primarily involved in the technical aspects of the expert system project, there are always others from the product laboratory,from quality control, and from the management ranks who can provide key input or support to the project. In our investigation of diagnostic applications in various manufacturing centers, we noticed that virtually all application areas specifically stated the following requirements of the expert system:

1.   All updating and editing of the knowledge bases should eventually be handled by the process experts directly rather than by knowledge engineers. After observing the frantic schedules of many process engineers and maintenance people, it was clear that those people would revolt if asked to learn a programming language or the complexities of a rule system. Instead, a specially designed environment was needed that could be learned in a few hours and would provide a natural interaction with the experts.

2.   The experts have to be able to read and understand each other's expertise. A complex manufacturing process may require a dozen process engineers and maintenance people. As in any large group, responsibilities of some people will change and others will be called in to fill the gaps. It is important for the expertise to be transferred smoothly as well. Therefore, the knowledge and diagnostic strategies contained in the expert system should be represented explicitly.

3.   The operators' interface to the system must be quick to learn and easy to use. We encountered many operators with little or no computer experience, and a few operators who had a tremendous fear of computers in general. In addition to a high degree of robustness, the delivered expert system must provide a sufficient amount of online instruction, and a minimum number of required operator actions.

4. Both the maintenance environment and expert system must be placed in close proximity to their respective users.

While this seems too obvious to even mention, we were still amazed at the dramatic increase in system use when the expert sys-

---

7   Waterman,D.A.,1986, A Guide to Expert Systems, Addison-Wesley Publishing Company, Reading, Massachusetts, 12-16.

tem was moved from an adjacent room to a spot just a few feet from the operators. This requirement may dictate that the expert system and maintenance environment be on separate computer systems. The KLUE environment was designed to satisfy these requirements, common to diagnostic applications in manufacturing centers. The following section describes the approach taken with KLUE, and then describes its software implementation.

The Knowledge Representation and Environment of KLUE

Overview

During the development of some of our early diagnostic expert system projects, the process and maintenance experts would often reveal what they called "fish bone" charts. These charts graphically represented their strategies for diagnosing problems by mapping symptoms and test results to the various potential causes of the problem. [These charts always displayed the symptoms and problem causes fanning out from some common starting point producing a skeletal structure similar to that of a fish!] The information contained in the fish bone charts was translated into a rule-based representation for use in the expert system. Although the expert system would work correctly, the translation to the rule system took time, and the resulting representation seemed to be more obscure than the original fish bone charts. Even after a considerable amount of training, the process experts did not feel comfortable or confident enough to maintain the rulebases.

In an effort to move our representation closer to that of the experts, we constructed an environment around the relatively simple concept of the decision graph. Figure 21-1 depicts a high level description of a decision graph. Basically, a decision graph consists of question nodes (e.g., Nodes A, B, C, E),potential problem nodes (e.g., Nodes D, F), null nodes (nodes filled with a large X), and answer links (each connecting a pair of nodes). A question node represents a question asked by the expert to the operator. The labels on the links leaving the question node provide the various choices available to the operator when answering a question. A potential problem node represents a partial diagnosis by the expert. Associated with each potential problem node is a likelihood factor (shown in brackets on nodes D and F) which represents the expert's confidence in the diagnosed problem given its location in the graph. Finally, a null node corresponds to a dead end path.

The Inference Engine (IE) in the expert system is simply a traversal of the decision graph, dictated by the operator's answers to

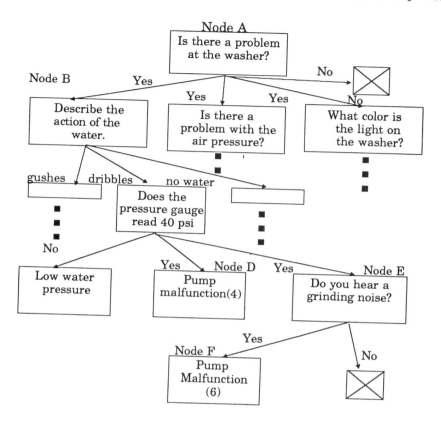

Figure 21-1   Example of a Decision Graph

questions posed by the system. Suppose the IE is using the decision graph in Figure 21-1 to diagnose a problem. IE will first pose the question in the root node (A), "Is there a problem at the washer?" If the operator answers "yes," then IE must follow all "yes" links from Node A. In this case, IE is directed toward three more question nodes.

After the operator has answered those questions, IE moves down to the next level for one of the questions. For example, if in response to the question in Node B, the operator answered that the water dribbled, then the next question posed by IE would come from Node C. Suppose the operator answered yes at Node C; the none of the "yes" branches leads to the potential problem of a bad pump (Node D). This potential problem, along with any others collected by IE, will be presented to the user after IE finishes traversing the decision graph. The other "yes" branch from Node C leads to another question, namely, whether the machine is making a grinding noise (Node E). If the operator enters no,that path ends. But if the operator answers yes, another instance of the bad pump problem is encountered (Node F). This last instance of the bad pump problem has a higher likelihood value than the previous instance, presumably due to the additional piece of supporting evidence. During the traversal,IE is maintaining a list of unique potential problems, along with their likelihood factors and network paths. When IE encounters multiple instances of a potential problem, only the information on the most likely instance is retained.

When IE completes one leg of the decision graph, it must then back up to the point where it can traverse another leg of the graph, and so on, until the entire graph is traversed. At that time, the entire list of potential process problems, ordered by likelihood, will be displayed to the operator along with appropriate operator actions. As mentioned earlier,the network path and likelihood factor is available for each potential problem and can be used in an explanation facility and in a "what if" analysis (rediagnosis using a new answer to an old question).

The KLUE expert system shell is made up of an environment for building and maintaining decision graphs, and an operator's interface. The maintenance environment allows process experts to operate directly on the decision graphs and the information contained in them. This environment satisfies our requirements for maintainability by the experts, and for an explicit knowledge representation; it will be discussed in greater detail in the next section. The operator's interface that comes with KLUE can be used by process experts to test changes made to the decision graph, or can be used by the operators to run actual diagnostics. On-line instruction and a simple control panel are always provided. The operator interacts with the expert system through mouse selection of objects on the screen.

It is important to note here that although KLUE provides both a maintenance environment and an interface to the expert system,the two facilities are largely independent. The maintenance environment builds knowledge bases; the expert system accesses those knowledge bases to perform the diagnosis (see Figure 21-2). It may be desirable to keep the process experts and operators on different machines entirely. A common scenario is one where the process experts work on some workstation with a graphics capability. Each process expert will "own" a unique knowledgebase. The knowledge bases built by the experts are then transferred to a single user or multiuser system where another program will execute the expert system. Of course, this runtime program can be in any language; the only requirement is that the runtime program contains the inference engine described earlier. In this scenario, access to the expert system is always convenient since it available from any terminal in the plant to any user with the proper authorization. Assuming the operator's

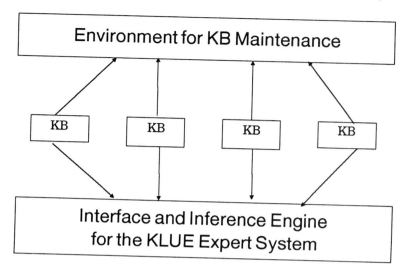

Figure 21-2   Functional Description of the KLUE Expert System Shell

interface is well designed and tested, the remaining system requirements are satisfied.

Implementation

Recognizing the need for a representation that closely matched the
mechanism used by experts in diagnostics led to the selection of a
semantic network as the representation used in KLUE. Once the representation had been selected, the process of implementation began.
Initial development of KLUE was accomplished on a Symbolics™ computer using the KEE™ development package.

In the development of KLUE, the requirements applied to the
selection of a representation were transferred directly to the selection of an implementation. The concerns regarding the updating and
editing of process knowledge required serious consideration of the
maintenance interface. Musen et al.[8] (1987) described the importance of the relationship between the underlying knowledge
representation and the design of the interface used for the maintenance of the knowledge in the expert system, indicating that a

8   Musen, M.A.,Fagen,L.M.,Combs,/d.M.,and Shortliffe,E.H.,"Use
of a Domain Model to Drive an Interactive Knowledge-editing
Tool," INternational Journal of Man-Machine Studies,
Vol.26(1), 105-121.

close parallel between the two can simplify the maintenance process. Since the basis of the system was a graph, and since the experts that were consulted on the project used graphical reference aids, it became apparent that the interface between the expert system and the maintainer should be graphical in nature. At the same time, the nature of a semantic graph, with information-containing nodes and control-producing links, was well-suited to the object-oriented programming paradigm.

The basis of an object-oriented program is a set of object classes, in which individual items are instances of one or more particular classes. Each object type is defined with a set of activities specific to that type. In KLUE, different classes of nodes can be created, each having its own specific action capabilities, with the creation of new classes and new actions in no way affecting existing classes. A user can select a node,then choose from the various activities specific to that node type, such as modifying the knowledge stored there or creating new descendant nodes. Previous research[9] demonstrated the usefulness of this paradigm in graphical implementations, further supporting its use in KLUE. The KEE development package was selected because it provided a powerful object system facilitating this type of implementation, but any reasonably capable object system could have been used.

In the KLUE design, the decision graph was formulated as a series of questions interconnected by links representing possible answers to those questions. The terminal nodes of the graph represent possible conclusions of the diagnostic process, again connected to upstream nodes by links representing answers to questions (see Figure 21-1). At any node or link on the graph,several activities are made available when that object is selected, with the specific activities available determined by the class of the object selected (see Figure 21-3). Construction of the graph structure is accomplished by the addition of a new node to a question node or the creation of a link between a question node and another existing node. As a graph is constructed, it is important that no cycles appear in any path through the graph. In KLUE, cycles are explicitly prohibited; at the time the connection between two nodes is to be created, KLUE searches the graph to ensure that the child node in the new connection is not"upstream" of the parent node of the new connection along any path in the graph. Also, every path should lead to some terminal node. To provide for the possibility that answers given in a diagnostic situation may not lead to a known conclusion, null nodes were added to the system. These simply terminate a path but provide no diagnostic information. Also, to provide for the possibility of unavailable or unclear data at the time diagnosis takes place, an option has been provided to allow marking of a question node so that a response option of

---

9   Eich,B.O.,1986,"An Object Oriented Graphics Library," Technical Report U1UCDCS-R-86-1273, Department of Computer Science, University of Illinois at Urbana-Champaign.

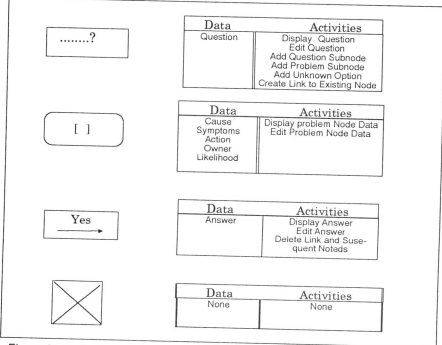

Figure 21-3   Data and Activities of KLUE Objects

"unknown" is provided. If, during inference, this option is selected, the system will simply continue inference through all possible paths from the question rather than selecting paths matching the answer provided.

Links in the graph provide much control in the system. Inferencing takes place by simple traversal of the graph along paths indicated by a match between the semantic label on a link and an answer to the connected question provided during the inference process. A significant activity is controlled at the link: the integrity of the graph inference process. Removal of a single node or link could create "orphaned" paths with knowledge present but inaccessible. For this reason, one significant activity centered at a link is the ability to delete information from the graph. Deletion requires selection of a link for deletion, and the deletion operation will then proceed down the graph from that link, deleting all nodes and links on paths connected to the selected link that do not also connect to nodes outside the current paths.

Note that nothing in the specification of the graph implementation requires that all answers to questions be provided by people. While systems developed using KLUE collect data through interaction with process operators, KLUE has also been used to develop an expert system that directly interfaces with a database containing process information, with the information extracted from the database providing the answers to the questions during inference. The answers need not be textual in nature; bitmaps have been used to provide operators graphical options for answering specific questions.

Finally, in developing diagnostic expert systems, one must always confront the issue of uncertainty. Several approaches to this have been taken, and there is no general agreement on the best implementation approach. In developing KLUE, it became apparent that in manufacturing processes, at least, the experts themselves seldom attempted to produce a single conclusion. They generally developed a list of potential conclusions, ordered by some heuristically generated weight that was determined by the particular information gathered, i.e., the path followed to that conclusion. Probability was not propagated; the relative likelihood of a particular conclusion was simply determined by its place in the graph. KLUE reflects our findings. Each conclusion node possesses a weight, which is an integer value between 0 and 10, that is provided at the time the node is added to the graph. It can be modified by the maintainer, but is a constant when inference begins. In this way, KLUE performs an exhaustive search of the problem space, and presents an ordered list of all non-null terminal nodes encountered.

The capabilities of this implementation have been proven. KLUE has been installed in manufacturing environments in 3M with noteworthy success. Maintenance responsibility has been transferred to the process engineers at the installation sites with a reasonably smooth transition. The KLUE system has also been used in the development of a nondiagnostic expert system, an extension of the design that was particularly interesting.

KLUE has now been ported to several computer systems in addition to the Symbolics, including SUN 3™, MicroVAX II™, and 386-PC workstations, all running the KEE package.

Conclusion

KLUE was developed to provide a tool for the construction of expert systems for diagnosis of manufacturing processes. In developing KLUE, it became apparent that a primary consideration when constructing expert systems must be the selection of a knowledge representation that most closely corresponds to the internal representation used by the experts in the domain. Similarly, the final implementation of the expert system should provide a maintenance environment in which the knowledge and control of the system closely parallel the mechanisms used by the experts to aid them in the use of their own expertise. In this way, the capture and maintenance of the knowledge can be facilitated, and the system becomes more easily verifiable, allowing the transfer of the maintenance responsibilities to the experts themselves.

In KLUE, the use of a graph structure as a diagnostic aid by the process engineers whose knowledge was to be captured led to the selection of a semantic network representation with an interactive graphical implementation. KLUE has been used to develop successful expert systems in several diagnostic domains, as well as in a nondiagnostic domain. This is not to say that the model used in KLUE will work in all domains, or even all diagnostic domains, but that if the representation used by experts in some domain corresponds to the KLUE model, then KLUE should provide the proper environment for the capture and maintenance of their knowledge.

In the past, selection of knowledge representations has generally been based on the ability of the representation to capture the knowledge. Since several representations have equivalent representational capabilities, the selection was often arbitrary. The development of KLUE has demonstrated that the selection of the representation to be used in an expert system should be based on the appropriateness of the representation to the domain, and the implementation should provide an environment that provides interaction that facilitates the use of the selected representation. The role of the knowledge engineer should be more clearly defined as one of selection and establishment of the most appropriate expert system environment for the domain in which the expert system is to function.

Author Biographical Data

Gerald Karel is a Supervisor in the 3M Hardgoods and Electronics Support Division. He received undergraduate degrees in Mathematics and Applied Statistics from St. Mary's College of Minnesota, and a M.S. degree in Operations Research from Stanford University. Before coming to 3M in 1985, he worked at AT&T Bell Laboratories in Naperville, Illinois in the area of telephone switching system performance. Since joining 3M, he has led the development of several expert systems in areas such as manufacturing, customer service, and sales support.

Martin Kenner is a Senior Software Engineer in the 3M Hardgoods and Electronics Support Division Laboratory, and has been employed at 3M for two years. He received his undergraduate degrees in math and physics from Dartmouth College, and is actively pursuing a PhD at the Universuity of Minnesota in Computer Science. His current research focuses on the integration of sensor information from multiple, non-homogeneous sensory systems for robotic navigation. Results of past research have appeared in *Perception and Pattern Recognition Letters*, as well as in the proceedings of several conferences.

Process Control and Planning

22

A Real-Time Expert System in the Area of Energy Management

Steven Silverman
Consolidated Edison
Co. of N.Y., Inc.

Alvin Shoop
Expert Systems Consulting Group
New York, NY

Introduction

The Consolidated Edison Company of New York, Incorporated (Con Edison) operates an electric utility system serving a 604 square mile area of New York City and part of Westchester County with a peak load in excess of 10,000 megawatts. The Con Edison bulk power system consists of both overhead and underground feeders in a voltage range from 69 kilovolts up to 500 kilovolts as well as generating units consisting of oil and gas fired steam units, aircraft and industrial gas turbines and one nuclear unit. The Con Edison electric system is interconnected with its neighboring utilities through common telemetry points.

The operation of this extensive, heavily used power system requires precise monitoring and quick response to any troubles that may develop on the electric grid. Con Edison relies on a sophisticated mainframe-based real time energy management control computer (SOCCS) to assist its operators in overseeing the operation of the electric system. SOCCS is also connected to all the other members of the New York Power Pool in a message network.

The SOCCS Host

SOCCS hardware consists of redundant Supervisory Control and Data Acquisition (SCADA) and Security Assessment (SAC) Gould/SEL 3287 computers. This four computer configuration as well as redundant disk drives, tape drives, etc. ensures continued real time operation even in the event of any critical hardware or software failure. Redundant configuration manager microprocessors and automatic failover software insures the reconfiguration of SOCCS in a minimal amount of time (less the 60 seconds) from the failure.

SOCCS supports our operators in the overall control and operation of both electric generation and transmission. SOCCS software provides automatic generation control, economic generation dispatch, transmission line security analysis, load management and a vast variety of additional functions necessary for the safe, reliable and economical operation of the Con Edison electric system. See Figure 22-1.

SOCCS CONFIGURATION

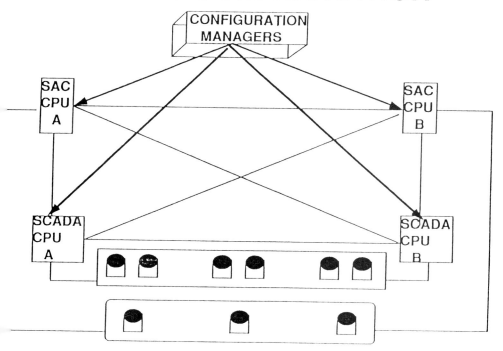

Figure 22-1   The SOCCS Redundant Architecture

The Real Time Data

SOCCS scans 69 Remote Terminal Units (RTUs) every two seconds, *updates in excess of 100,000 data points on 15,000 displays* on 28 color CRTs on a two-second or demand basis and runs all application programs. In addition SOCCS must log all operator actions, communications with the other New York State Utilities and the New York Power Pool as well logging and displaying to the operators of any unusual occurrences on the electric system.

The SOCCS Database

The SOCCS database consists of approximately 100 different point types. These represent some 12,000 analogs (floating point numbers representing feeder flows, generation levels, bus voltages, etc.) and 10,000 discretes (state information on breakers, disconnects, links, etc.). There are 1,200 feeders, 4,200 topology points (defining the electrical system configuration mapping) and other point types defining generators, loads, transformers and other data necessary to build the appropriate models necessary for the application software.

Most of the data points in the database are capable of being alarmed when an abnormal state occurs. In addition to recording these abnormal states, SOCCS also logs all operator actions such as data entry, limit changes, communications between Con Edison and the New York Power Pool, etc.

Even during operating periods free of any unusual occurrences on the electric system some 1,200 entries are logged in an average hour. During a system electric emergency, when the operators need the information the most, as many as 200 entries may be logged in a two second scan.

This large volume of alarms is not only cumbersome to the power system operators, since specifically needed information is imbedded in this large number of entries, but in certain cases may even slow down the response of the energy management system.

Overview of the SOCCS Alarm Advisor (SAA)

To provide Con Edison's system operators with a clearer, more comprehensible picture of the electric system, particularly during an emergency or abnormal condition, the SOCCS Alarm Advisor (SAA) was developed using expert system technology.

The SAA expert system monitors alarms and events observed by SOCCS. Alarms are generated by changes in the power transmission and/or distribution network topology due to breaker operations, measurements of electrical values which fall outside preset limits, telemetry failures from RTUs (Remote Telemetry Units) and various other conditions. Events reflect the return to a normal state of a previously alarmed entity, manual data entry by an operator, messages to and from other utilities, and other software messages. Alarms are presented on the display terminals and logged; events are only logged.

Telemetered data comes from 69 Remote Terminal Units (RTUs) scattered throughout the Con Edison service territory and is transmitted to SOCCS via telephone lines. Often faulty telephone lines cause all the data from a particular RTU (in excess of 200 data points) to "toggle" between the alarm and normal states at a very fast rate. These nuisance alarms and events are displayed, logged and then removed. Also, many of the software applications report an abnormal condition upon each cycle of its execution. It is clear that SOCCS need only to report this "repetitive" nuisance data once.

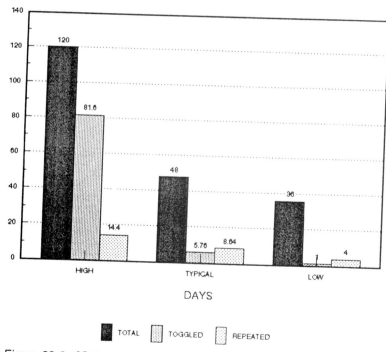

Figure 22-2   Nuisance Alarm Suppression

The first task of SAA is to filter the input stream to remove these nuisances since they can significantly clutter the input stream, making it difficult for the electric system operators to locate other relevant data. The use of SAA for this single task has provided substantial benefits to Con Edison. Figure 22-2 shows some results of this suppression.

The second task of SAA involves the analysis of the remaining "good" alarms to determine components of the electric system that might be out of service, alive on backfeed (energized from one end only) and various other conditions requiring corrective actions. The deduced statuses of these affected components are compared with available analog data to verify the analysis. In cases where discrepancies are found or insufficient analog information is available the analysis is propagated through a causal network. Finally, SAA provides recommended operator actions necessary to restore the electric system to a stable condition. These recommended actions are based upon Con Edison procedures and operator experience. The recommendations are encoded in the rules and take into account such facts as bus and feeder voltage classes, failed operator attempts to reclose switching devices, related alarms, etc.

SAA Real Time Requirements and Constraints

SAA's primary goal was to suppress nuisance data and to provide operators with a clear picture of a system incident along with the appropriate restoration actions. A basic constraint of SAA was that the expert system impose no measurable overhead on SOCCS processing. Also, because SOCCS is fairly well loaded in terms of the amount of software installed on the system, no SOCCS software was to be modified to support this expert system. Some other constraints on the design of SAA were:

- All data had to be transferred to SAA automatically.

- All output from SAA had to be on a separate crt screen and not the SOCCS crts.

The major requirement of SAA was that it perform all its functions in "real-time". In general, this meant that SAA had to keep up with the maximum expected data rate. The 69 SOCCS RTUs are scanned once every 2 seconds producing a maximum of 200 alarms in each scan. This maximum rate was confirmed with on line testing and a review of historical SOCCS data. Also, the operators expected an analysis and appropriate recommendations from SAA within 20 seconds of the detection of an incident. Once an incident is detected, SAA continues to monitor the electric system for 5 minutes during which time all new data is analyzed and could result in additional conclusions and recommendations. Because the system state may evolve non-monotonically, SAA has to "back out" deductions in order to respond to revised information.

SAA Architecture

SAA consists of four modules (see Figure 22-3). The first module resides on the SOCCS host and transfers the alarms and database information to the SAA host. SAA was developed, and is deployed, on a Texas Instruments Explorer II-Plus workstation. SAA is written in the Automated Reasoning Tool (ART), an expert system shell developed by Inference Corporation and Common Lisp. The link between the Explorer and the Gould is a "BUSLINK", a Direct Memory Addressing device developed by Flavors Technology, Inc. This device

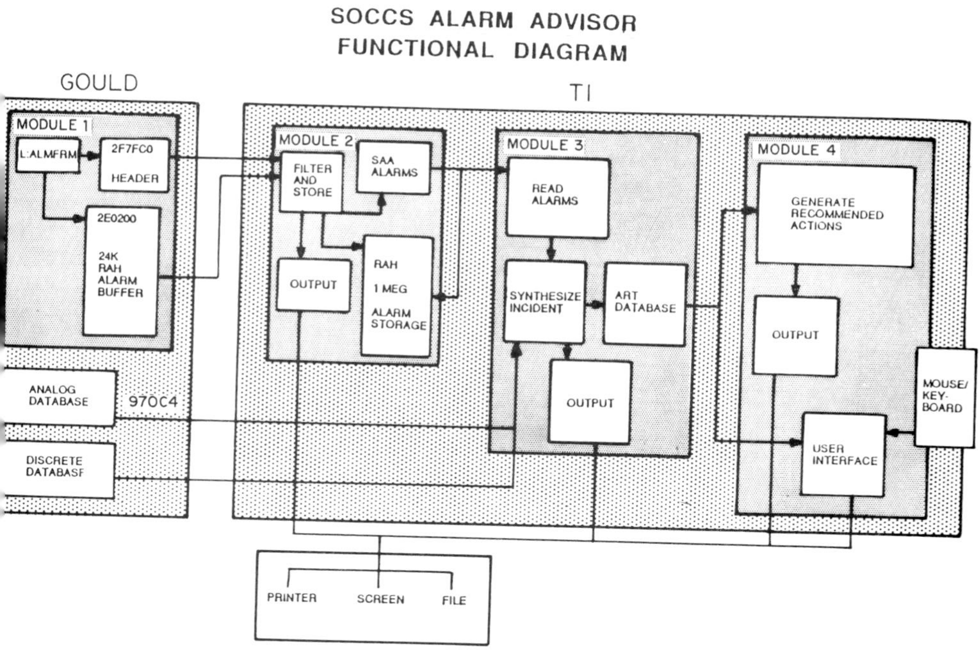

Figure 22-3 SAA Architecture

establishes a physical memory link between the computers. All process to process communication is managed over this link.

The initial SAA deployment was one-way from the Gould to the TI. Now that SAA has proven its worth we have begun plans to fully integrate SAA into SOCCS. In addition to displaying the SAA results on SOCCS, alarm and event processing will be offloaded from SOCCS to SAA thereby reducing utilization of SOCCS by as much as 12% during peak activity times. This has the benefit of extending the lifetime of the current architectures and delaying capital expenditures.

SAA Internal Models

SAA system design is based on the Three Model Methodology. These are physical, functional and temporal. The physical model establishes the actual connection in the network. In many cases, the physical model describes the dimensional or physical connection. The functional model describes how the pieces of the physical model relate to each other in a functional way, and the temporal model qualifies the evolution over time and/or the cause and effect of various observations **when the functional dependencies themselves are unknown.**

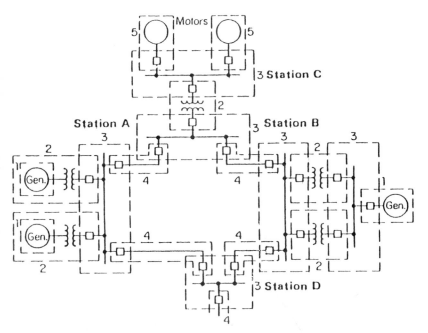

Figure 22-4   A Network and Its Relay Zones of Protection

SAA Physical Model: Here the physical model encodes the topological description of the electric power network. The physical model is automatically built from the data points in the SOCCS database. SAA constructs unique **objects** (also called frames or schemata) based on electrical **relay protection zones**. See Figure 22-4. Using these zones as a design paradigm is based on solid engineering principles. The benefit of this will be explained further. These objects, called "**clusters**" are defined as one or more entities bounded by discrete switching devices (breakers, disconnects or links). A cluster is also naturally bounded by generators, loads, and/or shunts which are located at the edges of the topology.

Any switching device or group of switching devices having a direct physical connection is considered a "**switch-group**". A switch group's status is determined to be open if any component device is open; otherwise it is closed. A switch group connects two clusters. Navigation and causal propagation from one cluster to another is along this common switch-group edge.

SAA Functional Model: The condition, or state, of a cluster is dependant on the states of the switch-groups at its boundaries and is determined by a set of cluster analysis rules. In addition, the functional model is used to propagate the effects of the cluster status to neighboring clusters. To implement the functional layer, two additional relations are used.

(1) A switch-group is **isolated** if it cannot provide power to its neighboring cluster. This is the result of all other switch-groups in the cluster either being open or themselves isolated. Switch groups connecting loads and shunts are, by definition, isolated.

(2) A switch-group is a **sink** if it connects to a load or shunt. Sinks propagate recursively, i.e. **if a closed switch-group is in a cluster containing a switch-group that is a sink, then all the other closed switch groups are also sinks.**

SAA Temporal Model: SAA must respond to changing conditions. Conclusions drawn at a certain instance in time may be radically altered upon changes to the original parameters. SAA employs a **truth maintenance system** to record logical dependencies between conditions and conclusions. If conclusions are based on conditions that are later retracted, the conclusions are automatically retracted. Thus, SAA maintains a consistent model of the network's state at every moment in time.

The temporal model also reasons about trends and previous history. The following SAA rules provide some examples:

- If a breaker opens, then closes, and then opens again within 90 seconds, then an automatic attempt to reclose the breaker has occurred, and a serious ground fault may exist, and do not attempt rapid restoration procedure.

- If a feeder overload condition preceded breaker operation, then check system loads before restoration, otherwise attempt a reclosure.

- If a feeder overload condition is followed by a Phase Angle Regulator hang-up alarm, then a tap run-away condition exists.

SAA Processing Phases

Identifying Nuisance Alarms: A set of rules identify if alarms or events are "toggling" or repeating. For example, If the rate of change of a discrete status exceeds four in six seconds, then the discrete status is considered to be toggling, and suppress discrete's status changes until a clear period of sixty seconds.

Similarly, alarms and events that repeat within the same or subsequent scans are also suppressed.

Determining Affected Clusters: Alarms pertain to elements located within the physical topology and model. SAA uses hash tables to determine those clusters containing elements involving alarms. These clusters are the initial set analyzed. Hash tables are also used to find the switch-groups located within a cluster; and for each switch-group, those switching devices contained within it.

Cluster Analysis: Functional model rules determine the state of the entities and breakers within a cluster. Some example SAA analysis rules are:

- If all switch-groups in a cluster are either open or isolated by adjacent clusters, then all entities in the cluster are out of service.

- If all switch-groups in a cluster are closed and not isolated, then all entities in the cluster are in service.

Within certain clusters, it is necessary to analyze the direction of current flow. Transformers, in particular, have primary and secondary sides depending upon the voltage level on each side. A breaker tripout on the primary side has a uniquely different effect than a secondary side tripout. Thus, clusters containing transformers, loads, shunts or generators all have different sets of rules.

Causal Propagation: Cluster analysis and causal propagation provide a complete description of the electric system's state (the scenario) at any instance in time. Causal propagation is implemented as an event declaration (as a fact assertion) in the clusters adjacent to the connecting switch group. For example, If only one switch-group in a cluster "c" is closed (all others are open or isolated), and the closed switch-group is not isolated from the adjacent cluster (it is seen as a possible source by its neighbor), then that switch-group is isolated within "c", and assert an incident in the adjacent cluster.

Implemented in a data-directed architecture, event declarations trigger initiation of cluster analysis rules for these adjacent clusters. The principal method of reducing search (i.e. stopping propagation through the clusters) is by analog confirmation, i.e., the SAA cluster analysis conclusions are supported by analog values. Thus, If analog measurements are inconsistent with the conclusions (analog denial), then create an incident in all adjacent clusters, otherwise do not create in the adjacent clusters and terminate analysis.

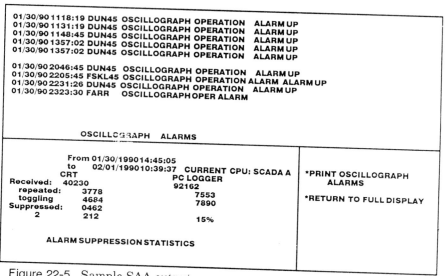

Figure 22-5   Sample SAA output

This rule does not mean that analog measurements are always accurate. The point is that if they agree with the conclusions then there is no reason to continue propagation. SAA will continue analyzing neighboring clusters until there is agreement. No qualitative penalty exists for doing this; the correct determination will eventually occur. The penalty is in computational analysis, i.e. time. As was mentioned before, the relay protection zone design is critical to the success of this approach. Since the zones are designed to isolate a fault, even in the event of a breaker failure, it is guaranteed that the search will not have to go very far before the correct analysis is done.

Recommending actions: Recommended actions have been implemented according to Con Edison's Operating Procedures manual and supplemented by senior system operators experience. These recommended actions include breaker failure recognition and action, transformer spare bank identification, and multiple feeder tripout procedures. Figure 22-5 shows a sample SAA output screen.

SAA Performance and Size

When an SAA run-time system is initialized, a separate filter process is forked, running at highest priority to read the alarm buffers at the same rate as produced by SOCCS (maximum of 200 alarms per 2 second scan). SAA's clock is synchronized with SOCCS' clock. After filtering nuisance alarms, analysis and recommendations for the remaining alarms are displayed on the operator's workstation within 20 seconds. SAA contains priorities to ensure that critical activities (e.g. reading the SOCCS buffers) occur before less critical activities (e.g. servicing operator requests for statistical summaries). Use of priorities was sufficient to achieve real-time performance in this application. When the topology of the electrical network is modified (e.g. adding a transformer or altering feeder connections), a new SAA system (with a new physical network and other changes) needs to be created. The total number of facts input to SAA from the SOCCS topology points is about 7,800 (3,900 entities and 3,900 discretes). SAA includes approximately 250 rules to create up to 2,000 objects (e.g. representing clusters), 100,000 facts and 75 distinct relations. This preprocessing phase takes between 1 and 2 hours to construct the datafiles necessary to build the delivery system. The SAA delivery system (real time system) contains approximately 100 rules, 4000 facts and objects, and uses about 40 relations. Twenty hash tables are created with a total of 25,000 entries.

Conclusion

Innovations

Innovations of SAA include:

- One of the first expert systems deployed in energy management.

- One of the first deployed expert systems that meet real time performance criteria in a time critical application.

- The use of physical, functional and temporal models to provide a complete and systematic analysis of alarms.

Criteria for Successful Deployment

Three criteria were critical in ensuring SAA's deployment and continued usage:

- SAA does not lose any alarm data, i.e., all alarms are processed and analyzed. Nuisance alarms are correctly identified and filtered.

- All information presented by SAA is correct, i.e., all data presented, all analysis performed, and all recommendations suggested.

- The information helps the operators in making correct decisions with regard to system restoration.

Payoff

Major benefits to Con Edison include:

- A significant reduction of alarms seen by operators (by suppressing nuisance alarms and by consolidating alarms).

- Improved response time of operators during incidents.

- Standardization of operator responses according to Con Edison procedures.

■   Identification and prioritization of maintenance tasks (focusing maintenance on the most severe problem areas identified by SAA).

In addition there were some unexpected benefits, including:

■   SAA identifies inconsistencies in the SOCCS database.

■   SAA generates reports useful to the operators and management, e.g., statistical summaries, logging reports and outage reports.

Development and Deployment Timeline

Development and deployment timeline of SAA is as follows:

September 1988	Project initiation.
September-October 1988	Functional specification
October-December 1988	High level design.
December 1988-January 1989	Kernel expert system.
January 1989	Gould-TI integration.
February 1989	Integration of SOCCS and SAA.
February-June 1989	Extensive enhancements.
June 1989	SAA beta version operational online.
July-September 1989	Extensive testing and system extensions.
September-November 1989	Final production system on-line testing.
November 1989	Acceptance testing of final production version.

SAA has been in operational use since June, 1989. The development lifecycle, including the beta period and final production testing, was 15 calendar months and involved 1 full-time expert system developer for the entire period. From the user community, the total manpower involvement was approximately 4-5 person months.

Future Development Plans

The SAA system has been so successful that plans are currently under way to enhance its capabilities. Some of the currently planned activities include:

- "Write-back" of SAA results to the SOCCS CRTs thereby allowing all operators access to its information.

- Downloading of SOCCS alarm and event processing to the TI thereby reducing valuable GOULD utilization.

- Providing a remote SAA screen to our Central Substation Control Center for their review of pertinent information.

Author Biographical Data

Steven Silverman has been responsible for database design, software design and expert system design at Consolidated Ediison. He received his Master of Science Degree in Mechanical Engineering in 1978 from Columbia University. He received his Bachelor of Mechanical Engineering in 1973 from City college of New York. He began work at Con Edison in the System Operation Department where he helped develop, test and deploy the energy management system.

Alvin Shoop is President of Expert systems Consulting Group. He is an independent expert system consultant with 10 years of experience. his specialties include real time monitoring and diagnosis and control applications. He received his Master of Science in Physics from the University of Massachusetts and an additional Master of Science in Computer Science from the University of California at Berkeley.

COOKER:
What's Happened Since?

Randy Bennett
Knowledge Engineer
Campbell Soup Company
Camden, New Jersey

Alan K. Carr
Director, Manufacturing Systems
Campbell Soup Company
Camden, New Jersey

Introduction

Forty-four years. Fourteen of them as the traveling troubleshooter for the massive cookers that keep soup production moving for Campbell Soup Company. Now all that experience is ready to walk out the door into retirement. The business problem facing a manager in such a situation is "How do you replace all of that experience and knowledge?" Usually, you don't, and hope for a "quick study" as a replacement.

This is exactly the problem Rubin Tyson, Director of Manufacturing Technology for Campbell Soup Company, faced in 1984. His answer was to encode the knowledge of production engineer Aldo Cimino into an expert system. So began an effort that resulted in COOKER, one of the most widely known of commercial expert systems.

Campbell Soup is the dominant firm in the condensed soup market; it is said that 97 out of every 100 households have one or more cans of Campbell soup in their cupboard.Campbell also holds the number one or two brands in its frozen, baked goods, beverage, and grocery business units.Annual revenues world-wide are around $6 billion.

In the United States food processing is a generally mature industry, despite periodic food fads. For the industry overall sales are growing only slightly faster than the population. Clearly the condensed soup component of the industry is mature.

Among the circumstances common to most mature industries and the food industry are narrow margins and competition on price. For this reason, production downtime on the hydrostatic "canned food product sterilizers", as cookers are more properly known, is to be avoided. In 1984 Campbell made soup at six locations in the United States, one in Toronto, Ontario, and one in Kings Lynn, U.K. Each plant had at least one hydrostatic cooker; a unit about 30 feet square, over 70 feet high and able to process tens of thousands of cans per hour.

COOKER

Any device over 70 feet tall and packed with equipment can be very difficult to diagnosis. It was not unusual to incur significant lost production time on each cooker malfunction.It could take at least a day for Aldo to arrive at the plant, before diagnosis and repair could even begin.Multiply several eight hour work shifts by tens of thousands of cans per hour, and you can imagine the Company's losses each time.

The pending retirement of Aldo Cimino threw the problem into sharp relief. Decades of expertise that had benefited Campbell were on the verge of walking out the door forever. After some quick discussions, it was decided to try to capture Aldo's knowledge in what was in 1984 a brand new and untried commercial technology; an expert system.

Two goals were quickly determined. The expert system would have to be able to replace Aldo Cimino (as much as that was possible), and it should be useful as a training tool for production and maintenance engineers. The target audience was maintenance and engineering personnel at the plant level. A team from Texas Instruments (TI) was brought on board to do the knowledge acquisition and programming necessary to create such an expert system.

The COOKER, as the expert system quickly became known, was developed using Texas Instruments' Personal Consultant, at the time a new product for TI. Personal Consultant, which has since been upgraded to Personal Consultant Plus, was a backward chaining expert system shell, with a limited forward chaining mechanism. While the knowledge base could be written in either English or LISP, LISP was used for COOKER. Development work was done on a TI

Portable Personal Computer, using 768 KB of RAM, a 10 MB hard drive, color display, and MS-DOS.

COOKER was developed by TI over a seven month period. Early on it was decided to include A-B rotary cookers to the knowledge base, in addition to the hydrostatic cookers.Together these account for the 151 rules in COOKER. About halfway through the development it was decided to add rules for each cooker's start-up and shut-down procedures to the COOKER rule-base. These require a separate start-up and shut-down file for each cooker at each plant. Once cued by the operator, COOKER calls the appropriate file when needed, and displays the required information.

Overall the COOKER project was judged a success. Copies of the Personal Consultant run-time module and COOKER knowledgebase were sent to the maintenance departments of each of the five plants, along with instructions on how to load and operate the system.

Although the user interface of COOKER is sufficient only to play "twenty questions" it found quick use among some of the plant's maintenance staffs. Even before Aldo left the company COOKER was being used in lieu of his personal consultations. However, as though to emphasize that an expert system can never fully replace an expert, just prior to his retirement Aldo found a cooker problem he had not encountered in 44 years of engineering. It was added to COOKER at once.

The project that resulted in COOKER was deemed a sufficient enough success to move forward on another manufacturing expert system. This one too was developed under contract by Texas Instruments.

The Next Expert System

Francis (Fran) Andriella was an engineer with 51 years experience with Campbell's manufacturing technology. By 1985 he had developed a profound depth of knowledge on the canning machines used in the soup process lines. Although Fran intended to "work forever" Rubin Tyson wanted his knowledge encoded in an expert system ASAP.

At the time Campbell was using a three-piece can for their soup products. This made the canner a very critical piece of machinery. In fact, a problem with sealing the cans would shut down an entire line.

Most problems could be handled by plant personnel, but serious or unusual problems required Fran's unique knowledge. These problems would wait either for Fran's personal visit, or for samples to be shipped to Fran in Camden, NJ. All the while, the process line could be idle.

It took five months, but once again TI's Personal Consultant had captured specialized knowledge of a Campbell production engineer. Like COOKER this system, called CLOSER, was written in LISP.

And, like COOKER, CLOSER was sent out to maintenance staff in plants producing canned soup.

The Next Six Expert Systems

In all Texas Instruments developed eight expert systems for Manufacturing Technology, none of which got the attention of COOKER. Each dealt with a specific piece of the machinery used in processing soup or producing cans. Each sought to capture specialized knowledge, accumulated by engineers over decades, to make it available to the maintenance staff on the job site.

CLOSER was followed by COIL LINE, an expert system based on the knowledge of engineer Jerry Crawford. COIL LINE took four months to develop, and helps diagnose problems with Littell Coil Lines. These coil lines are the first step in the manufacture of cans, processing raw coils of tin planted steel into flat sheets to produce can bodies and ends. FILLER was developed next, again using Fran Andriella as the expert. It also took four months to develop. FILLER helps diagnosis malfunctions in Pfaudler and syruper type machines machines used to fill the cans with soup.

Following FILLER a more complex project was undertaken, again with the assistance of Texas Instruments. Four separate expert systems (SHEAR, 314-AP PRESS, CURLER, and CAN ASSEMBLY) were developed using the knowledge of one engineer, Bob Baasch, plus information from operation and maintenance manuals for the machines, between August 1985 and May 1986. Using knowledge from manuals was a new approach at this time, but worked quite well.

SHEAR was developed to diagnosis malfunctions with the scroll shears used in can production. The scroll shear cuts the metal sheets passed from the Littell Coil Line into scrolled strips. These scrolled strips are then passed along for use in making both cans and ends.

314-AP PRESS helps in diagnosing problems with the press used to cut the can ends from the scrolled strips produced by the scroll shears. Smooth alignment is critical at this point, since all aspects of can production are controlled by corporate standards, and each step in the production process assumes all other materials are according to standard.

CURLER is used to diagnosis malfunctions with the double wheel curler used to form a curl on the can ends which is later used to form an airtight seal with the tops and bottoms. It accepts the scrolled strips, after ends have been punched out.

CAN ASSEMBLY concerns itself with the forming of the body of the can and closing the side seam. Unlike the other expert systems in this series, CAN ASSEMBLY uses graphical images to help with the diagnosis and repair. It can also create a file holding the recommendations of, and rationale for, the consultation to either be stored on disk or printed.

These expert systems were all developed under TI's Personal Consultant, all used knowledge bases coded in LISP, and all were developed on TI's Professional Computer. Altogether these eight expert systems were developed between November 1984 and May 1986. With each expert system a disk was mailed to the maintenance staff at each staff, along with instructions on operating the new expert system.

What Happened?

So ended the story four years ago. What has happened to the legendary COOKER, and its kin, since?

All found immediate use over the short term, but most often in the company's two oldest plants. COOKER, CLOSER, and FILLER were being used even before Aldo and Fran retired, saving thousands of dollars with each use. They were less used in the newer plants having more modern equipment. As a result the maintenance personnel in the newer plants seemed to lose awareness of COOKER and its kin, and the value to be derived from them.

The expert system world too has changed since these projects were begun in late 1984. Not only did TI enhance its Personal Consultant, but competing products gained creditability and power. LISP became less important as expert system shells began using English for their rules. And user interfaces became far more sophisticated than a "twenty questions" style interface permitted.

The business environment of Campbell Soup changed considerably as well. Over a two-year period the oldest plants, with the oldest equipment, were closed. And, a major effort in computer-integrated manufacturing (CIM) was mounted under the auspices of MIS and engineering.

With the CIM effort came a change in the plant computing environment. PC's running DOS, which is the environment for COOKER, FILLER, and CLOSER, became less important. The manufacturing control software now standard for Campbell plants runs under OS/2. So, any computer located on or near the production line will not have one of these expert systems available. All environmentally sensitive equipment, such as a computer, require special enclosures to protect them from daily steam cleaning and from the moisture of food processing. The expense of these enclosures tends to keep additional computers off the plant floor.

The effect of all this on the eight expert systems was pretty straight forward. Newer plants, using newer equipment, had fewer malfunctions and less need for COOKER and the rest. Since the distribution of the expert systems had been somewhat impersonal (a mailed letter, disk, and instructions), plant staff felt there was no one to call on for help in understanding and using them. Since the knowledge bases were written in LISP they could not be kept current by in-house staff.

With all of these changes going on the use of these expert systems had to be effected in some way. With these changes somehow COOKER, and its seven siblings, seemed to get lost in the shuffle.

However, as a result of the success of COOKER a new expert system was born: SIMON. Federal regulations require special handling of any processed food product in the event of a cooker malfunction. For years this was done by telephone. The plant contacted the correct department in Camden, who then interviewed the plant staff. Fast handling is critical, since some malfunctions do not effect food safety and the product can continue in processing if action is taken in time. Before SIMON was developed, it could take days to gather the data and make the decision.

SIMON, developed by the Process Safety department of the Campbell Institute for Research and Technology, automates the process. SIMON was written using the Aion Development System, which calls a database for product information and some FORTRAN programs for numerical analysis. Based on information about the malfunction, mathematical analysis, and rules based on Federal regulations and Campbell standards, SIMON produces a product disposition report.

Following a year of use on a central mainframe connected to each canned soup plant by telephone, SIMON was ported to PC's and distributed to quality assurance staff in each plant. The product disposition report is sent via modem to Process Safety in Camden after the report's conclusions have been acted upon. Michael Mignogna, Director of Process Safety, estimates that SIMON saves millions of dollars each year.

Future Plans

It is ironic that these pioneering expert systems are now"old" technology. COOKER, for example, requires the operator to key in answers to questions, which may require taking readings from gauges on the machine, as well as observation of the operation of the cooker. Operating the expert system must take place off the plant floor, away from the equipment. But the memory of benefits from COOKER remains strong within Campbell. Currently the Maxton, NC plant is in the midst of replacing an existing soup line with a flexible line, able to process multiple soup products, with processing speed controlled by the demands of the equipment further up the processing line. This CIM system is being put into place by the CIM department of MIS. At meetings engineering personnel have asked if some form of AI, specifically COOKER, can be used to help maintain the flexible line.

Toward that end some design goals have been written up for the new COOKER expert system. These take into account the changes to the expert system world, to the Campbell soup plants, and to the availability of knowledge technologies in general.

For example, the new COOKER will fit into the existing plant computing environment, meaning it must run under OS/2. A general CIM goal, which the new COOKER must observe, is to avoid "islands of automation" on the plant floor. The CIM architecture mandates OS/2 running on a model 50 or 70 PS/2 computer.

The new COOKER should have access to signals from equipment sensors. This will allow it to obtain as much information as possible on-line, as opposed to operator entry. This should help avoid operator error in data entry. But, it should also speed up the diagnosis. It is possible that some malfunctions could be recognized by the new COOKER before the operator is aware of them.

Given that the new COOKER will be connected on-line it should operate as fast as is possible. This suggests some sort of compiled knowledge base. This would allow the expert system to react to sensory data as fast as possible. With 600 cans per minute being processed through a cooker speed is essential.

In addition to these design goals two "nice-to-have" goals are under consideration. One is an automatic link to SIMON, so that a malfunction detected by the new COOKER could initiate SIMON, leaving SIMON to notify its operator that a malfunction has occurred. This would speed the analysis SIMON does, probably saving more product in the event of minor malfunctions.

The second is a link to some sort of hypertext/hypermedia environment. This could allow diagrams, pictures, pages from original plans or manuals, or even video images to be captured as hypertext cards. These could then be linked to the expert system in the appropriate rules. So, if the new COOKER is asking about the condition of the chain drive, which is hidden behind a guard plate, the new COOKER could display a diagram instructing the operator how to remove the guard, or show a video sequence demonstrating the best and safest method of removal.

Conclusion

The final picture is somewhat mixed. COOKER, CLOSER, and FILLER are still available to plant maintenance personnel, but with the departures of the people involved in their development they are sitting in limbo. The other expert systems developed at about the same time are likewise in limbo because in-plant can production been changed.

Still, COOKER was enough of a success that additional expert systems were developed, SIMON being the most direct link to COOKER. As the CIM project in the Maxton plant shows, plant engineers want an updated COOKER created. All of this activity led MIS to create a position dedicated to overseeing the development of expert systems, as well as other knowledge technologies.

And, of course, COOKER continues to bring world-wide fame to both Texas Instruments and to Campbell Soup Company. Perhaps this fame will be continued in the next few years as the new COOKER, a sort of SON-OF-COOKER, is developed as a real-time expert system.

Author Biographical Data

Randy Bennett is the Knowledge Engineer in the Management Information Systems department of Campbell Soup Company, at the corporate headquarters in Camden, NJ. He is responsible for developing AI and Knowledge-Based Systems, connecting these systems to corporate information systems, establishing corporate AI/KBS standards, and supporting users across the corporation. He has been a worker in a rural anti-poverty program and a management consultant before becoming a programmer. He joined Campbell in 1983, and has been dedicated to AI/KBS since 1989.

Alan Carr is the Director of Manufacturing Systems in the Management Information Systems department of Campbell Soup Company, at corporate headquarters in Camden, NJ. He is responsible for designing and implementing Computer Integrated Manufacturing systems connecting plant floor control and monitoring systems to corporate information systems, as well as the MIS AI function. He has worked as an Electrical and Software Engineer in the Engineering Research and Development department at Campbell, where he designed hardware and software for process control and data acquisition.

24

CABPRO Case Study

MIchele Van Dyne
IntelliDyne
Kansas City, Missouri

Robert Schaefer
Allied Signal, Inc.
Kansas City, Missouri

*Operated for the U.S. Department of Energy by the Allied-Signal, Inc.,
Kansas City Division, under contract No. DE-AC04-76DP00613.*

Introduction

Allied-Signal Inc., Kansas City Division, is a prime contractor for the Department of Energy. It manufactures plastic, mechanical, and electric/electronic components. Because of the low volume of any one part produced compared to traditional manufacturing operations, the Kansas City Division atmosphere is much like that of a job shop. Additionally, there are few continuous production processes, with most manufacturing being discrete production. As part of the Nuclear Weapons Complex, the Kansas City Division strives to use state-of-the-art techniques in both manufacturing and productivity improvement. The Kansas City Division requires a significant engineering staff. Therefore, any system that can improve engineering efficiency or accuracy has a great effect on the bottom line.

Artificial intelligence was recognized as an important technology by Allied-Signal in the early 1980's. It promised to provide an automation solution to some of the more knowledge intensive activities of the engineering staff as opposed to the algorithmic procedures in other departments.

Because AI appeared to be a valuable tool for solving engineering problems, initial activity in AI was focused on process planning. Presently there are several efforts in process planning at the Kansas City Division, in addition to those in quality engineering, and some activity in failure analysis and diagnostics.

The AI effort at Allied-Signal, Kansas City Division, began in the early 1980's. Since that time, several projects targeted at process planning were undertaken, including the expert system, CABPRO, which is the topic of this case study.

CABPRO

CABPRO (Cable Processor) is targeted as a tool for process engineers in the multi-wire cable fabrication area of KCD. While the number of end users is low, approximately one dozen engineers, the time saved per user is potentially extremely high, thus justifying the pursuit of the expert system. Also, development by-products such as an object-oriented database and process characterization were identified as useful for other analysis work.

CABPRO is a rule-based expert system which produces work directions used to fabricate multiwire cables. Creation of these work directions, called travelers, is a significant part of the engineering process-planning function. Writing travelers now require a large amount of manpower to do important, but repetitive, work. CABPRO's objective is to reduce the manpower effort in process planning, thus freeing engineers for more challenging assignments. The CABPRO system creates traveler text for a defined subset of cable configurations. A prototype was developed initially to demonstrate the feasibility of using expert system technology to automate the engineering process-planning function for the manufacture of multiwire cables.

The original major objectives for beginning the CABPRO project were: (1) To create accurate, consistent and timely work directions which would help eliminate many work direction errors; (2) To produce a first draft set of work directions the same day a design is received; (3) To capture and store knowledge of experienced process engineers to reduce the effects of engineers retiring, transferring, and leaving; (4) To gain Artificial Intelligence development experience in a manufacturing domain.

Multiwire cable fabrication is performed by human operators following written instructions which document and communicate the processes used in cable assembly. Cables are manufactured in low volume and come in a multitude of different designs, so that many different cables are in production at one time. In addition design changes often require new written instructions. In this environment, manufacturing seldom becomes completely familiar with specific cable designs. Therefore, work instructions for each cable design must be accurate, detailed, and current.

The written instructions that the operator follows are called travelers.

They "travel" with the cables to whatever workstations and operators are needed during assembly. All materials, components, tools, machines, and any additional information needed for fabrication must be explicitly defined for each operation. Whether the operation is done by hand (such as twisting wires to improve flexibility of branches) or with the aid of machines (such as transfer molding to insulate connectors), each discrete action is defined in the traveler.

A typical multiwire cable traveler might consist of 80 to 100 pages defining 60 different manufacturing operations. The writing of these travelers represents a significantly large part of the process-planning function. To select which operations apply to a specific cable design, to determine what sequence of those operations will produce the best quality cable at the lowest cost, and then to compose the traveler takes an average of 40 engineering manhours. Looking at the number of manhours involved in the writing of travelers, the benefits of automating the process are obvious.

By using expert system technology to automate the writing of travelers many manhours can be saved. Travelers can be created with greater consistency and accuracy. The time required for process-planning can be dramatically reduced, freeing engineers for more challenging tasks. It is important to note that a side benefit of using expert system technology will be that the knowledge of experienced cable process engineers is saved in a rule-base and may be used for future reference.

Process planning begins by review of a cable design to identify features and assign manufacturing activities. A cable with three connectors would require an activities for soldering wires to each connector. Therefore, "solder wires" is an activity assigned to each connector. The engineer knows that once connectors have been soldered, they must be insulated to prevent, among other things, exposure to moisture. Therefore, "encapsulate connector" is another activity assigned to each connector. After all activities have been determined the engineer then decides how to sequence those activities for the process plan. Sometimes natural constraints are used in the decision-making. For example, a connector cannot be encapsulated before wires are soldered to it. At other times decisions are based on design parameters and general knowledge of cable fabrication techniques. In either case, the engineer must select words to communicate the process plan to the manufacturing operator in the form of a traveler.

Process planning is a multi-step process that follows a hierarchical pattern the initial planning stages are at a macro "activity" level, followed by planning at the "operation" level and finally planning on the "task" level. Figure 24-1 shows this hierarchy graphically. "Activities" are general processes in cable fabrication that are mapped to specific cable features. For example, soldering and encapsulation are general processes that are mapped to a specific cable feature connectors. "Operations" are more specific sets of actions grouped under activities that can be performed by one operator at a single workstation. Mold preparation, material preparation, cure, and clean-up are some examples of operations that are grouped under the activity encapsulation. At the micro level are tasks —discrete actions grouped under operations. Some tasks associated with the material preparation operation include: weigh material A, weigh material B, combine and mix materials.

Significant progress has been made in this cable domain for communicating process plans through what is called a "standard traveler." The standard traveler consists of words that have been agreed upon by both engineering and manufacturing to describe the various tasks needed for cable assembly. Standard words (tasks) for approximately 300 operations have been documented. These 300 operations are in a library of available resources that engineers can consider when forming a process plan.

Organization of the standard traveler takes advantage of the activity-operation-task hierarchy. An operation is the basic unit of the standard traveler and may be up to three pages of text with optional illustrations. Tasks are represented as distinct paragraphs in the text. All operations for a given logical activity are grouped together for easy reference.

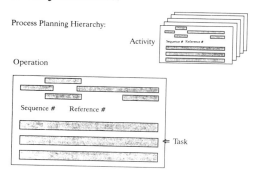

Figure 24-1   Process Planning Hierarchy

When an engineer has conceptualized an assembly plan, specific operations (or pages) are selected from the standard traveler and tasks (or paragraphs of text on the page) are customized to fit the specific cable design. Before implementation of CABPRO, the actual pages of text that made up the standard traveler were kept in a file cabinet and engineers gathered the pages, in sequence, and customized the text by hand. CABPRO can do this faster, more accurately, and more consistently. CABPRO automates this process planning function, by beginning with the cable design, and from that, producing the entire traveler required to produce the part.

CABPRO consists of the user interface, database, and database interface, rule base, inference engine, and traveler output sections. This is illustrated in Figure 24-2.

It was developed under the AEGIS operating system in the Apollo Engineering Workstation environment, and currently runs under their System 5 Unix operating system.

The purpose of the user interface (UI) is to provide a comfortable means of entering all design information necessary for process planning, then to organize that information into a format usable by the expert system. The UI is perhaps one of the most important modules in CABPRO, since it is the only part of the system seen by the user. If the UI did not present a quick and simple method of interaction for the user, CABPRO would not have gained user acceptance.

A typical CABPRO user would be a process engineer who may or may not have cable-domain-specific knowledge. To start a session the engineer must have a drawing of the cable design. No advance review of the drawing is necessary (although it may be helpful). Input is made through keyboard entries or simple menu selections using a mouse. The UI does a minimum amount of error checking for things like improper data types. Help is available in the form of pop-up windows, and help files containing, for example, specifications defining approved fabrication methods. See Figures 24-3 and 24-4 for user interface examples.

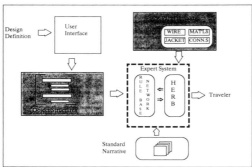

Figure 24-2   CABPRO Architecture

In the prototype version, when the user was finished entering all of the information about a cable, the data was formatted into a Flavors structure that was recognized by the Lisp-based expert system. This Flavors file, called the cable description file (CDF), was created using a token replacement procedure. For improved efficiency, the production version uses Lisp structures to store the data, rather than Flavors objects.

The CABPRO prototype is a rule-based expert system that incorporates a forward chaining strategy in the inference engine HERB to select and sequence manufacturing operations necessary for the assembly of multiwire cables. Figure 24-5 illustrates the architecture of the expert system portion of CABPRO.

HERB (Hierarchical Expert rule-base system) is the expert system shell used to develop the CABPRO prototype. It supports multiple rule-bases, allowing the developer to define and manipulate a rule-base stack. Rule-bases can be pushed onto and popped off of the stack. Rules in HERB are typical IF-THEN statements having a set of IF conditions and THEN actions. HERB matches conditions with known facts and data with IF portions of rules, then selects rules to

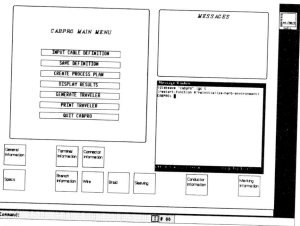

Figure 24-3   CABPRO User Interface: Main Menu

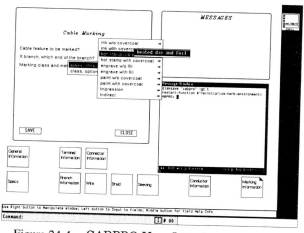

Figure 24-4   CABPRO User Interface

be executed. When a rule is chosen, the THEN actions (usually Lisp commands and/or HERB data base manipulation commands) are executed. Five different conflict resolution strategies are offered to break any ties that arise when more than one rule matches. HERB continues the match-select-apply cycle until current facts and data no longer match any rule. Control is then passed back to a Common Lisp environment.

As shown in Figure 24-5, CABPRO rule-bases comprise a hierarchical network operating at three levels for process planning. Sixteen rule-bases are grouped into the categories illustrated. At the top level, a rule-base is used to decide what rules will be needed to solve the next set of problems and to push and pop the appropriate rule-bases as they are needed. Activities are selected and sequenced at

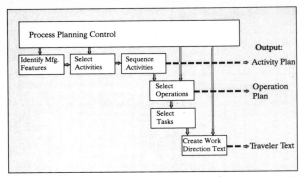

Figure 24-5   Hierarchical Rule-Base Network of CABPRO

one level, then, at a lower level, operations and tasks for each activity are selected and mapped to specific standard traveler operation text pages and paragraphs, respectively. A final rule-base takes care of miscellaneous functions like printing the traveler.

All rule-bases use the age-directed conflict resolution strategy. The age-directed strategy ensures that rules with conditions matching the newest facts and data in the HERB data base are executed first, except rules with a condition designating a GOAL. If the keyword GOAL is used to begin a matching condition of a rule and if all other conditions of that rule match, that rule is given special consideration and executed first. This strategy fits the cable process-planning problem well because of the goal/subgoal problem solving-method used by expert engineers in process planning.

Cable fabrication is done in two stages, a preparatory stage and an assembly stage. In the preparatory stage, all pieceparts are gathered, cut to length, identified, and cleaned. In the assembly stage, the cable is actually formed and pieced together. To take advantage of this constraint, separate rule-bases select preparatory and assembly activities.

The sequencing portion of process planning is perhaps the most difficult because of the variety of choices. At the same time, cable fabrication procedures offer several constraints to simplify sequencing. For example, all wires must be soldered to the connectors and all solder joints must be visually inspected before connectors can be encapsulated. These constraints are exploited by the rule-base design. Three rule-bases are dedicated to sequencing activities: one for preparatory activities, one for an assembly starting point, and one for the remainder of the assembly activities.

Nine cable fabrication activities can be selected by the CABPRO prototype. A separate rule-base exists for each activity and each rule-base selects operations for that activity. As operations are selected, a list is created that contains reference numbers mapped to standard traveler text pages in a file. Once all of the operations and tasks have been selected and sequenced, the operation list is passed to a program that prints out a copy of the traveler if the user so desires.

New or revised cable designs are received from the design agency every month, and CABPRO is designed to handle approximately 60% of these (only round-wire polyurethane cables are addressed). A typical process plan requires about 40 hours of engineering time to determine and assemble, while CABPRO can do this function in under 2 hours, only about 30 minutes of which is requires user interaction.

In acquiring the expert knowledge for CABPRO, interviews were conducted with expert process engineers, producing 50 60 hours of audiotape. The audiotape was transcribed into outline format, and was assembled into a notebook which is still available for review. What is somewhat unusual about the knowledge acquisition for this project was that multiple experts were used, despite cautioning against that approach by expert system literature. This actually created a more robust knowledge base, where methods of problem solving were verified by different experts, and areas of expertise possessed by one individual were incorporated that may otherwise have been missed.

There are presently two people assigned to CABPRO on a part-time basis. Because a good deal of their time is now spent on interface issues with traditional software, not much maintenance has been done on the original rule base, and not much has been required. Although cable designs are relatively dynamic, the procedures for assembling those cables and the process decisions required appear to be more static.

Figures 24-6 and 24-7 show an example rule used in CABPRO and the definition of an object.

Apollo Engineering Workstations were used as the hardware platform, primarily because of their capability and secondarily because of the availability of this hardware across the user community. An internally developed expert system development environment (HERB) was used because of its fit with the problem area. HERB is a hierarchical, Lisp-based tool, and the nature of process planning is hierarchical.

HERB is an inference engine/expert system development environment produced by Keith Hummel of KCD. It supports multiple rule-bases, forward and backward chaining, pattern matching, multiple conflict resolution strategies and the execution of Lisp code at any point in a rule. The multiple rule-base feature was attractive not only to modularize programming but also to take advantage of the inherent hierarchical nature of process planning. Rules are entered in typical IF-THEN form and require a working knowledge of Lisp. The internal HERB data base supports FLAVORS representations as well as any Lisp data structure.

Common Lisp was chosen as the language to use from within the rules of the CABPRO system because of its ties to HERB and FLAVORS. FLAVORS was used to define data structures designed with information analysis work in the prototype version. Lisp structures now contain the data.

DEFRULE "*Solder Wires to Connector*"
    IF
          (GOAL select assembly activities)
          (GOAL select solder wire activity for P1)
          TEST
          search CDF to locate connector-termination for P1
          match connector-termination value with "*cup*"
          search CDF to locate conductor type to be
              terminated at P1
          match conductor type with "*wire*"
    THEN
          add (solder P1 to the selected activities list)
          remove-fact (GOAL (select termination activity for P1))

Figure 25-6   Example CABPRO Rule

(make-instance 'connector
        :connector-designation "*P1*"
        :connector-type "*rack-and-panel*"
        :connector-termination "*cup*"
        :connector-insert-type "*standard*"
        :connector-part-number "*355652-02*")

Figure 25-7

C was chosen as the language for both input and output programming, outside the realm of HERB. It is used to create output for the traveler text file.

Lisp windows are used for the user interface. In the prototype version, an operating system windows management tool was used, however, this did not allow for sufficient flexibility. Lisp windows provides facilities for multi-window management, mouse and/or keyboard input, and the ability to incorporate user-defined help screens and help files.

There are 750 rules in the CABPRO production system, and 50,000 lines of code. About 85% of this code is Lisp, with the remainder being C. Lisp was used for internal procedural work, while C was used for input/output. CABPRO is to be interfaced with an SQL database system residing on the networked Apollo workstations. There is a patent pending on the software developed for this interface. Other interface work also remains to be completed. A plant-wide Automated Traveler System (ATS) is used by all departments for the editing, maintenance and control of travelers. The requirements for interfacing with that system are under investigation.

Conclusion

The criteria used for measuring performance of the prototype version was the comparison of results from CABPRO to process plans developed by engineers. Three cables currently in production that contained features known to CABPRO were chosen for comparison.

One cable consisted of three connectors and two branches while the other cables were very similar to the example detailed above. Design data was entered into CABPRO and in all cases the CABPRO-generated process plans were equivalent to existing plans. Two cables similar to our example with two connectors and one branch were chosen because they had similar designs but different process plans. Their designs differed in total length and number of conductors in the branch.

In addition to accuracy, the CABPRO prototype proved a significant time savings in using an expert system to help engineers write travelers. Time required to input data for the test cables averaged 15 minutes and execution time averaged 7 minutes. For an engineer to perform the same tasks requires approximately 2.5 hours. Actual time comparisons may not directly transfer from the prototype to a production system but the promise of significant man-power reduction benefit is encouraging.

User acceptance of the system is always a concern. Therefore, a continuing effort is made to improve the user interface. Additional error checking, error recovery mechanisms, and additional help information are items for incorporation. Adding windows with graphics to display help was identified as a primary target of effort. As a prototype, CABPRO successfully demonstrated the time savings that were projected in the initial justification. As a production system, no measures have been made, however, it appears to be continuing to save large amounts of time when used. Future efforts on CABPRO include trimming the expert system down to only handle process planning, and let other systems handle database, printing, revision tracking, text manipulation and editing.

Author Biographical Data

Michele Van Dyne is the owner/operator of IntelliDyne, an expert system consultation and development company based in Kansas City, Missouri. She has worked in the area of artificial intelligence since 1985, primarily in the manufacturing and telecommunications industries. Michele has a BA in psychology and an MS in computer science. She worked on the CABPRO project during her employment at Allied-Signal Aerospace, Kansas City Division, from 1985 to 1988. Michele also serves as the chairman of the steering committee for the Kansas City area expert system interest group, ESKaMo.

Robert M. (Mike) Schaefer is a staff engineer at Allied-Signal Inc., Kansas City Division. He has worked with expert systems since 1985 in the process planning area, and had four years of process planning experience prior to that. Mike is the project leader for the CABPRO project, and has a BS in mechanical engineering. He also serves as the treasurer for the Kansas City area expert system interest group, ESKaMo.

25

Expert System
for Automatic Generation
of Printed Wiring
Assembly Process Plans

David Liu
Liu and Associates
Granada Hills, California

Introduction

The research problem at hand is automatic generation of printed wiring assembly (PWA) process plans. It is an industry-wide problem that has yet to be successfully addressed in its entirety. This difficulty arises out of the nature of process planning—it may be conceived of as art, as opposed to a formalized engineering discipline. This research embraces the entire spectrum of process planning, i.e., process characterization, product definition, producibility analysis and machine programs generation.

Process planning is a highly personalized activity which requires a great deal of insight obtainable through experience. The style and quality of the process plan characterize the level of expertise of the process plan creator. An individual's preferences can substantially bias the end result.

Typically, a process plan reflects the creativity, skill and knowledge of a variety of manufacturing disciplines. Process plans for similar parts, however, are created each time by different individuals, thus introducing planning inconsistences which may confuse the assembly worker. Inconsistent process plans (although each may be an excellent plan when evaluated on its own merits) also cause parts to

348

flow in different sequences during assembly, thus making scheduling much more difficult.

Process planning consists of interpreting a design, selecting appropriate operations, sequencing correctly the operations, selecting the type of equipment, and detailing the specific work instructions at each operation. Process plans can differ in appearance and content, depending on the skill level of the intended reader (operator or assembler).

The simplest form of a process plan is a route sheet, which contains the proper sequence of operations. Operation instruction sheets and pictorial planning books are both illustrative types of work instructions. The operation instruction sheets are more technical in nature. The illustrations in an operation instruction sheet resemble that of a blueprint, while a pictorial planning book is similar in style to a cartoon book. Obviously, the most precise and elaborate work instructions are those that instruct machines, e.g. numerical control (NC) programs for automatic insertion equipment.

Due to the magnitude and complexity of the task, the majority of today's process planning systems are computer aided process planning (CAPP) systems. Computer aided process planning systems assist the process planner in carrying out the task by guiding him/her through the manufacturing steps via an interactive session on the computer. But the process planner still performs the drawing interpretation function, thus providing the feature recognition capability that is missing in most computer aided process planning systems.

Once the part description is made available—in the form of part features—to the computer aided process planning system, an

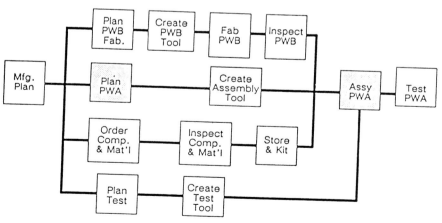

Figure 25-1   Scope of Printed Wiring Assembly Manufacturing

extensive amount of manual editing is required. The editing process is an open forum which offers the process planner an opportunity to freely modify the process plan. But since both the interpretation of drawings and editing of instructions are highly personalized activities, it is difficult to achieve consistency in process plans created by different planners, even for the same part.

Today there is an increasing trend toward total automation of process planning, i.e., automated process planning (APP). Yet, existing automated process planning systems are still in their experimental stage and largely unrefined; the systems developed to date produce only simple skeletal process plans, i.e. route sheets. In order to create more highly detailed process plans, a great deal of additional input and editing are required to augment these skeletal plans.

The difficulty in building an automated process planning system capable of providing complete and comprehensive process plans can be attributed to the magnitude of the task, the complexity of the problem and the dynamic nature of today's manufacturing environment. Because today's manufacturing environment demands more robust and flexible planning systems, those systems need to be more intelligent.

Quality decision-making requires accurate and timely information; even though the knowledge of the planner can be mimicked, it is not entirely clear what information is essential—and therefore what information must be made available—to the decision-making process. The research will thus, also define the essential information required for automated process planning.

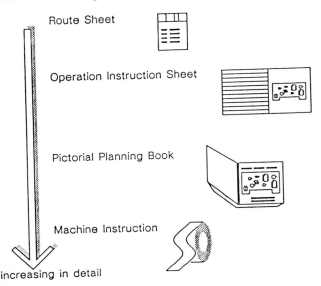

Figure 25-2   Level of Detail in Process Planning

During the current transition from the Industrial Age to the Information Age, manufacturing is shifting from being capital intensive to information intensive. Facing ever mounting economic and competitive pressures, American manufacturers have been forced to rethink how information can best be harnessed to support the decision-making process; in the process, automation and integration have become key words in manufacturing technologies.

In today's global marketplace, low overseas production costs, in combination with increasing demands for a wider variety of products and shorter product life cycles, have compelled domestic manufacturers to reassess the strategies and technologies which have enabled them to be competitive. The future success of manufacturers will require low cost production of quality products and increased flexibility in response to rapidly shifting market demands.

Low cost production necessitates containing and minimizing soaring support costs. Support costs account for as much as 50 percent of the production cost in some sectors of the manufacturing industry. Support cost is the cost of non-touch labor effort contributed by white-collar workers towards the production of a good. Generally, these support activities entail some sort of decision-making process.

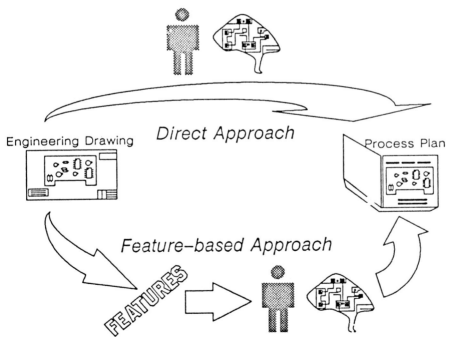

Figure 25-3   Approaches to Automated Process Planning

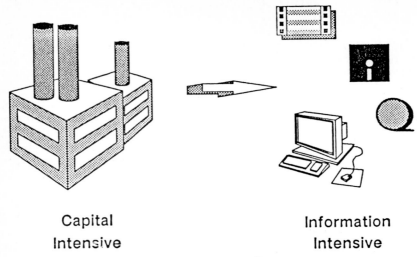

Capital   Information
Intensive   Intensive

Figure 25-4   The Changing Manufacturing Environment

Minimizing support costs, while also maintaining a high level of decision-making capability, requires advanced computing technology. Automated decision-making can significantly reduce the cost of manufacturing and, at the same time, consistently provide quality decisions.

Manufacturing relies on a broad spectrum of decision-making processes. The two most prominent, planning and scheduling, fall into the general category of manufacturing planning. Manufacturing planning encompasses long-range planning (production planning), intermediate planning (manufacturing resource planning and process planning) and short range planning (scheduling). Much of today's manufacturing research focuses on methods to automate intermediate and short range planning.

Suitability of Knowledge-Based System

Since little formalism has been introduced to enable the application of engineering principles to process planning generation, the automatic generation of process plans would to be an ideal application of knowledge-based systems (KBS). Knowledge-based systems are suited for solutions which are ill-defined and for decisions which seem ad hoc. By using an expert knowledge base, the reasoning of expert planners can be captured and re-used.

Knowledge-based systems can embody facts, logic and rules which mimic the reasoning of an expert planner. Manufacturing knowledge is used to deduce the required manufacturing operations based upon an interpretation of the product description and the analysis of "fit" to the capability of the equipment. The product description includes:

- The physical data

- Functional specifications

- Geometric information

- Special manufacturing instructions

- Inspection criteria

The word fit infers that the product is within the constraints of the equipment. The capability of the equipment includes descriptions of its functions and its operational limits.

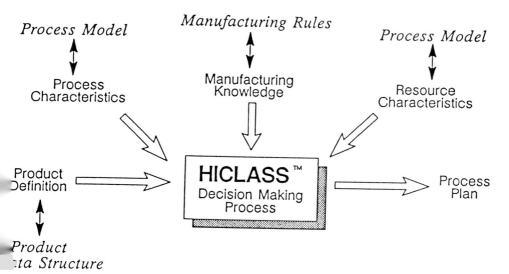

Figure 25-5   Knowledge Base to be Developed

Research Objectives and Goals

This research addresses the need for an industry-wide endeavor that attempts to apply knowledge-based systems to process planning for PWA assemblies. Its objectives are to validate the suitability of knowledge-based systems for automatic generation of PWA process plans and to explore the limits of automated process planning via knowledge-based systems.

A further objective of this research is to develop a generic framework for characterizing electronic assembly processes, equipment capabilities, and product data definition requirements, and to apply this framework in a prototypical knowledge base.

Scope of Research

This research is not intended to develop expert system shells or to extend manufacturing knowledge regarding PWA. Expert system shells and PWA techniques are certainly worthy subjects for intensive research, but I do not attempt to treat them here. The intent of this research is to apply, in a functional and demystified manner, the capabilities of knowledge-based systems to the task of automated process planning.

To exercise the knowledge base, the HICLASSTM Software System is used. HICLASS is a network-oriented, rule-based knowledge-based system developed by Hughes Aircraft Company.

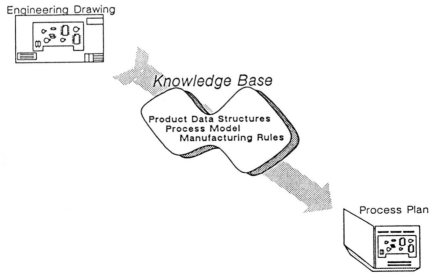

Figure 25-6   The Goal of this Research

The Hughes PWA Application

The Hughes Process Planning Environment

Under the traditional manual process planning system — which is still the predominant mode of operation at the Tactical Product Operations Division of the Electro-Optical and Data Systems Group, Hughes Aircraft Company — planners must sift through engineering drawings, notes, and other relevant data in order to prepare a paper planning package. A paper planning package is comprised of a routing sheet and pictorial manufacturing instructions. The pictorial manufacturing instructions appear in the form of a picture book that has both textual explanations and hand-drawn illustrations.

This traditional manual method of creating and maintaining process plans is very time-consuming and labor intensive. To create a process plan for an average printed wiring assembly (PWA) at the Tactical Production Operations Division, approximately 160 man-hours (or four man-weeks) are required. These paper planning packages must also be manually maintained and updated to reflect frequent engineering and manufacturing process changes. The entire manual process is subject to errors which must later be manually corrected.

Additionally, given the same engineering data, different planners are likely to create differing planning packages, depending on their level of expertise and past experiences. The manual method also lacks an automated (electronic) link between engineering and manufacturing which could aid not only in the generation of process

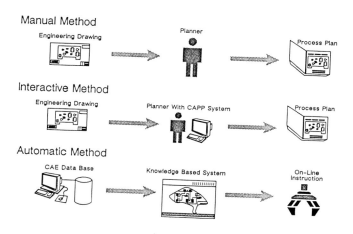

Figure 25-7   Evolution in Methods of Process Planning

planning, but also in providing automated producibility analysis for the design engineers.

The Motivation

The PWA process planning application is the driving force behind the continual refinement of the Hughes Integrated Classification (HICLASSTM) Software System. At the Tactical Product Operations Division, a variety of PWAs are produced at low to medium volume. Thus, instead of automating the actual assembly process, it was determined that more benefit could be derived by automating the "white collar" (i.e. support) activities. HICLASS is the vehicle used for carrying out the mission to automate PWA process planning.

HICLASS Rules

While HICLASS I was being developed, industrial engineers worked to translate manufacturing logic into HICLASS rules. The resulting rule base and data base were termed the Hughes PWA application, also known as the HICLASS PWA network. The Hughes PWA application automates the generation of instructions for printed wiring assemblies, with the intention of reducing the manufacturing implementation cycle time.

After a re-codification of rules from the HICLASS I syntax to the HICLASS II syntax — HICLASS II rules were much more compact and expressive than those of its predecessor — the rule base was reduced by a ratio of 6:1. In the Hughes PWA application, two types of rules were identified:

■ Inference rules

■ Data modeling rules

Inference rules refer to the group of application specific logic codified in the HICLASS syntax. Inference rules exclude rules which are used to define the application's data models. Data modeling rules refer to the group of rules used to define and construct the meta-structure of the application. Terms such as conceptual structure (viewing it from the philosophical perspective) and working memory (viewing it from the computer programming perspective) are often used synonymously with the term meta-structure (viewing it from the information modeling perspective). All of them refer to the frame/slot definitions and external data access methods of the application.

The rule base now contains over 440 unique "inference rules" which are partitioned into approximately 120 rule sets or nodes. A majority of the inference rules concentrate on formatting and painting the graphics for the process planning packages. Ten percent of the inference rules are dedicated to graphics generation. An additional 40 data modeling rules are used to define the meta-structure. Since HICLASS supports the concept of sharing common knowledge—some rules, nodes and sub-networks make multiple appearances in the application—the resulting effect of the total rule base is much greater than the sum of its parts.

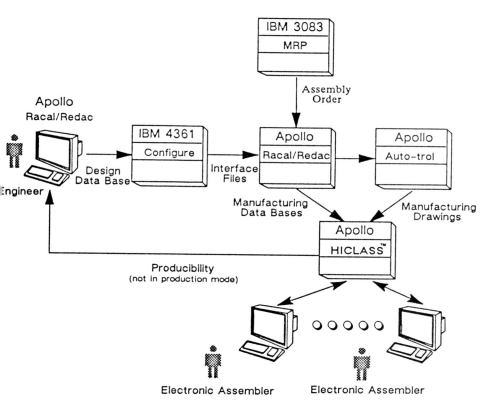

Figure 25-8   Hughes Printed Wiring Assembly Application

Information Requirement

In the Hughes PWA application, the computer aided engineering (CAE) data bases for a PWA design are electronically transferred to Manufacturing. The transferred information contains data that describes a specific PWA design.

The bill of materials contains the part numbers and descriptions of components comprise the PWA. Component physical data includes component reference designators, each component's pin 1 location and its orientation. Special graphics are IGES (Initial Graphics Exchange Specification) formatted files containing information such as board outline, pad patterns, tooling holes, component bodies and coating areas. Gerber plot data are the drill patterns of the printed wiring board (PWB) along with the aperture of the holes. Net list is a list of all the connections (between pins) in the board. The reference information contains pertinent manufacturing information such as maximum component height, maximum lead protrusion, and board thickness. Engineering notes contain drawing notes that conform to a special syntax designed to ease the process of computer interpretation. The component electrical data and the net list are not being used by the Hughes PWA application at this time.

A master component library and manufacturing data bases are the other primary sources of information for the Hughes PWA application. The master component library supplies information such as component dimensions, body material, lead material, lead thickness, number of pins and other intrinsic characteristics of the components. The information that manufacturing itself supplies are:

- Component list for automatic insertion

- Component list for semi-automatic insertion

- Kit location

- Tooling information (including graphics)

- Packaging information (including graphics)

- Graphics for special illustrations (e.g. side view explosions)

If the assembly planning package generated by the application is not producible, or if it does not meet quality assurance (QA) requirements for some other reason, the approval cycle begins again by submitting an "engineering change request" so that changes can be made and a new CAE data base issued. When the final package is approved by quality assurance, it can then be downloaded to the factory floor where the assemblers can view it electronically on a workstation as opposed to viewing the traditional paper planning package.

HICLASS PWA Network

The actual generation of process planning for the Hughes PWA application is implemented using the HICLASS PWA network (herein known as the PWA network in this section). The PWA network is a modular and multi-level network. The primary or top level of the network is a logical mapping of the Tactical Product Operations Division's manufacturing process flow into the HICLASS network notation. The PWA network is updated and enhanced in order to reflect changes in the flow layout and the factory capabilities.

The implementation of the PWA network emphasized the creation of graphical assembly work instructions to be displayed on the factory floor. Thus, the majority of the rules are biased toward planning formats as opposed to process characterization or resource characterization.

In order to ease the translation of the industrial engineers' perspective from manufacturing process flow diagrams to a HICLASS network, two types of nodes were identified:

- Process nodes

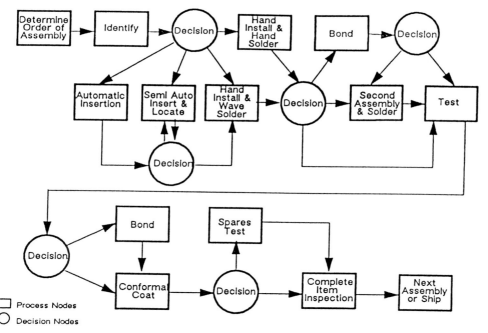

Figure 25-9   Primary Level of the HICLASS Printed Wiring Assembly Network (PWA)

■ Decision nodes

Process nodes refer to those nodes which contain rules relating to actual manufacturing operations. The rules in a process node deal with how an operation is to be performed. There is at least one process node associated with each manufacturing operation. Decision nodes, on the other hand, refer to those nodes which contain rules relating to selection of the manufacturing operations. The rules in a decision node determine when an operation is appropriate. Each process node has only one exit path while each decision node has at least one exit path.

The division of process nodes and decision nodes is artificial. There is nothing intrinsic within HICLASS to impose or enforce this separation. It is strictly done for the convenience of the industrial engineers. Internally, HICLASS processes these two types of nodes in identical fashion.

Modus Operandi

The way that the Tactical Product Operations Division implemented their HICLASS PWA network was to have the process nodes contain inference rules and facts used to produce work instructions. Although each wiring board might require a different sequence of assembly instructions, the types of operations and the instructions for each specific process (represented by these nodes) are similar for all boards. Every board, for example, must be identified by either a stamp or stencil, but all boards for which stenciling is chosen will have similar instructions for that operation. In the decision nodes, inference rules determine the next logical path to traverse in the PWA network for a given PWA.

The PWA network is executed for each and every unique PWA. PWAs are uniquely identified by their assembly number and the associated configuration number. Execution of the PWA network begins with a set of nodes which define the network's data models, after which it knows where and how to retrieve external data.

The HICLASS Conceptual Structure Manager, previously known as the Experience Table in HICLASS I and II, then locates the correct data—much of which is in the files created from the CAE data base—and loads it into the meta-structure. It is this data which is unique for each PWA and which drives the execution of the network. The data enables the inference engine to find a logical path through the network.

Once the initial meta-structure is set up, the logic evaluation of process planning generation begins. The first process node is entitled "determine order of assembly". This node contains rules which use information about the wiring board—such as characteristics of its hardware and components—to make decisions such as what components must be installed first, what parts are to be automatically inserted or what types of soldering might be required. An example of one such rule is the following:

> FOR EACH hardware WHERE height IS GREATER THAN 2
> DO the second assembly hardware reference designator list
> IS EQUAL TO UNION (the second assembly hardware list,
> hardware reference designator).
> END

This rule looks at the "height" attribute (slot) of each entry (instance) in the "hardware" frame of the meta-structure. For each one that has a value greater than two, it adds the value of the "hardware reference designator" for that instance to an existing list of second assembly hardware. The reference designator is a unique slot value which identifies the particular instance of the hardware frame under consideration.

Later in the PWA network execution, an inference rule within a decision node will look at the value of this "second assembly hardware list". If the set is not empty, it will use the "reference designator" values in it to locate other information (slot values) for the specific instances of the hardware frame, and the rules will use that data to generate instructions for the installation of the second assembly hardware.

The traversal path through the PWA network will eventually flow through the final two process nodes which generate instructions for the inspection and shipment of the completed PWA. One final example is a rule from the PWA network used to determine how a wiring board will be packaged for transportation:

> IF bare board width IS GREATER THAN 4
> AND bare board height IS LESS THAN 6
> AND bare board length IS LESS THAN 6
> THEN
> exterior cover # IS "35-256"
> interior cover # IS "35-2580"
> tote box # IS "35-2573" pouch # IS "35-6370"
> tag holder # IS "35-2594"
> END

This simple inference rule from the PWA network illustrates how knowledge about routine decisions, such as the packaging to be used for a certain type of board, can be encoded directly into the rules when appropriate. If information such as the material handling part numbers above were to change frequently, it could be included in a data base instead of within the rules. This complies with the general philosophy of being data-driven, thus, allowing the rules to remain generic and defer the decisions until the data is available.

Once this data (exterior cover #, interior cover #, tote box #, etc.) has been determined, it can be processed through either the Route Sheet Generator (generates routing documents) or the Rule Driven Graphics Generator (generates graphical assembly work instructions). A reason for having the data printed on the route sheet is to explicitly indicate on the "traveler document" the material handling part numbers needed to package the boards prior to transporting them.

Network Organization

The PWA network divides the PWA manufacturing process into a number of smaller problems, i.e., nodes. Since there are approximately 120 individual nodes in the PWA network, nested sub-networks are used to collapse some of these nodes into logic modules according to functional areas. Domain experts in each particular functional area are responsible for their portion of the PWA network.

Although these major sub-networks have been identified, many nodes within these sub-networks have not yet been implemented. The degree of implementation will be discussed as each sub-network is described.

The meta-structure module brings together all the information required for decision making during the network execution. Once the

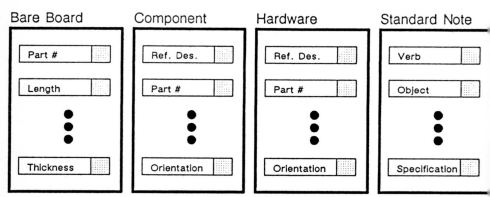

Figure 25-10 The Meta-Structure of the PWA Network

information is loaded into the meta-structure and instances of frames are created, information is easily accessible in an intelligent manner. Even the information loading process is highly data-driven. The PWA network has a "flat file" node that assigns the appropriate file names to assist in the loading of the meta-structure.

The meta-structure of the PWA network is comprised of a set of distinct objects: bare board, components, hardware and standard notes. Each object has a list attributes, each attribute is contained within a slots. A bare board, for instance, have attributes such as part number, length, width and thickness. A component have attributes such as component reference designator, component part number, length, height, width, location, orientation, pad span, description, body type and number of pins.

Hardware possesses attributes similar to those of a component. Hardware consists of less standardized types of components such as relays, connectors, transformers and power transistors. A special frame is used to capture the relationship of hardware which is attached to the PWB via other hardware. For example, screws, nuts and washers are sometimes used to attached a connector onto the PWB.

A standard note possesses attributers such as verb, object, object modifier, specification, verb modifier, material and comment. These attributes describe portions of a note on an engineering drawing which are recognizable by a special HICLASS sub-network called the Standard Notes Parser. The notes on an engineering drawing are an important part of the product definition, because they determine the processes and methods to be used.

A node named "determine order of assembly" is placed at the beginning of the network to make most of the major decisions regarding routing and assembly process selection. Instead of the traditional usage of rule-based systems — where decisions made later can affect decisions made previously, causing the system to make multiple passes through the rule base — the Hughes industrial engineers decided that it would be more straightforward to make the major decision regarding routing and assembly process selection up front.

Possible choices for the assembly processes are automatic insertion of dual in-line packages, automatic insertion of variable center distance devices, semi-automatic insertion, pre-wave solder assembly, post-wave solder assembly or manual assembly with hand soldering. The general preferences are for components to be automatically inserted and for PWAs to be wave soldered whenever feasible. As a rule of thumb, automatic insertion implies that the wave solder operation will follow. In addition, the order of preference is that automatic insertion is done prior to semi-automatic insertion, which is done prior to manual insertion.

A list of component reference designators—reference designator is the unique identification of each component on the PWA—is created for each of the assembly processes. In the event that an assembly process has an empty list, it means that this particular PWA does not need to be processed through that operation. These lists are created by applying manufacturing logic to determine whether there is a fit between a component and a process. For example, a rule uses temperature sensitivity or size criteria to choose whether to place a component in the pre-wave solder or the pose-wave solder list. A list of component reference designator is not, however, created for the automatic insertion operations; rules for assigning components to the auto-insertion machines have not been developed as part of the PWA network. Therefore, the PWA network does not determine auto-insertability nor does it generate machine instructions for the automatic insertion machines. Instead, the PWA network takes the pre-programmed machine instructions as input. It is assumed that when a component is being called out by the machine instructions, the component will always be automatically inserted.

State of Implementation

Overall, the state of implementation is that the PWA network is operational for manual operations. The rules for the manual assembly process have been fully implemented, i.e., the "hand insertion and hand solder" and "hand insertion and wave solder" sub-networks are complete. Since the ultimate goal of the PWA network is to provide assembly technicians with work instructions displayed on color terminals prominently positioned along the manual assembly area, the majority of the logic implemented deals with formatting and painting the graphics for the process planning packages.

The formatting rules provide a means of maintaining uniformity throughout and among the various process planning packages. Formatting rules are responsible for creating operation numbers, external document references and the following graphics displays: material-handling caution notes, components' polarity orientations, static sensitive caution notes, color keys for parts status and packaging instructions.

Formatting rules are used to generate or piece together illustrations showing the PWA as it is to be assembled at various stages of the assembly process. A running record is kept of which components have already been inserted and which components are yet to be inserted. To illustrate the building up of the PWA, the appropriate component graphics are generated, along with the component part numbers, kits locations, caution notices and special graphic instructions. In addition, formatting rules scale and rotate components as well as Initial Graphics Exchange Specification graphics files.

During pre-wave solder assembly, components are first installed, then hardware is installed. Post-wave assembly follows this sequence. Only three (3) types of hardware are handled in pre-wave solder assembly: connectors, relay sockets and power transistors. Of the three, only the logic for installing connectors has been implemented. The existing PWA configurations demanded that the connector installation logic be implemented first.

Formatting rules compile both a top view and an exploded side view of the connector installation. The exploded view consists of the connector itself and its associated hardware (nuts, washers, screws, etc.). These views are not drawn automatically, but rather, are manually drawn via a computer aided drafting system.

In addition to the assembly instructions, the PWA network also generates inspection instructions. Component polarities, placement and existence are checked prior to wave solder to minimize post-wave solder rework. An illustration of the PWA is generated to show the proper insertion of components and their proper polarities.

The second-stage assembly and post second-stage sub-network have not yet been implemented either. The implementing of the various sub-networks is dependent upon the installation schedule of the shop floor computer equipment. The areas with availability workstations and color display terminals was implemented first.

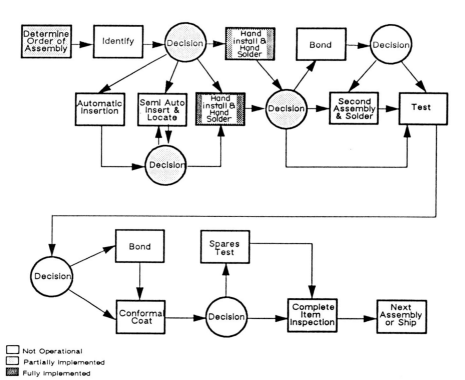

☐ Not Operational
☐ Partially Implemented
▨ Fully Implemented

Figure 25-11   State of the PWA Network Implementation

Integrating Application Networks

A primary HICLASS network can also take advantage of other HICLASS networks by joining one as a sub-network. These sub-networks may modify or expand the meta-structure established by the primary HICLASS network. The PWA network included several of these sub-networks, for example, the HICLASS Standard Notes Parser network 48.

The HICLASS Standard Notes Parser network is used to extract information from the notes on an engineering drawing. Each different note is stored as an instance of a "standard engineering note" frame in the meta-structure of the PWA application, and components of the note, such as "verb", are saved as slots within the frame. The decision node can then include a HICLASS rules similar to the following:

> IF ANY standard note's verb WHERE THE verb IS EQUAL TO "BOND"
>   THEN GO TO "bond".
>   ELSE GO TO "conformal_coat".
> END

This rule will search for any engineering notes stored by the HICLASS Standard Notes Parser network which include the verb "BOND". If any are found, it will proceed to the process node labeled "BOND". If the verb "BOND" is not found, as would be the case if the design engineer had decided that a bonding process was not required, the network would skip the "BOND" and continue with the "CONFORMAL COAT" process node.

Benefits of the PWA Network

The PWA network, as opposed to the manual planning method generates a consistent and error-free planning package for each PWA. If the inference rules are well constructed, the resulting package also reflects the most efficient and cost-effective method of assembling a particular wiring board.

Currently, using the PWA network to generate an average process planning package takes approximately 100 hours, a time savings of one and a half weeks off the design-to-manufacturing cycle. This includes the time required for transferring the CAE data base, modifying the knowledge base, executing the PWA network, screening planning, QA verification, and repeating the cycle when necessary.

More importantly, Hughes has been able to realize the benefit of a paperless environment for both its process planning and factory floor display activities as a direct result of the PWA network.

The Hughes PWA application involves much more than merely automating and integrating existing procedures and information systems. The success of its implementation did not rest soles upon technical innovations, but rather also involved efficiently controlling and utilizing information, and changing attitudes about how manufacturing tasks should be performed.

Limitations

A current limitation of automatic process planning generation using HICLASS, as compared to a planner, is the lack of ability to infer certain part features directly from the graphical information stored in the CAE data base. To circumvent this shortcoming, part features are explicitly incorporated into the graphics entities of the CAE data base so that they may be referenced from within the PWA network.

Conceptually speaking, the network is a body of knowledge containing know-how on every phase of the PWA manufacturing process. However, at the current state of implementation, the PWA network does not cover the entire spectrum of PWA manufacturing processes. Thus, the Hughes PWA application is not 100 percent generative for all PWAs. New conditions may exist which require additional rules to be written or old rules to be modified. Furthermore, a person needs to review the generated planning package to ensure its validity by checking for missing and incorrect results.

Research Results

Overview

The results of the research are embodied in the development of an automated process planning (APP) knowledge base (KB) for printed wiring assemblies (PWA) which is presented in this chapter. This consist of:

1. A conceptual framework for process modeling

2. A software program (written in HICLASS) which is capable of: Auto-insertability analysis
Generation of insertion instructions

The conceptual framework for process modeling provides a general model for describing the entire PWA manufacturing process. The software program contains sufficient knowledge, about both the product and the process, to generate automatically insertion instructions as demonstrated in five test cases. The software program takes into account the following facets of process planning:

- Process characterization
- Product definition
- Producibility analysis
- Planning generation

More precisely, the software program carries out the following tasks:

1. It accepts a feature-based description of the PWA design as input

2. It analyzes the PWA design for auto-insertability

3. It generates the insertion instructions for the axial and dual-inline package (DIP) automatic insertion equipment, as well as for the sequencer/verifier.

In the course of developing this knowledge base, the state-of-the-art in automated process planning has been extended. However, many more, as yet, unchartered tasks have also been uncovered.

The resulting knowledge base of this research is a network of rules called the Research Network. The research network is expressed as a software program which is partitioned into the following major segments:

Each segment is represented by a sub-network of rules. Each serves a critical function within the spectrum of automated process planning.

The development of the four major segments (or software modules) is based on detailed conceptual analyses of the tasks they represent. The product definition segment had been adequately addressed by state-of-the-art developments in process planning—such as the Hughes PWA application—and the resulting concepts were adopted here. An important part of the present research was devoted to the development of conceptual models for the other three segments, with particular emphasis on process characterization. The scope of producibility analysis and planning generation is quite broad; thus, because of its emphasis on printed wiring assembly, this research focuses on automatic insertion operations. Furthermore, the framework of the knowledge base for the automatic insertion operation is generic enough to exemplify other insertion methods such as semi-automatic, manual and robotic, as well as various placement operations.

In completing the spectrum of automated process planning, work instructions must be generated. Hence, the output of the Research Network is insertion programs which drive the automatic insertion equipment. However, before the Research Network can generate the insertion programs, an evaluation for auto-insertability is conducted to determine which, if any, of the components on the PWA are appropriate for the automatic insertion operation. Process characterization is crucial to the task of producibility analysis and planning generation.

Going Beyond Surface Knowledge

Although the state-of-the-art developments in PWA process planning have dealt with issues of process characterization, the knowledge that is being embodied in those state-of-the-art systems is primary surface knowledge. Thus, for instance, processes are generally modeled after process flow diagrams, as shown in figure 25-11. The correlation of one process with another exists only in the sense of sequential flow.

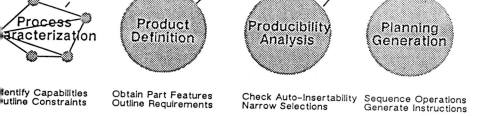

Figure 25-12   Research Network Organization

Although sequencing is important, it is, nevertheless, only surface knowledge; that is, knowledge which is adequate for describing the flow of parts through the factory, but inadequate to explain why the parts must flow through the particular sequence. Because surface knowledge contain no underlying (deep) reasoning, it is difficult to extract generalizations based on such knowledge which would allow the development of a generic set of process planning rules.

In order to achieve a higher level of sophistication in automated process planning — a system that is capable of supporting distributed and hierarchical decision-making — knowledge base with a more in-depth understanding of the processes is required, one which capture the intrinsic nature of the processes involved. The development of such a knowledge base represents a significant part of this research.

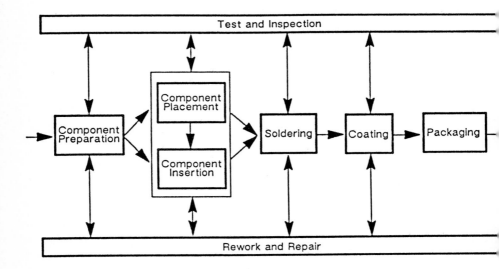

☐ Process
➤ Material Handling

Figure 25-13   PWA Process Flow

Process Characterization

Process characterization traditionally appears in the form of a process flow diagram. The objective of the process characterization segment of the Research Network is to develop a new concept for systematically describing a richer set of semantics than merely simple sequential order. Hence, this research produces a new conceptual framework for process characterization. This new conceptual framework is the result of the following steps:

- Literature survey on PWA manufacturing

- Synthesis of the characteristics of each process/operation

- Development of a new conceptual model for process characterization

- Validation of the model by walking through a few scenarios @t1-1st par = This new process model uses a network of process elements and relations, as explained below.

Process Elements

To capture the intrinsic nature of the processes, each process can be decomposed and characterized in terms of its constituent elements. However, these elements can themselves be processes, operations, equipment and material. Therefore, the term process elements is used to describe categorically both the processes and the entities related to processes.

Although the terms process and operation are often used interchangeably, this research refers to a process only as a combination of lower level processes or operations. An operation can consist of lower level operations (i.e., tasks), but not processes.

This "characterization by successive refinement" approach is similar to a popular technique used in group technology 5 (GT) called classification. By applying the classification technique, the following PWA process taxonomy (see figure 25-14) has been derived.

Although a process taxonomy is informative, it still does not provide enough information about each process/operation to describe relationships among the process elements or to generate work instructions. In addition to knowing the decomposition of a process/operation, the salient features of each process element must also be captured. Salient features include information on the process element's capabilities, constraints, advantages and disadvantages, methods of application and preferred part features.

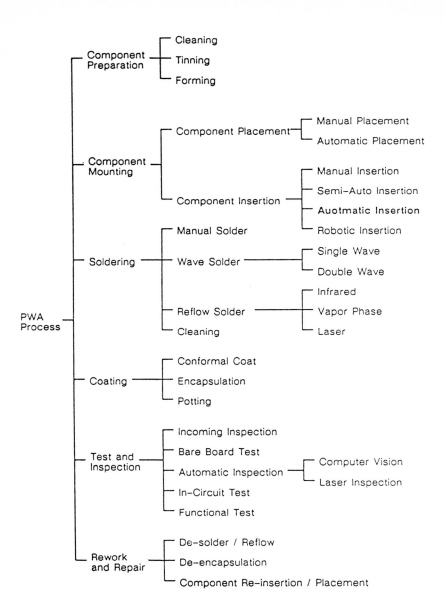

Each node in the taxonomy tree is a specific type of process element, i.e. a process or an operation.

Figure 25-14   The PWA process Taxonomy

Process Element Relations

An understanding of the relationships among process elements is of paramount importance because they provide insights into the dynamics of PWA manufacturing. The relationship among the process elements is termed the process element relation.

Among the numerous possible process element relations, six may be identified as fundamental. Their definitions are as follows:

Consist of
> "non-restrictive" membership relation. If X consists of (A,B,C), then X may be replaced by any one or any combination of A, B and/or C.

Choice of
> A "mutually exclusive" membership relation. If with X you have the choice of (A,B,C), then you may substitute either A, B or C (only one) for X. It is a special case of consist of.

Involve
> If X involves (A,B,C), then X = A u B u C, i.e. X always requires all the elements (A,B,C) under all circumstances.

Followed by
> A timing relation, i.e. If X is followed by A, then X must precede A. The followed by relation is assumed in the involve relation. The order (or flow) is the default sequence of the elements.

Need
> A material-requirement relation. If X needs A, then X requires a material called A in order to complete the process X.

Use
> A methodology relation. If X uses A, then X requires an action called A in order to complete the process X. It is generally paired with at least one need relation.

These six fundamental relations are sufficient to characterize the entire PWA process. Furthermore, the relations that are identified among process elements may be generalized to cover a broad spectrum of similar processes because they are built on top of fundamental relations.

Process characterization is accomplished by combining process elements and process element relations. By doing so, this research blends the traditional use of group technology classification of processes along with the semantic relations of artificial intelligence (AI) to form a new type of process model, a "network of process elements and relations".

Modeling Taxonomies

One might ask whether there is a correlation (or semantic equivalence) between this network of process elements and relations model and the process taxonomy. Or to broaden the question, is there a correlation (or semantic equivalence) between this type of model and group technology type of decision trees? The answer is "yes" and lies in two of the fundamental relations: choice of and consist of. Indeed, figure 25-15 shows how the consist of and choice of relations are to be used in conveying the semantics of a process taxonomy. For instance, the figure explains that the component preparation process (see top of second column on figure 25-15) consists of three operations: cleaning, tinning and forming. In addition to the consist of relation, figure 25-16 also depicts the use of the choice of relation. In the soldering process for example, there is a choice of four operations: manual solder, wave solder and reflow solder.

Modeling Component Mounting

Since the two most often discussed processes in PWA manufacturing are component mounting and soldering, these two processes will be used to illustrate how the "network of process elements and relation" process model is constructed.

Figure 25-16 is a graphical representation of the component mounting process. The figure explains that component mounting consists of component placement and component insertion, meaning that one or both processes may be appropriate, depending on the types of components involved. In turn, component placement is decomposed into bonding and placement. The involve relation shows that both steps are required.

In addition to the usual membership relations, figure 25-16 depicts several interesting relations for the following processes/operations:

■ Bonding

- Dispensing

- Sequencing

Bonding requires a pair of relations—the use methodology relation and the need material-requirement relation—to convey fully the essence of the process. Bonding is accomplished by using an attachment technique which needs a bonding material. There is a choice of bonding materials available: epoxy, acrylic, cyanoacry or pressure sensitive tape. The option is to choose one out of the four.

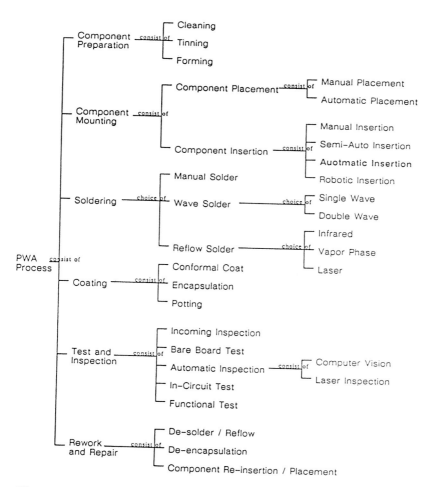

Figure 25-15 The PWA Process Taxonomy Modelled as a Network of Process Elements and Relations

The dispensing operation is required whenever epoxy, acrylic or cyanoacry is used. There is also a choice of dispensing methods available: pneumatic, screen or pin transfer. The dispensing operation, regardless of the method of application, must be followed by a curing process. Curing provides the opportunity for the physical properties of the bonding material to change, thus creating a strong bond.

The sequencing operation is interesting from the standpoint that it is only applicable for certain operations; namely the automatic insertion of axial or radial components. Since sequencing can be either a standalone operation or an integral part of the automatic insertion operation, the involve relation can be used to demonstrate the flexibility of its definition. Furthermore, it demonstrates that there is not always a 1:m relation, but there can also be an m:1 relation.

Modeling Soldering

The graphical representation of the soldering process is quite similar to that of component mounting. Except for the "dispense" process element, there exist no common process elements between soldering and component mounting. However, the process element relations and

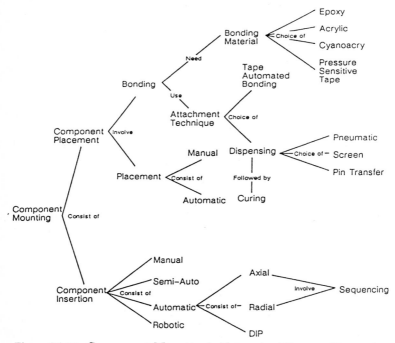

Figure 25-16   Component Mounting's Network of Process Elements and Relations

their uses are identical. Based on these observations, it can be concluded that these six fundamental relations are sufficient for representing a wide variety of manufacturing processes.

The network of process elements and relations for soldering provides a revelation in the use of the need-use relation pairs. The process of soldering has multiple material requirements; both flux and solder paste are needed in order to complete the process. Additionally, both flux and solder paste need to be dispensed. Except for restrictions which are stated in the definition of these fundamental relations, there are no conceptual (or implentational) limitations on how they are to be paired, combined and used.

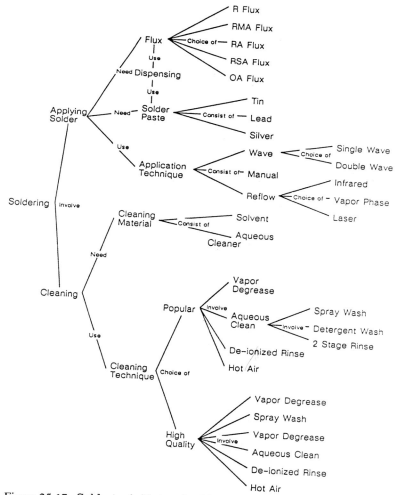

Figure 25-17   Soldering's Network of Process Elements and Relations

Process Characterization via Frames

In actually implementing process characterization, frames are used to represent both the process elements and process element relations. There is a process element frame template and a process element relation frame template. Each individual process element and each process element relation is modeled as an object. Therefore, a "network of process elements and relations" is a body of objects (process elements) connected to one another via other objects (process element relations).

The attributes (or slots) of the bare board frame are conceptually categorized into two groups: identification and construction. The slots of the identification groups are required to uniquely identify the process element. The slots of the construction group describe what the process element is and how it is related to other process elements.

The groupings of slots are strictly conceptual, serving to communicate and clarify the purpose of the slots. HICLASS syntax and semantics do not support the notion of grouping of slots. Therefore,

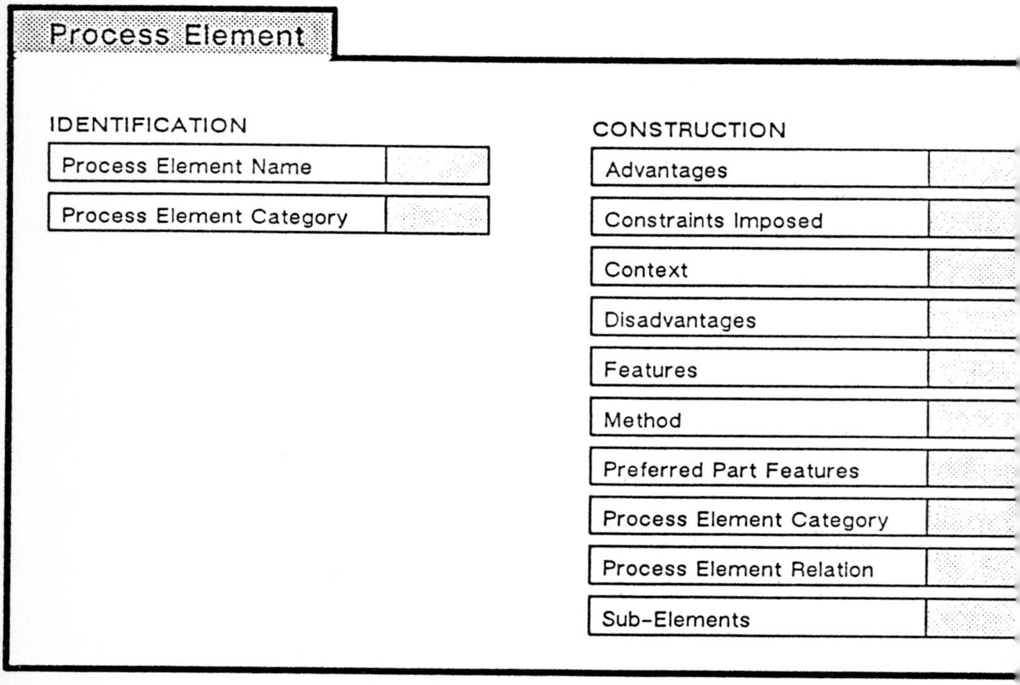

Figure 25-18   The Process Element Frame

grouping of the slots is a part of the research result, but does not play a role in the actual implementation of the knowledge base.

HICLASS uses a node to represent a frame and a rule to represent a slot. Therefore, in the Research Network, there is a node name "process_element" to represent the process element frame and it contains the following rules:

- Process element name

- Advantages

- Constraints

- Context

- Disadvantages

- Features

- Method

- Preferred part features

- Process element category

- Relation and sub-elements

Each rule in the process element frame defines a slot and contains information about what type of data it expects hold. For example, the "process element name" DEFINE rule takes on the following form:

> {RULENAME: process element name}
> {This rules defines the key slot of the process element frame. Process elements are any process characterization objects.}
>
> DEFINE
> data type IS EQUAL TO "STRING"
> default IS EQUAL TO" ".

The first few statements which have braces ("{" and "}") around them denotes that they are merely comment statements and they do not participate in the definition of the process element slot.

In a DEFINE rule, the name of the rule is also the name of the slot. The first statement of this DEFINE rule informs the system to expect only data values which are of type "string". The second statement initializes the default value of the slot. The default value is assumed by every instance of the frame. When an instance receives an explicit assignment of a new value, the default value is overwritten for that particular instance.

The "process element name" rule is the first slot of the process element frame. According to the HICLASS syntax, being the first slot makes it the key slot of the frame. The significance of a key slot is that it uniquely identifies each instances of a frame. HICLASS ensures that within a frame, the key slot value of each instance is unique.

In the Research Network, the complete set of process element instances are codified in a rule name "add process elements" which belongs to a node name "init process element". Although every instance of the process element frame has been codified, only the automatic insertion process elements belong in the normal execution of the Research Network.

These automatic insertion process elements are separately codified in a rule named "add implemented process elements" which also belongs to the "init process elements" node, and which is written as follows:

{RULENAME: add implemented process elements}
{This rules create process element instances and initializes the slot variables}
ASSERT
  Implemented process element list IS EQUAL TO ("automatic insertion",
                                "axial insertion", "DIP" "sequencing").
  {create the 'automatic insertion' instance by assigning with a new key slot value}
EXECUTE ADD FRAME INSTANCE ("process_element", "process element name", "automatic insertion").
{initialize the slot variable for this instance}
FOR EACH process_element WHERE process element name IS EQUAL TO "automatic insertion" DO
advantages IS EQUAL TO
   LIST ("faster....", "higher precision", "higher yields", "mature technology").
   constraints IS EQUAL TO "dedicated to 1 type of leaded components".
   context IS EQUAL TO "component insertion",
    disadvantages IS EQUAL TO
   LIST ("larger lot size", "longer set-up time").
   features IS EQUAL TO "highly refined equipment",
   method IS EQUAL TO "want_to_check_auto_ins".
   preferred part features IS EQUAL TO EMPTY.
   process element category IS EQUAL TO "operation".
   relation and subelement IS EQUAL TO LIST ("consist of", "axial insertion", "radial insertion", "DIP insertion").
  END.
{create the 'axial insertion' instance by assigning with a new key slot value}
EXECUTE ADD FRAME INSTANCE ("process_element", "process element name", "axial insertion").

Product Definition

Product definition traditionally appears in the form of an engineering drawing. The objective of the part definition segment of the Research Network is to develop a method for systematically describing and capturing engineering drawing information in a computer interpretable format. The description must be sufficiently detailed so that it may enable the software program to automatically generate process plans.

The product definition of the Research Network performs the following basic functions:

- Define the basic constituent frames

- Obtain the PWA design information from external files

- Create the frame instances from the PWA design information

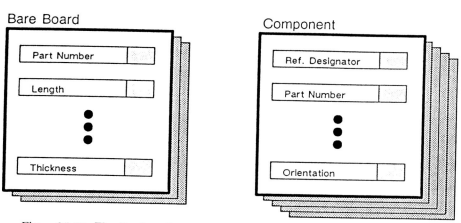

Figure 25-19 The Product Deffinition Frames

Basic Constituents Frames

The definition of a PWA must be described and captured in terms of its basic constituent elements. The basic constituent elements of a PWA are a bare board and the components (which include hardware such as brackets). Each constituent element is then further decomposed into its attributes.

Product Definition via Frames

In the actual implementation, the attributes of the constituent elements are represented as frames. They conform to a pre-defined structure which resembles a template. There is one template for each type of frame. Since there are two types of frames in the product definition segment, there are two corresponding frame templates.

Each individual bare board and its components are modeled as an object. An object appears as an instance of a frame template. All the instances of the same frame template belong to the same "class" and contain exactly the same slots. Of course, the content or the value of their slots may differ.

The slot values of the instances are obtained from external files or data bases whenever they are available. If the slot values are not directly accessible from external sources, then there are rules in the knowledge base attempt to derive the information based on other facts that it possesses. If there exists insufficient known evidence to support automatic deductions, the knowledge base relies on default values or interactions with the expert.

Although a PWA may have smaller boards that plug into it, usually, there is only one bare board per PWA. Thus, there usually need be only one instance to the bare board frame. The attributes (or slots) of the bare board frame are conceptually categorized into five groups.

The slots of the identification group are required to identify uniquely the printed wiring board (PWB). The slots of the construction group describe the physical form and composition of the printed wiring board. The slots of the electrical group define electrical function and the connection type used for the printed wiring board. The slots of the application group are germane only in the context of the printed wiring assembly. the slots in the presentation group are used to hold information which is essential in composing graphic displays.

Since HICLASS uses a node to represent a frame and a rule to represent a slot, the "bare_board" node represent the bare board frame and it contains the following rules:

- Part number
- Item number

- Board length
- Board thickness
- Board width
- Hole dimensions
- Maximum lead protrusion
- Maximum component height

Only a subset of the slots—those which are essential to support the decision-making process for automatic insertion—shown in figure 25-19 are implemented. In addition to defining the data type of the slot, each rule in the bare board frame also contains instructions on how to

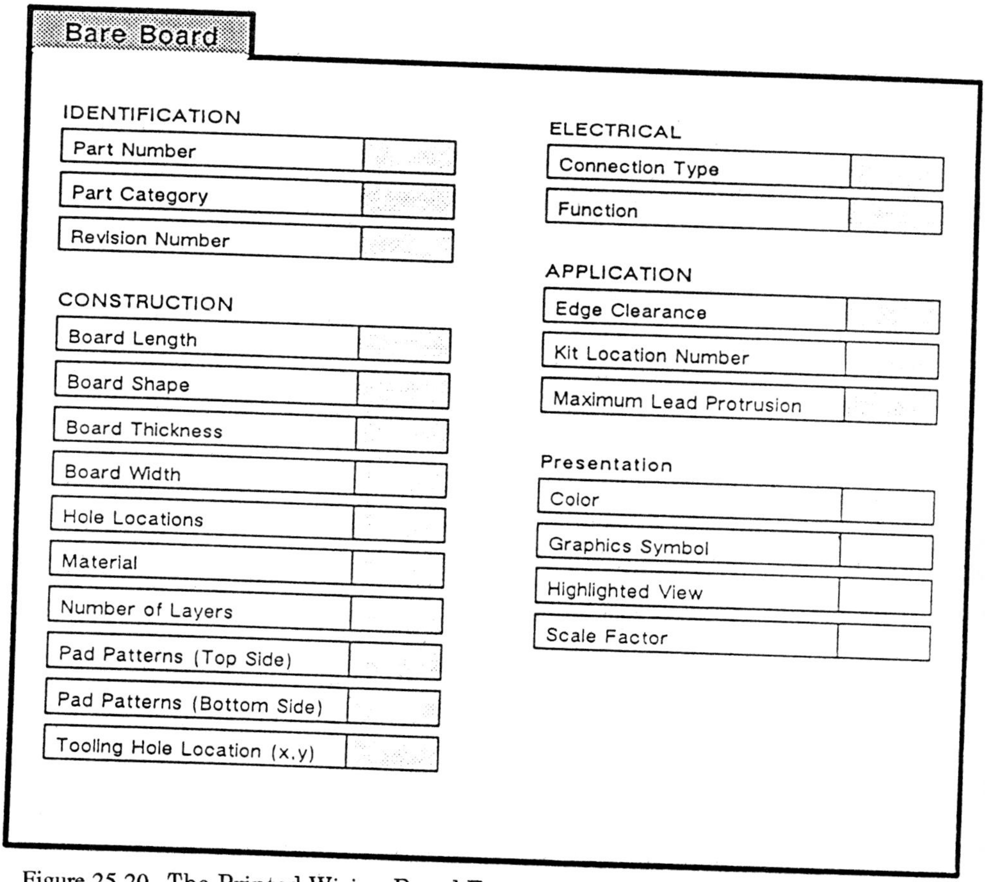

Bare Board

IDENTIFICATION

Part Number

Part Category

Revision Number

CONSTRUCTION

Board Length

Board Shape

Board Thickness

Board Width

Hole Locations

Material

Number of Layers

Pad Patterns (Top Side)

Pad Patterns (Bottom Side)

Tooling Hole Location (x,y)

ELECTRICAL

Connection Type

Function

APPLICATION

Edge Clearance

Kit Location Number

Maximum Lead Protrusion

Presentation

Color

Graphics Symbol

Highlighted View

Scale Factor

Figure 25-20 The Printed Wiring Board Frame

locate and access the information that is required to fill the slot. For example, the "part number" DEFINE rule takes on the following form:

{RULENAME : part number}

{This rule can either defines the key slot for the bare board frame or it can define the 'part number' slot of the component frame. In the case of bare board, the part number is selected from the BOM file. For component, the part number is selected from the CPD file.}

DEFINE

   data type IS EQUAL TO "STRING".
   default IS EQUAL TO " ".

IF CURRENT FRAME () IS EQUAL TO "bare_board" THEN
   {in the context of the 'bare board' frame}
   select string IS EQUAL TO "SELECT part num {STRING, 11,15}
      WHERE desc{STRING, 27, 20} = 'PRINTED WIRING BOARD'
      OR desc {STRING, 27,3} = 'PWB'".

   storage string IS EQUAL TO type of bom.
   location string IS EQUAL TO location of bom.
ELSE
   {in the context of the 'component' frame}
   select string IS EQUAL TO "SELECT part num {STRING, 6, 15}
      WHERE ref des {STRING, 24, 5} = '[reference designator]'".
 storage string IS EQUAL TO type of cpd.
location string IS EQUAL TO location of cpd.
END
   search string IS EQUAL TO select string.
   storage type IS EQUAL TO storage string.
   location IS EQUAL TO location string.
   computer IS EQUAL TO host computer.
   db password IS EQUAL TO file pass.
   computer password IS EQUAL TO computer pass.

The "part number" rules is share by both the "bare_board" node and the "component node", thus, the part number slot appears in both the bare board frame and the component frame. The role that part number serves is different within each frame. It serves as the key slot for the bare board frame and it is only a regular slot in the component frame.

   The location of the information that is required to fill the slot and access method also differ depending on which frame the part number rule is serving. For the bare board frame, the part number information is stored in the bill of material file and it is accessed by matching 20 characters starting in position 27 of the bill of material file with the strings "PRINTED WIRING BOARD" or "PWB". If a match is found, a new instance of the bare board frame is created and the key slot (i.e., the part number slot) of this new instance obtains its value from 11th to 26th characters of the matching record in the bill of material file.

The bare board's attributes groupings are generic and may be used for categorizing the slots of the component frame. Although the slots themselves differ from one another (because the frame templates differ,) the grouping of the slots is the same as shown in figure 25-21.

Physical Relationships

The physical relationships among the constituents (i.e., board layout) are an integral part of the product definition. Physical relationships

Component

IDENTIFICATION

| Part Number |
| Part Category |

CONSTRUCTION

| Body Length |
| Body Thickness |
| Body Weight |
| Body Width |
| Lead Diameter |
| Lead (original) Length |
| Lead Pitch |
| Leads Quantity |
| Lead Thickness |
| Lead Type |
| Packaging Type |

ELECTRICAL

| Connection Type |
| Function |
| Polarity |
| Static Sensitivity |

APPLICATION

| Reference Designator |
| Coordinate (x,y) |
| Feed Type |
| Head Sensitive |
| Kit Location Number |
| Lead Clinch Angle |
| Lead Length below PWB |
| Lead Span Length |
| Mounting Method |
| Offset (x,y) |
| Orientation |
| Preferred Processes |
| Required Processes |
| Soldering Method |
| Tinning |

Presentation

| Color |
| Graphics Symbol |
| Highlighted View |
| Scale Factor |

Figure 25-21   The Component Frame

between the bare board and the components, between the components themselves and, in the case of sub-assemblies, between components and their supporting elements (e.g., screws, washers, nuts, etc.) are all important factors.

The physical relationship resembles a taxonomical representation of the bill of material. In addition, there are logical relationships that can be superimposed on the top of the physical relationships. There logical relationships include common sense knowledge such as the fact that the screw must go through the washer and is used to fasten the part onto the bare board.

In the actual implementation, there are rules in the knowledge base that are written especially for handling both the physical relations and logical relations. The physical relations are calculated whenever they are needed. Calculations are necessary because the needed information is not directly available; however, the needed information can to be derived from what is available. The logical relations are handled via scripts, which have predetermined expectations and procedures on to how to react to events in a prototypical scenario.

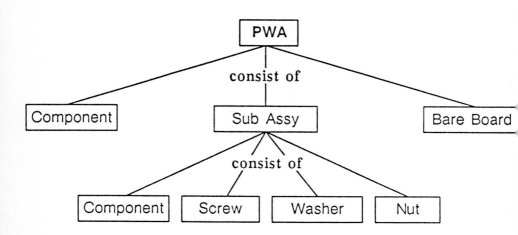

Figure 25-22   Physical Relationship Among Constituents

Frames and Taxonomies

This research uses frame-based representation for part definition while most of the recent development work in part definition continues to reply on a traditional group technology approach, i.e. creating classification via taxonomies. Is there a correlation between the frame-based representation and the taxonomical representation? Yes, the frames provide a structure for holding values while the taxonomies provide a structure for classifying the possible values which can go into the slots.

It is conceivable that the Research Network can utilize demons—procedures activated by the changing or accessing of values of the slot variables—to serve as the mechanism for providing the means of achieving semantic integration between frames and taxonomies. The taxonomies can be coded as rules, the rules can then be associated with a node, and in turned, the node can be attached to a slot and serve as a demon for that slot. The demon can ensure that only "proper" values are stored in the slots and provide the "inheritance" behavior that normally accompanies a taxonomy.

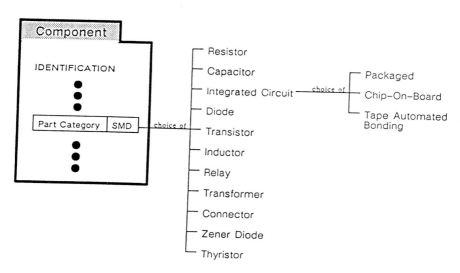

Figure 25-23 Semantic Integration of Frame and Taxonomy

Planning Generation

The objective of the planning generation sub-network is to complete the spectrum of automated process planning by automatically generating work instructions. To make this possible, the Research Network must sufficiently embody the necessary process characterization knowledge and product definition information. The planning generation sub-network automatically produces the insertion programs that drive the axial lead component inserter, the axial component sequencer/verifier and the DIP inserter. In addition, it produces a VCD summary report and a DIP summary report.

The two nodes in the Research Network that are responsible for generating the insertion programs are "generate_axial_reports" and "generate_dip_reports". The "generate_axial_reports" node contains the following rules:

- "init axial reports parms"

- "make axial list for ai"

- "gen vcd summary report"

- "calc seq of axial for ai"

- "gen vcd insertion report1"

- "gen vcd insertion report2"

- "gen vcd insertion report3"

- "gen seq instruction report"

The "init axial reports parm" rule initializes the temporary variables that are used in the generation of the axial insertion and sequencing instructions. The "make axial list for ai" rule creates a list of the auto-insertable axial components sorted in quantity order. Since the USM-DynaPert axial leaded component inserter provides a Variable Center Distance (VCD) between component leads, the word VCD is often used instead of axial. The "gen vcd summary report" rules generates a listing of the component-to-channel assignments. The "calc seq of axial for ai" rule prompts the user for the insertion sequence of each of the auto-insertable axial components.

Due to the limitation on the maximize size of a rule, the logic for generating the axial insertion reports are separated into three rules: "gen vcd insertion report1", "gen vcd insertion report2" and "gen vcd insertion report3".

These rules embody the following procedures:

1. Establish the orientation of the polarized parts

2. Sort the parts according to quantity, then according alpha/numeric order

3. Assign station numbers

4. Establish the sequence of insertion

5. Rotate the printed wiring board 90o

6. Repeat 2 through 5 for the second axis

After completing these procedures, the Research Network generates a report such as the one shown on figure 25-24.

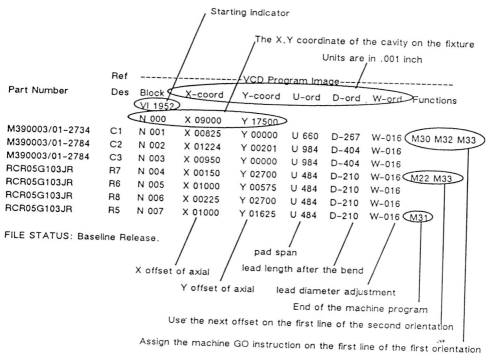

Figure 25-24 Axial Insertion Instruction Generated from Test Case #4

The machine program for the axial lead component inserter is generated first, followed by the generation of the machine program for the axial component sequencer/verifier. The machine program for the sequencer/verifier verifies the electrical values of axial components and sequences them in reverse order to their insertion. Generally, the throughput of the sequencer/verifier is much greater than that of the inserter. Hence, it is important to maximize the efficiency of the inserter, which is the pacing rate.

The sequencer/verifier's capability to process components in the reverse order is assumed to be sufficiently fast so that it does not hamper the rate of insertion. If it is not, then the rules in the Research Network need to be changed to generate the machine program for the sequencer/verifier first, then reverse the order for the inserter.

After the Research Network generates the machine instructions for axial leaded component sequencing and insertion, it proceeds with the processing of the DIP components. The "generate_dip_reports" node contains the following rules:

- "init dip reports parms"

- "make dip list for ai"

- "gen dip summary report"

- "calc seq of dip for ai"

- "gen dip insertion report1"

- "gen dip insertion report2"

The "init dip reports parm" rule initializes the temporary variables that are used in the generation of the DIP insertion instructions. The "make dip list for ai" rule creates a list of the auto-insertable DIP components sorted in quantity order. The purpose of the sort is to place the most frequently used components nearer to the insertion mechanism, i.e. the center channels of the insertion machine. By doing so, the travel time of the insertion mechanism is minimized. The "gen dip summary report" rule generates a listing of the component-to-channel assignments. The "calc se of dip for ai" rule prompts the user for the insertion sequence of each of the auto-insertable DIP components.

Again, due to the limitation on the maximize size of a rule, the logic for generating the DIP insertion reports are separated into two rules: "gen dip insertion report1" and "gen dip insertion report2". These rules embody the following procedures:

1. Establish channel numbers

2. Establish the sequence of insertion

3. Establish x,y location

4. Establish offset coordinates (from tooling hole)

5. Assign machine functions

After completing these procedures, the Research Network generates a report such as the one shown on figure 25-25.

The only part of the machine program generation procedure that requires human interaction is the sequencing of insertion. The ideal insertion sequence is one that minimizes the movement of the insertion head of the machine, this is accomplished by calculating

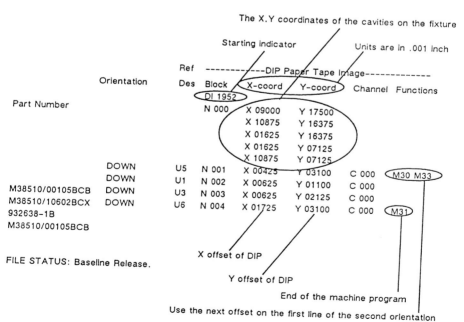

Figure 25-25 DIP Insertion Instruction Generated from Test Case #4

the minimum amount of travel distance between each component that is to be mounted. This, of course, describes the classical "traveling salesman problem". Since the implementation of an algorithm to solve this problem would add little value to the essence of this research, the sequencing is left as an external input to the Research Network.

Conclusion

The original approach for verifying the machine program was to compare it with the existing machine programs that were manually generated. Unfortunately, the parts possessing sufficiently complete information in the design data base did not have accompanying machine programs, including the five test parts that were chosen. This highlights the importance of electronically capturing enough information during the design cycle so as to enable automation of down stream processes in manufacturing.

As an alternative for verifying the machine program, the machine programs were reviewed as to the correctness of their syntax in accordance with the documentation. Furthermore, machine programs from similar parts were used as a basis for comparison and visual inspection. Using both of these validation activities, it is reasonable to conclude that the generated machine programs are correct.

The research objectives are to validate the suitability of knowledge

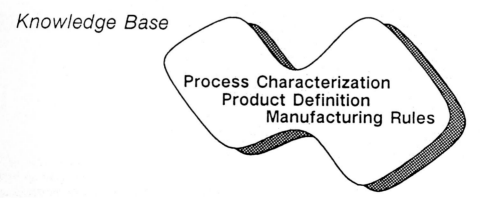

Figure 25-26   The Research Result

based systems for automatic generation of PWA process plans and to explore the limits of automated process planning via knowledge based systems. A further objective of this research is to develop a generic framework for characterizing electronic assembly processes, equipment capabilities, and product data definition requirements, and to apply this framework in a prototypical knowledge base.

The result is a successful development of a prototype knowledge base (KB) that met those objectives. It is clear that within the scope and assumption of this research, knowledge-based systems are an appropriate vehicle for exploring the limits of automated process planning. This prototype knowledge base embodies an overall process model of the entire PWA manufacturing process, the production definition of a PWA, and manufacturing rules specific to the automatic insertion operation. A new model for process characterization has been developed which can serve as the cornerstone to the successful implementation of the automatic generation of work instructions in general.

Exercising this prototype knowledge base with HICLASSTM Software System demonstrates that it is sufficiently robust since it is able to generate a detailed machine programs for the USM-DynaPert automatic insertion equipment. Since machine programs are the most detailed form of process plans, the research provides adequate evidence to conclude that automated process planning for PWA is feasible at any level.

The Research Network also reveals that process planning can be performed modularly (by operation) and dynamically (on an as needed bases). Providing this added flexibility supports the advanced concepts of distributed and hierarchical decision-making in factory management. Currently, because process plans are generated all at once and months in advance of board assembly, they do not take into account current conditions on the factory floor. Being able to generate, in near real-time, process plans allows the decision of committing manufacturing resources to be deferred to the latest possible moment, thus providing for maximum flexibility in the utilization of manufacturing resources.

Another obvious benefit of automated process planning lies in productivity enhancement. By embodying manufacturing know-how in a knowledge base, the entire process planning cycle is shortened because a majority of the support activities can be automated. Furthermore, the decisions that are made are more consistent. In short, automated process planning offers an opportunity for cheaper, faster and better manufacturing of PWAs.

References

1.Beck, T. J. "Surface Mount Soldering Techniques". Assembly Engineering. May 1987. pp. 40-42

2.Bao, H. P. and Reodecha, H. BRCODE: Electronic Part Coding System, Technical Manual. Raleigh, North Carolina. North Carolina State University, 1986.

3.Brachman, R. J. and Levesque, H. J. Readings in Knowledge Representation. Los Altos, California. Morgan Kaufmann Publishers, Inc., 1985.

4.Chryssoluris, G. M. and Sack C. F. Process Plan Data Interface Study, Final Report. Arlington, Texas. Computer Aided Manufacturing - International, Inc. August 1986.

5.Capillo, C. "How to Design Reliability Into Surface - Mount Assemblies". Electronic Packaging & Production. July 1985. pp. 74-80.

6.ELECTRI.ONICS. "1986/1987 ELECTRI.ONICS Desk Manual". ELECTRI.ONICS. June/July 1986.

7.Ford, D. "The Managerial Side of SMT". Circuits Manufacturing. March 1984. pp. 62-70.

8.Hastie, W. M. "Repairing Surface-Mount Boards". Circuits Manufacturing. March 1984. pp. 56-60.

9.Lapin, P.J. "Surface Mount Technology: A Market Impact Analysis Based Upon End-User Attitudes". PCFAB Market Measures Survey. January 1985. pp. 90-91.

10.Liu, D. "A Case Study of the HICLASSTM Software System, A Hughes Expert System". Proceedings of Information Technology for Manufacturing. Troy, New York. June 1986.

11.Loh, L. M. Development of An Expert System for the Auto-Insertability of Electronic Components on Printed Wiring Boards. Unpublished Master Thesis. Los Angeles, California: University of California, Los Angeles. August 1986.

12.Mangin, C. H. "Managing Automated Electronics Assembly". Assembly Engineering. May 1986. pp. 20-25.

13.Patel, A. Development of Production Rules for Automatic Assembly of Printed Circuit Boards at "Hughes" El Segundo Plant, California. Unpublished Master Thesis. Los Angeles, California: University of California, Los Angeles. Fall 1984.

14.Schwartz, W. H. "Chip-on-Board: Shrinking Surface Mount". Assembly Engineering. March 1987. pp. 32-34.

15.Smith S. F., Fox, M. S. and Ow, P.S. "Construction and Maintaining Detailed Production Plans: Investigations into the Development of Knowledge-based Factory Scheduling Systems". The AI Magazine. Volume 7. Number 4. 1986. pp. 45-61.

16.Zucherman, M. I. "A Knowledge Base Development for Producibility Analysis in Mechanical Design". Proceedings of ULTRATECH. Long Beach, California. September 1986. Vol. 1. pp. 2-15/2-36.

Author Biographical Data

David Liu is a principal of Liu & Associates, an independent software vendor of medical/dental and general accounting software systems. He is past president and founder of Ellison S^3, one of the largest value-added resellers of UNIX workstations in Southern California. He is also past president and founder of C-ATLAS, an independent software vendor of artificial intelligence and hypermedia-based software. While a manager of CIM at Hughes Aircraft Company, he managed a multi-year, multi-million dollar project which integrated design and manufacturing. This project resulted in the successful and efficient implementation of an AI-based paperless factory which was the first system of its kind in the world. Many still consider this system to be one of the finest implementations to date.

26

The USS Blast Furnace Advisor

Mike Stehura
USS
Pittsburgh, Pennsylvania

Introduction

Labeled as either dead or dying in the early 1980's, the steel industry in the United States needed to make drastic changes in order to survive. These changes not only included the need to produce products more economically, but more importantly to satisfy the consumer demand for higher quality steel products.

In response to these changing needs, United States Steel Corporation, long the major steel producer in the United States, began an aggressive and often painful plan of restructuring and diversification. Compelled by changes in the world steel market, U. S. Steel discontinued product lines and facilities with marginal profitability. At the same time, it began to move into the energy field, with the purchase of Marathon Oil Company and Texas Oil & Gas Corp. These changes have resulted in today's USX Corporation, with energy, steel and diversified segments structured to meet the changing needs of today's society.

But restructuring alone would not guarantee the survival of USS, the USX steel division. The company also needed to renew its commitment to its customers. Today, product quality and customer service are the major concerns within the operational departments of USS. This commitment to quality and service is one of the main forces within USS pushing the company towards the use of new problem solving techniques.

Blast Furnace Advisor

In the late 1980's, as the promise of technologies such as expert systems became a reality in the business world, USS began piloting its use in selective operational areas. One of the first areas targeted was blast furnace control.

The making of molten iron in a blast furnace has often been termed a 'scientific art', where the delicate balance of chemistry is dependant on a knowledgeable and talented team of employees. It was in applying some consistency to this 'art' of furnace operation that USS saw the potential for the use of expert systems technology.

The blast furnace process consists of charging iron ore (in pellet form), coke and fluxing materials into the top of the furnace and blowing heated air in through the bottom. Within the furnace, the heated air ignites the coke and produces a continuous combustion to reduce the iron pellets into molten iron, often referred to as hot metal. The fluxing materials used in the furnace, such as limestone and dolomite, act to restrict other elements like sulfur, silicon and alkali from combining with the hot metal. The flux combines with these elements to form a fluid waste product called slag, which can easily be removed from the furnace.

The blast furnace itself is a huge cylindrical structure. A large furnace may be over 100 feet high and 45 feet in diameter, capable of producing over 8,500 tons of molten iron per day. As materials are charged continuously into the top of the furnace structure, they begin to undergo a series of reactions as they slowly travel down the blast furnace stack. The entire reduction process for a particular charge of material requires about 8 hours. Then, at the bottom of the structure, or hearth, the furnace is tapped to remove the hot metal and slag. Most smaller furnaces are tapped once every two to three hours, while a large furnace, with the proper metal removal facilities, can be tapped continuously. Once the hot metal and slag have been removed, metallurgical tests are performed to measure and record the chemical properties of the products. The hot metal, still in molten form at a temperature exceeding 2650 degrees, then begins its journey to other facilities for conversion into steel and eventually to finished products for USS customers.

The goal of USS blast furnace operations is to control this massive chemical reaction, consistently producing the high quality hot metal needed in the steelmaking process to satisfy customer orders. For a number of reasons, the achievement of this goal is often quite difficult.

Much of the variability in the output of a blast furnace results from the variability in its material input. The raw materials charged into the furnace are sampled and analyzed on a regular basis. However, due to the massive volume of material charged into the top of a blast furnace, exact measurements of the chemical makeup of each component material is impossible. Thus, any programs which attempt to track furnace operations, can at best 'guesstimate' what has actually been charged.

Another cause of variability in the furnace has to do with the process controls available to the operator. Controls such as pressure, moisture, blast temperature, wind rates and blast enrichments affect the furnace reactions in certain ways. For example, all of the process controls just listed are available to the operator and could be used to help regulate the temperature of the hot metal. Using the proper controls in response to a particular set of conditions, is essential for maintaining a smooth furnace operation.

The most critical, and sometimes least predictable, of these sources of variability is the furnace operator himself. Here is where the 'art' of furnace operations comes into play. Based on his knowledge and experience, an operator must balance the quantity of materials charged into the furnace, using the available process controls, in order to continuously produce hot metal of the proper quality.

Due to the nature of the movement of material through the blast furnace—recall the 8 hours required to completely reduce the raw materials—operators must know what is occurring at all levels within the furnace. Should an operator overreact to, or fail to recognize, a problem situation, it may take the better part of a day to correct the problem and return the furnace to an optimum condition. So it is extremely important for operators to make the right adjustments at the right time.

Over the years USS built systems to help operators in this task of furnace control. Modeling systems were put in place for use as predictive tools, to assist in balancing the material and energy input and output of the furnace. Other systems gathered and stored data from various points in the process, all with the intent of aiding the operator in furnace control. Today, the USS Blast Furnace Advisor ("Advisor") works as an extension to these systems. Using information from the existing data collection system, the Advisor was designed to go beyond the capabilities of the systems which modeled the pure chemistry of the furnace. The Advisor makes use of the skill, judgement and expertise of the best USS experts in furnace operations. It was designed to help blast furnace operators recognize existing or potential furnace problems and recommend corrective operator actions.

The development of the Advisor began as a pilot project to evaluate the viability of expert systems and expert systems building tools within USX. Groups participating in the development included: Plant Process Systems, which served as the initial source of furnace operations expertise, and USX Corporate Technology Assessment. The pilot was conducted in conjunction with an apprenticeship program provided by a firm experienced in expert systems development.

Based on early interviews, the development team concluded that sufficient data for the Advisor was available from existing computerized systems. Data collected at the plant site in a DEC network was transmitted to the USS scientific computing mainframe as input to the existing furnace modeling system. By directing the output from this existing model back to the plant location for use in the expert system, the development team satisfied the bulk of the data requirements.

Conceptually, the furnace operator's job is very straightforward. For the furnace to operate at a consistent level, certain measured data elements needed to remain within acceptable ranges, known as the furnace aims. As recording systems reported elements out of aim, the operator noted trends and combinations of unacceptable values. Based on knowledge of the chemical reactions in the furnace and knowledge of past problems, the operator would take corrective action.

As the development team began to understand the application area and the environment, an evaluation of expert system building tools was conducted. The application was obviously data driven and needed to execute in the plant's DEC network. VAX OPS5 was chosen as the development language, using Digital's Forms Management System (FMS) as the user interface tool and custom FORTRAN code for data file interfaces. The system would run under the standard VMS operating system, initially as an option on the blast furnace operator's menu. Since its installation, dedicated terminals have been set up to display the Advisor's output continuously.

Development began slowly, focusing more on the concepts of expert systems rather than on the task of knowledge gathering. Educational sessions were held with the expert to explain and position the technology. Afterwards, a standard question and answer interviewing process was begun to work through possible problem situations facing the furnace operators. These interviews were generally held in an office setting. Although valuable process knowledge was gathered by talking through theoretical situations, a clear picture of the expert's problem solving methodology was still not apparent.

To change the pace of development and hopefully gain more insight, the interviewing sessions were moved from the office environment to the actual furnace control center. In this setting the tools and techniques used by the expert in problem solving were easily seen. This also helped direct the sessions toward working through the analysis and resolution of actual problems which had recently occurred at one of the furnaces. As prototypes of the application were built and extended, the development sessions became more and more productive, since new situations and rules could be tested and analyzed quickly.

Internally the Advisor, mimicking the expert, works in a two step process: find problems and recommend actions. Chemical analysis data gathered from the raw material samplings, along with the output from the furnace modeling system, are used to initiate the Advisor.

To 'find problems', the Advisor uses approximately 70 forward chaining rules to : read input, determine furnace aims, calculate trends, identify problems by comparing actual data to aims, and finally to qualify the severity of the identified problems. In the Advisor's 'recommend actions' phase, approximately 60 forward chaining rules look for combinations of problems or undesirable trends. The Advisor can also look ahead to anticipate a certain class of problems. The rules are grouped and prioritized by the expert's classification of furnace problems into categories of severity. Advice concerning critical problems appears first on the Advisor's list of recommendations. The number of recommendations displayed to the furnace operator is dependant on the number of distinct problems identified.

The Advisor functions in both batch and interactive modes. In batch mode, the Advisor outputs the current furnace conditions and its recommendations to log files. The logs serve as a historical tracking of the Advisor and provide a means for the development team to periodically monitor and verify the recommendations made by the system.

The Advisor's interactive output provides operators with a simplified view of trending information and raw data from the mainframe modeling system. The trending of data is important in the furnace application, since operating problems tend to build gradually and corrective actions taken by operators may need hours to take effect. The tendency to oversupply data on an output display is resolved in the Advisor by making less pertinent data viewable using function keys. Although this data was available prior to the Advisor, the new display eliminated the need for operators to locate numerous trending reports for comparison to the current furnace conditions.

The new data displays also served to lead the operators to the Advisor's main objective—providing expert operational recommendations. A recommendation consists of a single line of text (up to 79 characters) alerting the operator of an unusual condition or recommending that a certain action be taken. While concise communication of the condition or required action was the aim of this display, the reasoning behind the recommendation was also known to be invaluable. This reasoning is available to the operators through a special function key labeled 'why'.

Also important to the development team was a method of evaluating and extending the expert system. Rather than relying on purely verbal feedback to enhance the system, the Advisor was programmed to accept input from its users. This feedback, entered in a computerized suggestion box, has been used to pinpoint errors or shortcomings in the Advisor. These suggestions are reviewed by developers and the expert and used to develop new rules and enhance the Advisor's useability.

At the conclusion of the development, the expert termed the Advisor a 'technological success'. Then he wondered if anyone would ever use it.

Reaction to the Advisor by furnace operators has been mixed. As the system was phased into the operation's environment, questions about the technology and the Advisor's position in the day to day decision making once again needed to be addressed. Some looked at the Advisor as a threat, others considered it a nuisance. Since the initial sponsorship for this project came from outside of the furnace operations area, gaining the support of furnace management was critical to the success of the Advisor. Little progress was made achieving acceptance for the system until operations management at the furnace began to make some commitments.

Slowly, furnace management has allowed different aspects of the Advisor to be phased into the operation. Initially it was used to replace the manual tracking, trending and control of less critical aspects in the process. More recently, as the Advisor has gradually become a source of documentation for the Standard Quality Practices at the furnace, it has been accepted as a credible repository of operations expertise.

The issues addressed by the Advisor were those of consistency and accuracy in USS blast furnace operations, both in problem identification and actions taken. The Advisor is intended to lessen the consequences of inexperienced operators and also challenge the veteran operators who have a "I've done it this way for 20 years" attitude. Early in the project cycle, the development team realized that the Advisor, like any advisor, would need a certain period of time to build confidence with its users.

Furnace management does not want operators to stop thinking and blindly follow any system's direction. USS believes that the time spent allowing the Advisor to grow into a peer-like role in the blast furnace area will insure that it will be a reliable, useable source of operation's expertise for many years to come.

Note : Other USS blast furnace locations are currently developing blast furnace expert systems patterned after the Advisor. These locations are utilizing the Aion Development System in a PC environment.

As stated, the development of the Advisor was the USX/USS pilot project using expert systems methods and tools. Pilot projects are meant to be learning experiences. Looking back at the project, USS has learned much about the technology of expert systems, but more importantly we have learned about our company and the effort and people needed to bring about significant change.

Many of the texts and presentations on expert systems technology talk about finding champions for expert systems projects. Our experience with the development of the Advisor has shown us that simply finding a champion is not enough; the *proper* champion must be found. To really effect change in a particular business area, the push must come from within the business area itself. Finding that individual at the proper level in the operational organization is critical.

Our experience has also shown us that the selection of the proper expert to participate in the development process is not such an easy task. A broad range of blast furnace expertise exists within USS. The right expertise for a particular project, is that expertise which will be used and supported by the target location. The Advisor project began with a single expert. Today's Advisor is the blending of expertise from at least 4 individuals with blast furnace knowledge. The lesson -involve management from the target organization in the process of locating expert knowledge.

Conclusion

Based on our experience with this project and the potential seen in the use of expert systems, we have begun to position this technology as part of the total USS application development strategy. When considering which organizations in the company would develop expert systems applications, two options were raised: the centralized 'SWAT' team approach—or—the use of traditional local MIS development staffs. As we examined our own company and culture, we saw a close relationship between USS systems organizations and the businesses they serve. Our business priorities and systems priorities have grown to be one and the same. Centralizing and specializing expert systems development would certainly damage the business/systems relationships we have built.

As we attempt to integrate expert systems into the USS environment on a larger scale, development groups are being established within our traditional MIS structures. This strategy will allow business units within USS to focus the use of new technologies, like expert systems, in appropriate areas as we strive, not to advance technology, but to meet our commitments of product quality and customer service.

Author Biographical Data

Mike Stehura, as a member of the USX Technology Assessment group, was the lead developer of the USS Blast Furnace Advisor. Mike, a graduate of Digital Equipment Corporation's Business Fellowship in Artificial Intelligence, is currently working in the USS Application

Support Group in its Pittsburgh Service Center as a senior planner for expert systems applications.

27

A Hierarchical Planning Knowledge System Applied to Manufacturing Processes

Aldo Dagnino, Ph. D.
Alberta Research Council
Calgary, Alberta, Canada

Introduction

This paper presents a hierarchical knowledge-based system for generative process planning (GPP), employed to assist in the production of printed circuit boards (PCBs) at Northern Telecom, Canada. The objective of the computer programmed manufacturing and assembly planning system (CPMAPS II) is to determine the manufacturing operations, resources and times required to assemble components in a printed circuit board. The knowledge-based system has been developed in collaboration with experienced process planners in different production areas of the company. CPMAPS II is based on a hierarchical planning model. This hierarchical architecture determines a process plan based on two levels of planning. The first level, determines a plan taking into consideration the design constraints applicable to the PCB. The second level planning takes into account "workplace" constraints that impact the plan. Presently, CPMAPS II is capable of interacting with Northern Telecom's manufacturing databases and determines process plans to assemble "through-hole" and "surface-mount" electronic components on the board. CPMAPS II is a hybrid knowledge-based system that incorporates frames, object-oriented programmed methods and a sophisticated rule-base and knowledge-based structures that allows easy maintenance. The system has been deployed in one plant and its performance is being tested and will be incorporated in a global corporate process planning environment. CPMAPS II has been developed in KEE/LISP and appropriate database interfaces have been built to obtain information from the company's manufacturing databases.

Previous Work Developed in Process Planning Systems

Process planning in manufacturing environments involves deciding how to use the capabilities of the shop floor to produce a given part. The input to the process planning activity includes information about product design and shop floor capabilities. Design engineers supply a detailed product specification along with a description of each part to be manufactured. Production engineers supply information about the entities and capabilities of the manufacturing facility including machines, tools, fixtures and human operators functions. The final plan provides information about the sequences of operations, the resources and the parameters required in the manufacturing process. There are a number of process planning issues that can be identified. Two of the most obvious are cost reduction and quality control. Manufacturing engineers are always striving to meet low cost goals as well as high quality standards for their products.

Berenji and Khoshnevis (1986) identify two main categories of computer-assisted process planning systems. The first category refers to systems that retrieve standard plans of coded parts and generalize these plans to similar parts, and it is referred to as variant process planning (VPP). This approach requires that all the existing process plans in the plant be summarized and coded so that similar part families can be classified. Some examples of these systems include the CAPP system (Tulkoff, 1978), Miplan of Organization for Industrial Research (Schaffer, 1980), and AUTOPLAN (Tempelhof, 1980). The second category of computer-assisted planning systems include systems capable of creating a process plan for a new part by synthesizing the process information and decision logic employed in the system. This approach is known as generative process planning (GPP). In generative process planning the descriptions of the parts, the manufacturing processes, the machines, the tooling the human operators, and the logic of the manufacturing process are stored in the computer, and the system develops a new process plan. GPP systems make no reference to previous plans.

Artificial intelligence (AI) is a very promising approach for the development of generative process planning systems. By using AI techniques, usually in the form of knowledge-based systems, the computer can look at the attributes of a product and by using decision rules, it can decide exactly how to manufacture a part. Knowledge-based systems can be defined as special-purpose computer systems that are "intelligent" and contain large amounts of "high quality" specific expert knowledge related to a particular domain (Waterman, 1986).

Despite the great potential for process planning knowledge-based systems few organizations have acquired this technology. Most of the developed systems are in the experimental phase. Some of these systems include the following: Technostructure of machining (TOM) (Matsushima et al., 1982) is a knowledge-based system that concentrates on hole-making processes and considers one hole at a time; GARI (Descote and Lathomb, 1983) is a system in which the plan generation proceeds through successive refinements; Proplan (Phillips et al., 1984) concentrates on the problem of integrating a CAD system with a knowledge-based system for generating process plans. The hierarchical and intelligent manufacturing automated process planner Hi-Mapp (Berenji and Khoshnevis, 1986) is an automated process planner for manufacturing employs AI techniques to determine the operations needed to manufacture a part. Ford Motor Co.'s direct labor management system (DLMS) (Johnson et al., 1989), is a multi-phase knowledge-based system for process planning. The knowledge in the DLMS environment is described in terms of a frame-based description language (FDL), which is a variation of the FL language described by Brachman and Levesque (1984). Kempf (1988) provides a comprehensive overview of the variant, generative, and AI approaches to process planning. Tempelhof (1980), Freedman and Frail (1986) explain the importance of AI techniques in process planning.

This paper presents a hierarchical generative process planning system employed in an electronic manufacturing environment. This system employs knowledge-based techniques to develop process plans for the manufacturing of printed circuit boards. Different features of the system are presented in Dagnino (1990a; b) and Crystall et al. (1990).

Manufacturing Processes Associated with the Production of PCBs

After a PCB has been designed, several manufacturing processes are required to build the printed circuit board. These processes include: (a) prepare components and board; (b) component assemblies; (c) wave soldering; (d) face plate and special assemblies; (e) test and repair. Most of the effort during the creation of a process plan for manufacturing a PCB, goes into the component assembly stage. Traditionally, the information required to develop a process plan for component assembly has been collected manually from the neutral design file, manufacturing databases, components databases, and CAD systems. Process plans to assemble components in a PCB are developed manually by the manufacturing engineers and take several days. At this point it is convenient to introduce two definitions.

Definition 1. Workunit represents an abstraction of an entity which possesses production capacities and production capabilities (Harapiak, 1990). Examples of work units include a manufacturing plant, a manual assembly line, a robotic cell, an assembly robot, an insertion machine, etc. As can be observed, there are several hierarchical levels of workunits consisting of large sets of workunits and subset workunits.

Definition 2. An operation is defined as an activity that must be performed on an object or one of its components, to change its characteristics, during the course of its manufacture (Harapiak, 1990).

There are several stages that a process plan for assembly of electronic components on a PCB must meet. These stages include the following:

1. Identify the component types that will be assembled on the PCB.

2. Determine the operations required to assemble the electronic components, based on their characteristics and the design constraints of the board.

3. Assign "generic" resources to perform the manufacturing operations previously identified. These "generic" resources refer to resources that are available in the corporation or the "universal workunit".

4. Determine which workunits are capable of performing the operations identified in stage 2.

The above stages of a process planning system are complemented by an optimization stage in which resources are efficiently allocated to the manufacturing operations previously identified. This optimization stage is not included in the scope of CPMAPS II because there are already several optimizing algorithms in the company.

The four stages of process planning described previously have been implemented in CPMAPS II in a hierarchical planning architecture which constitutes a framework for developing process plans at Northern Telecom. There are two levels in this hierarchical planning structure. The first level in the hierarchy, or "generic process plan level", maps engineering constraints, product characteristics, and design constraints of the product, to generic resource capabilities in the corporation. Stages 1, 2, and 3 are performed at the generic process plan level. The second level of the planning hierarchy, or "workunit process plan level", refines the plan obtained at the first stage and adapts it to the particular capabilities of the workunit being considered.

The two levels of the planning hierarchy can be accessed following a "top-down" approach or a "bottom-up" approach, depending on the type of information required. To illustrate these approaches consider an example where a process plan must be formulated considering the corporation which contains a set of several plants and a particular plant A which is member of this set.

Following the "top-down" approach, the corporation will develop a generic process plan, which corresponds to the first level of planning. At this level, it will obtain information about the options of operations that must be followed to manufacture a PCB mapping engineering and design constraints and product characteristics to a pool of capabilities spread among all the plants. The corporation refines the generic plan by developing the workunit plan.

With the information obtained at the second level planning, the corporation will be able to determine which plant is the most efficient to manufacture the PCB.

Conversely, following the "bottom-up" approach, plant A will develop a generic process plan, that will provide the possible options in operations and resources to be employed to manufacture the PCB. Based on this information, plant A will refine the process plan based on its particular constraints. These constraints include equipment, plant layouts, resources, manufacturing cells, component substitutions, and plant policies.

Figure 27-1 shows a diagram of the hierarchical process planning architecture implemented in CPMAPS II and its inputs and outputs. The input at the generic process planning level are the neutral design database (NDDB) and the integrated and engineering database (IEDB). These databases contain the design and engineering information for all PCBs and their components. The generic process planning module is a knowledge system that employs the information stored in NDDB and IEDB and produces a generic process plan for a particular PCB. This generic plan is stored in a file which together with a rates database and a component substitution database, serve as the input for the workunit process planning module. The rates database contains information about the micromotions and times associated with all the operations performed in the workunit. The substitute components database contains information about substitute components that are employed in the manufacture of a PCB due to changes in suppliers. The workunit process planning module is a knowledge system which produces a process plan particular to the capabilities of a workunit. The generic and workunit process planning modules will be explained below.

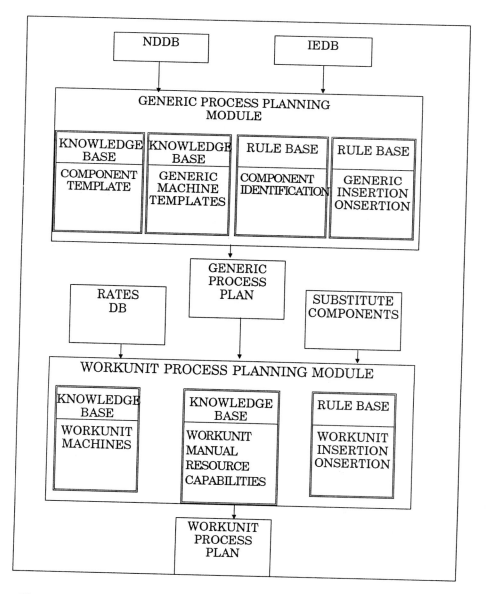

Figure 27-1   Hierarchical Process Planning Structure Implemented in CPMAPS II

Architecture of CPMAPS II

CPMAPS II is a GPP knowledge system that assists the manufacturing engineers at Northern Telecom to determine the assembly types required to insert "through-hole" and "surface-mount" components in a PCB. CPMAPS II has been designed to map the hierarchical planning model explained in Section 2.0. The system is divided into two distinct knowledge system modules that generate a generic and workunit process plans respectively, as shown in Figure 27-1. The generic process planning module employs the information stored in the neutral design and integrated engineering databases (NDDB and IEDB) to identify electronic components to be assembled on a PCB, to generate the generic manufacturing operations and to assign general resources to perform the operations. This module contains four subsystems: the COMPONENT TEMPLATES and the GENERIC MACHINE TEMPLATES knowledge bases, and the COMPONENT IDENTIFICATION and the GENERIC INSERTION ONSERTION rule bases. The workunit process planning module of CPMAPS II employs the generated generic process plan file, and the rates and substitute components databases, to refine the plan. This module contains the following four subsystems : the WORK UNIT MACHINES and the WORKUNIT MANUAL RESOURCE CAPABILITIES knowledge bases, and the WORKUNIT INSERTION ONSERTION rule base. The following section explains in more detail the functions and structure of the two planning modules and their respective subsystems.

CPMAPS II Planning Modules

CPMAPS is a knowledge system that has been built employing the Knowledge Engineering Environment (KEE) shell developed by IntelliCorp, and several modules written in LISP have been developed. The system employs rules, frames, objects, methods, active images, icons, active values and other object oriented programming features. The following sections describe the generic process planning and workunit process planning modules of CPMAPS II.

Generic Process Planning Module

The generic process planning module of CPMAPS II contains several subsystems. These sub-systems are explained below.

1. COMPONENT.TEMPLATES Knowledge Base. The electronic components that are inserted in the printed circuit board are

represented as objects. There are two approaches followed for the representation of these components in CPMAPS II. In the first representation, component templates that the system employs during its reasoning cycle, are stored in the COMPONENT.TEMPLATES knowledge base. These templates are employed to validate the components that are evaluated for assembly on the board. The second representation refers to the components that are assembled on a particular bare board, which are referred as component instances. Component instances and their characteristics are stored in NDDB and IEDB which are read by CPMAPS II. These components are compared to the template components stored in the knowledge base during the development of the process plan. The COMPONENT.TEMPLATES knowledge base contains the characteristics of the different types of components that can be assembled on a board. Presently, CPMAPS II supports more than 200 component templates. These component templates are sub-divided into surface-mount and through-hole components.

Table 27-1 shows the structure of one component template in the knowledge base.

2. GENERIC.MACHINE.TEMPLATES Knowledge Base. The automatic assembly machines employed to populate a board and their capabilities are represented as objects in this knowledge base. This knowledge base is divided into through-hole insertion machines and surface-mount assembly machines. Presently 16 different types of assembly machines are included in this knowledge base, and represent the general capabilities of the corporation. Table 27-2 shows an example of the structure of a machine frame.

3. COMPONENT.IDENTIFICATION Rule Base. CPMAPS II starts the process planning activity by recognizing and categorizing the component instances to be assembled on the board by comparing them to the component templates. With the design information of a PCB and the characteristics of the component instances, the

COMPONENT.NAME:	GAS FUSE
COMPONENT.CLASSIFICATION:	ELECTRONIC.COMPONENT.
IDENTIFICATION:	THROUGH.HOLE
COMPONENT.SHAPE:	AXIAL
COMPONENT.TYPE:	TWO.LEADS

Table 27-1   Frame in the COMPONENT TEMPLATES Knowledge Base

```
MACHINE.NAME:                  DIP.INSERTION.MACHINE
ELECTRONIC.COMPONENT.TYPE:     TWO.ROWS.PINS
MAXIMUM.PIN.COUNT:             16
MINIMUM.PIN.COUNT:             6
POLARITY:                          1,3
```

Table 27-2   Frame in GENERIC MACHINE TEMPLATES Knowledge
Base

COMPONENT.IDENTIFICATION rule base recognizes the component instances to be assembled on the board and categorizes them according to the component templates.

4. GENERIC INSERTION ONSERTION Rule Base. This rule base contains the rules needed to identify the generic operation required to assemble the component instances in a PCB. A generic operation is defined below as follows:

Definition 3. A generic operation is defined as an activity that must be performed on an object or one of its components, to change its characteristics, during the course of its manufacture, and it is independent from the capabilities of a particular workunit. In the assembly of through-hole components, for example, the generic operations include: insert with standoff, insert without standoff, prom blasting, VCD auto-insertion, etc.

The GENERIC INSERTION ONSERTION rule base takes into consideration design and engineering constraints, product and component characteristics, and generic resource capabilities, to define the generic manufacturing operations. Table 27-3 displays the form of a rule in this rule base.

Workunit Process Planning Module

The input for the workunit process planning module of CPMAPS II consists of the generic process planning file, the rates database and the substitute components file. The workunit planning module consists of three subsystems which are explained below.

1. WORKUNIT MACHINES Knowledge Base. This knowledge base is structured similarly to the GENERIC MACHINE TEMPLATES knowledge base. The difference is that it contains the machines and capabilities of a specific workunit.

```
((RULE.INSERT.IN.SOCKET.TWO.ROWS.PINS
  (IF (( ?COMPONENT.INSTANCE IS IN CLASS
       COMPONENT.INSTANCES) AND
       ( THE MANUAL OPERATION OF ?COMPONENT
       INSTANCE IS MANUAL.INSERTION) AND
       ( THE PREDECESSOR OF ?COMPONENT.INSTANCE IS
       SOCKET) AND
       (THE MOD OF ?COMPONENT.INSTANCE IS NO) AND
       (THE COMPONENT.NAME OF ?COMPONENT.INSTANCE
       IS PROM) )
       THEN
  (CHANGE.TO
       (THE MANUAL.OPERATION OF
       ?COMPONENT.INSTANCE IS
       INSERT.COMPONENT.IN.SOCKET)
       USING RULES.DETAILED.OPERATIONS))))
```

Table 27-3

2. WORKUNIT MANUAL RESOURCE CAPABILITIES Knowledge Base. This knowledge base contains the frames of the manual operations that are performed in a workunit. Each manual operation frame represents the operations required to assemble a component instance based on the generic operation identified in the generic process plan. Table 27-4 shows a manual operation frame.

3. WORKUNIT INSERTION ONSERTION Rule Base. This rule base relates the results obtained at the generic process planning level to the particular machine and manual resource capabilities of the workunit, and develops a more refined process plan. This rule base employs the workunit manual resource capabilities templates

OPERATION.NAME:	INSERT.STANDOFF
COMPONENT.TYPE:	TWO.LEADS
MANUAL.OPERATION.DETAIL:	((LEAD.FORMING,
	CUT.CLINCH))
MANUFACTURING.CELL:	((CELL.A , CELL.B))
OPERATION.RATE:	((RATE.XX2A,
	RATE.XX2B))

Table 27-4   Manual Operation Frame

level to the particular machine and manual resource capabilities of the workunit, and develops a more refined process plan. This rule base employs the workunit manual resource capabilities templates and specific workunit machine templates to provide the detailed process plan.

An Example

To illustrate the manner in which CPMAPS II determines a process plan following the hierarchical planning structure presented in Section 2.0, an example concerning only one component instance will be given. Figure 27-2 shows a computer screen where the characteristics of a RESISTOR component instance to be inserted in the PCB are analyzed by the system. The top three level slots represent the design characteristics of the component which are stored in NDDB and IEDB and are read by CPMAPS II. In our example, the instance has a body diameter of 0.2 units, a body length of 0.55 units, a lead diameter of 0.03 units, is regular insertion and not a modified insertion, an insertion span in the board of 0.8 units, a wattage of 0.5 units, has a plastic casing, and its predecessor is the board. At the first level planning, CPMAPS II determines that the RESISTOR is a through-hole electronic component, and it is an axial component with two leads. It determines that the generic operation is an electronic insertion and that the generic machine that can be employed is a VCD machine. Moreover, the component can be inserted manually with no standoff.

At the second level planning, CPMAPS II takes into consideration the capabilities of plant A. Based on these capabilities, it is determined that VCD machines 1 and 2 in the plant can be employed to insert the component instance. Additionally, the component can be inserted manually and the two operations required are a machine lead forming and insertions cutting and clinching. These operations are performed in the stock room manufacturing cell and in the assembly line with their respective associated rates.

If further information is required about the micro-motions and times associated with the operations identified at the workunit planning level, the system access the rates database. The information obtained will give details about these micro-motions and their times.

This example shows only one component instance to be assembled on the board. Typically, there are approximate 150 component instances to be inserted in the board and CPMAPS II develops a process plan for all of them.

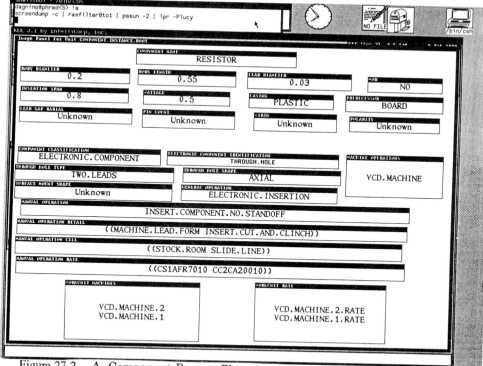

Figure 27-2　A Component Process Plan

Conclusions

This paper presented a review of a hierarchical generative process planning knowledge system for assembling components in a printed circuit board. CPMAPS II is a system that assists the manufacturing engineers in determining the assembly operations, resources, times and costs associated with the assembly of through-hole and surface-mount components in a PCB. The system is capable of interacting with the company's manufacturing databases to gather design data. The main benefits that CPMAPS include: (a) reduction in process planning time; (b) increased consistency in determining the process plans for assembling components across the corporation; (c) development of an accurate system to relate times and micro-motions associated to assembly types; (d) development of more effective costing methods during the assembly process; (e) some intangible factors such as: improvement in the manufacturing activities, portability of the system to several plants, easy maintenance of the system, modularity of the system, and better communication between manufacturing and design.

Acknowledgements

Special thanks are given to Paul Harapiak from Northern Telecom Canada Ltd. for his contribution in the development of a hierarchical planning structure for CPMAPS II. A special recognition is given to the efforts that Patrick Feighan and Mark Brinsmead from the Alberta Research Council and Mike Hoogstraat from Northern Telecom put during the development of the CPMAPS II modules. The completion of this project would not have been possible without the collaboration and good disposition of Judy Beaudoin, Cathy Bowden, Mike Dressler, Begum Fazel, Claire Evancio, Paul Goldsmith, Phyllis Gow, Tom Malaher, David Anderson, Isabel Race, and Tony Tse from Northern Telecom Canada, Ltd.

References

1. Berenji, H. R. and Khoshnevis, B. (1986). Use of Artificial Intelligence in Automated Process Planning. Computers in Mechanical Engineering, September, pp. 47-55.

2. Brachman, R. J. and Levesque, H. J. (1984). The Tractability of Subsumption in Frame-based Description Languages. AAAI-84, Proceedings, Austin, Texas.

3. Chrystall, K., Dagnino, A. and Feighan, P. (1990). A Robotic Planning System. The Sixth CASI Conference on Astronautics to be held in Ottawa, Ontario. November 19 21.

4. Dagnino, A. (1990a). CPMAPS: A Generative Process Planning System. Third International Symposium On Artificial Intelligence: Applications of Engineering Design, Manufacturing and Management in Industrialized and Developing Countries, to be held in Monterrey, Mexico from October 22-26.

5. Dagnino, A. (1990b). The Computer Programmed Manufacturing and Assembly Planning System: CPMAPS. AAAI SIGMAN Manufacturing, Planning and Control Workshop. Boston Ma., July 29 to August 3.

6. Descote, Y. and Lathomb, J. C. (1983). Gari: An Expert System For Process Planning. Solid Modelling by Computers. New York: Plenum Press.

7. Freedman, R. S. and Frail, R. P. (1986). OPGEN: The Evolution of an Expert System for Process Planning. The AI Magazine. Winter, pp. 58-70.

8. Harapiak. P. (1990). CAPP II System Overview. Northern Telecom Internal Technical Report.

9. Johnson, W. P., Woodhead, R. and O'Brien, J. (1989). The Ford Motor Company DLMS. AI Expert, August, pp. 42-52. 10. Matsushima, K., Okada, N. and Sata, T. (1982). The Integration of CAD and CAM by Application of Artificial Intelligence. Annals of the CIRP, Vol. 3, No. 1. 11. Kempf, K. (1988). Artificially Intelligent Tools For Manufacturing Process Planners. Intelligent Manufacturing. Proceedings from the First International Conference on Expert Systems and the Leading Edge in Production Planning and Control. The Benjamin/Cumings Publishing Company, Inc., Menlo Park, CA.

12. Phillips, R. H., Zhou, X. D. and Mowleeswaran, C. (1984). An Artificial Intelligence Approach to Integrating CAD and CAM Through Generative Process Planning. Proceedings of the ASME International Computers in Engineering Conference. Las Vegas, Nevada.

13. Schaffer, G. (1980). GT via Automated Process Planning. American Machinist. May, pp.119-122.

14. Tempelhof, K. (1980). A System of Computer-Aided Process Planning for Machine Parts. New York, Elsevier North-Holland.

15. Tulkoff, J. (1978). CAM-I Automated Process Planning (CAPP) System. Proceedings of the 15 Numerical Control Society, Annual Meeting and Technical Conference , Chicago.

16. Waterman, D. A. (1986). A Guide to Expert Systems. Addison-Wesley Publishing Company, USA.

Author Biographical Data

Dr. Dagnino is an Associate researcher in the Applications group of the Advanced Technologies Department at the Alberta Research Council. His background in Industrial Engineering and Artificial Intelligence provides him with the expertise to develop advanced manufacturing systems. Currently, he is involved with projects that explore the possibilities of applying Artificial Intelligence technology in advanced manufacturing systems and the integration of these systems to other manufacturing processes in place. These projects are sponsored by large Canadian manufacturing companies such as Northern Telecom Ltd. Dr. Dagnino received his Ph.D in systems design engineering from the University of Waterloo in 1987. He is the author of numerous articles and technical reports as well as the course developer and seminar leader for many courses in the area of AI.

Design

28

Expert Systems
for Engineering Design
and Manufacturing

Gavin A. Finn
Stone & Webster Advanced Systems Development Services
Boston, Massachusetts

Introduction

The migration of expert systems from the research domain into practice has followed a path that has three predominant characteristics:

- a focus on smaller, more manageable problems rather than larger, more general problems,

- a concentration on the application, as opposed to the underlying software technology, and

- use of conventional software and hardware environments, in contrast to the early use of specialized languages and computers.

With regard to engineering design, as with most domains, it is important to view the application of expert systems in the context of the above three principles. Many early attempts at applying expert systems in the design process subscribed to the first principle (narrow problem domain) but fell short in the areas of application-orientation and use of conventional computing environments. (Examples of these early design applications include integrated circuit and VLSI design.) The use of such systems in practice, however, is minimal.

It has become clear that in order to successfully deploy an expert system in a production environment the system must exhibit all three characteristics. Rather than attempt to solve the highly complex problem of design synthesis (that is, the generation of a complete design from scratch) it is more practical to define smaller parts of the design problem, and to implement design advisors that concentrate on these well-defined sub-problems. The availability of commercial expert system tools has enabled the expert system development process to be mostly oriented to defining and representing the design knowledge and problem-solving strategy. The use of such tools for design problems allows developers to adhere to the second principle.

Most engineering designs involve a variety of analysis and design software and databases. Until very recently, expert systems for design problems have been developed, and some implemented, outside of the existing design software and hardware environment. Most importantly, much of the design process involves the use of Computer-Aided-Design/Computer-Aided manufacturing (CAD/CAM) systems. Expert systems for design have hitherto been deployed on stand-alone hardware, not connected to or integrated with the CAD/CAM system. This distinction has inhibited design related expert systems from becoming more prevalent in real-world engineering and manufacturing organizations.

Stone & Webster Engineering Corporation is a large engineering, design, construction, and consulting company, whose primary business is the provision of engineering, design, and construction services to all industries. Stone & Webster's Advanced Systems Development Services Division is responsible for providing advanced computer systems to clients and for internal operations. Included in these advanced systems are expert systems applications, advanced 3-dimensional and 2-dimensional graphics for CAD/CAM applications, and software for integration of different application systems. Recognizing the need to integrate expert system applications into the existing engineering computing architecture, and specifically into the CAD/CAM environment, Stone & Webster developed an integration system (STONErule) for expert systems and CAD/CAM systems, allowing for the development and implementation of a wide variety of expert systems applications directly within the CAD/CAM application.

The need for integration of expert systems with CAD/CAM

The need for companies to be more competitive is more important now than ever before. In order to improve competitiveness, companies must use their resources as effectively as possible. The most valuable resource in any design, engineering, or manufacturing environment is the knowledge of the people in that organization. The application of expert systems in the engineering, design, and manufacturing environment is one way to leverage the knowledge of the company's experts and to distribute this knowledge to a variety of people.

Design is a highly iterative process. Designs are reviewed, modified, and reviewed again at every stage of the design process, from conceptual design, to detailed design, to manufacturing planning, even through manufacturing. In order to reduce the number of iterations, and the time that it takes to complete the design-review-modify cycle, it is desirable to apply critical expertise from a number of perspectives at design time. This concept is known as "concurrent engineering", and is a technique for improving the efficiency of design by using as much multi-disciplinary knowledge as is available at the earliest opportunity in the design cycle. The application of expert systems for manufacturing review during the design process is an example of concurrent engineering, because the designer is made aware of the impact of design decisions on the manufacturing process while these decisions are being made! the most visible result of the application of expert systems in this manner are shorter time to market, improved accuracy and consistence, and improved product quality.

The engineers and designers who use CAD/CAM systems should expect to have expert system applications integrated directly within the CAD/CAM environment. Having to use another terminal, or a separate computer for the expert system session is a time-consuming and inefficient process. In addition, much of the information needed by the expert systems is directly available in the CAD model, but in a separate expert system environment, entering geometry or model attribute data into the expert system is redundant. The best method to communicate between CAD/CAM and expert systems is to execute the expert systems on the same hardware as the CAD/CAM system, and to communicate dynamically via direct data exchange in real-time.

Applications in Design and Manufacturing

A wide variety of design and manufacturing related subjects can be addressed using an integrated expert systems approach. In general, applications may fall into the following broad categories:

- Design Review
  One of the most time-consuming activities in the design process is the design review. Reviewers are typically experienced engineers or specialists, who provide a subjective analysis of the

proposed design. Often, this analysis is predominantly qualitative - that is, the review is not an analysis of loads, stresses, fluid flow, or other numerical parameters. A qualitative review generally deals with issues that relate to the reviewer's experience and judgment.

Many design reviews relate to rules that the company or organization has put in place for engineering and design standards, drafting standards, engineering specifications, etc. Often, these rules are written down in manuals, and must be examined by the designer or engineer after every design (or during the design process.) Similar to these standards may be project-specific standards and specifications, which are easily represented in the expert system rule format, and can become immediately available to provide specific project constraints and direction to engineers and designers.

- Drawing Review
  Once the design has been completed, a drawing is usually created. Since the drawing is produced for a particular purpose or function, it is important to determine if the required information is conveyed, and that the information on the drawing is correct. While this process is not always based on expertise, it typically involves a checklist, or a series of consistency rules. For example, some organizations have specific requirements for notes and dimensions, manufacturing notation, drawing number and revision control. Expert systems to provide advice regarding these issues could be used in a batch mode (after the drawing has been created) or in an interactive mode (to assist the designer at the time that the drawing is being developed.)

- Design Assistance
  Much of the process of design can be structured as a series of modular pieces, the combination of which provides a complete design method. Typically, people who are expert at some of the pieces, such as material selection, fastener selection, equipment selection, geometric configuration, or human factors review, are consulted by the designers when a design decision has to be made. These types of specialized design problems are ideal expert systems candidates, and can be effectively applied in a modular, incremental fashion.

In most design organizations, many experts contribute to any given design. The use of expert systems as "pockets of knowledge" can be extremely effective at helping to provide this important resource at design time!

- Parametric Design
  In some cases, designers and manufacturers generate highly repetitive designs. The process of using the same shapes and configurations, but changing the dimensions on a case-by-case basis is known as "parametric design". In order to optimize this process, expert systems are highly effective at helping to generate the inputs (dimensions, load requirements, etc.) for the parametric design methods. (An integrated approach, such as the STONErule method will actually allow for automated, parametric design, if the application warrants this. In this method, the expert system would be used to determine the parameters for the geometry and configuration, and STONErule would actually call the requisite procedures for creating, or modifying the design model directly.)

- Manufacturing Review
  Once a design has been generated, it typically undergoes a manufacturing review. This review is usually performed after the engineering design has been completed. The result can be an iterative process of changes due to the manufacturing review, and subsequent design-review cycles. By implementing expert systems for manufacturability review, machining and manufacturing tolerance specification, these important manufacturing issues can be accounted for at design time. Therefore, the time that it takes to generate an acceptable design and manufacturing process plan is significantly reduced.

- Manufacturing Tooling Selection/Process Planning
  The selection of manufacturing tools (such as cutters, dies, etc.) can be a complex process, often involving the use of judgment and experience in deciding on the best tool for a given application. Some problems have a clear, definite solution (a single cutter size is available, for example) while others require choosing the best tool from a list of acceptable alternatives. A knowledge base that uses basic selection rules (table lookup and selection) in addition to heuristic rules for optimal tool selection can be an extremely powerful decision aid to engineers and manufacturing specialists. In addition to helping to optimize the process of tooling selection, corporate preferences and physical constraints (such as availability of certain tools at specific locations) can be inherently accounted for.

The use of the best tools for a manufacturing process is one key element in manufacturing optimization. Another important process is the effective planning of the manufacturing sequence, and the details of how the part/assembly is to be made. This manufacturing process planning also involves knowledge of how to schedule the process to minimize tool changes, operator/machinist actions, and manual operations. A knowledge base which advises manufacturing engineers on the best process plan for a part or assembly would incorporate rules about efficiency in tool usage, machining, and practical manufacturing issues (such as manufacturing/ machining rates, scheduling, etc.)

- Maintenance and Troubleshooting
  Many organizations use their CAD/CAM models to help in the maintenance and operations functions for facilities, machinery, and other equipment. Schematic diagrams, electrical and wiring diagrams, flow sheets, and 3-D models are all useful tools for maintenance and operations. Expert systems for troubleshooting applications have proved to be highly successful for improving the accuracy and efficiency of the maintenance and diagnostic processes. By integrating these expert systems with the graphical models, accurate diagnostic and repair information is displayed to the user in an interactive manner. For example, in a diagnostic session for an electrical system, the wiring diagram can be used as the medium for displaying the failed element (the part can be highlighted on the screen) and for illustrating procedures for repair.

In summary, for applications that involve the use of expertise, rules-of-thumb or heuristics, judgment, and experience (as opposed to closed-form algorithms), the expert system approach offers the following significant benefits:

- ease of maintenance

- non-programming development environment

- incremental development and enhancement

- natural-language rules, easily readable by end-users

- focus on the application, not the program

- overall savings in cost and time

Integration Architecture

In order to provide an environment in which expert system will be actually be used by engineers, designers, and manufacturing engineers, STONErule was designed to seamlessly integrate expert system applications directly into the CAD/CAM environment. Simply stated, STONErule allows CAD/CAM users to run expert system applications from within their CAD systems, as if the expert system applications were a part of the CAD system itself. The expert systems are developed as if they were to be run in a stand-alone (or consultative) manner, exactly as all other expert system applications would be developed. No changes are made to the knowledge base or to the CAD application to facilitate this integration.

There are three primary modes of interaction between the user, the CAD/CAM system, and the expert system:

1. question and answer interaction (consultative mode) directly on the CAD workstation screen, without having to switch sessions or terminate or suspend the CAD system session. In this mode, the consultation appears exactly as it would if the user were executing the expert system on a stand-alone terminal, except that the consultation appears on the CAD screen.

2. transfer of CAD model data directly from the CAD system to the expert system, without asking the user to type the answer. Some of the information that the expert system needs is available in the CAD model directly, and can be accessed by STONErule and bound to expert system parameters automatically.

3. use of expert system results to update the CAD model directly. In addition to displaying the results of the expert system session on the screen, it is sometimes appropriate to change the model to reflect the advice of the expert system. STONErule allows for direct model updates (adding, changing, or deleting model geometry and attributes.)

In most production applications, a combination of all of the above modes of interaction is necessary in order to achieve an optimal expert system consultation.

STONErule operates on IBM mainframes, and ties the mainframe CAD/CAM systems CATIA and CADAM to the Aion Development System expert system shell.

The user can invoke a STONErule/ADS expert system application from within any CAD/CAM function, at any stage in the CAD session. When the user invokes STONErule, the CAD/CAM session is not suspended - the STONErule programs initiate an expert system consultation in parallel with the CAD/CAM session. The ADS task is run in the CAD/CAM region, thus establishing an extremely efficient communications and execution environment. In fact, the user does not actually see ADS at all.

The knowledge base is developed as if the user were to enter data in the standard manner for ADS expert systems. Therefore, any parameter that is "user sourced" (requiring a user input when ADS reaches that point in the reasoning path) will automatically appear as a question directly on the CAD/CAM screen! No changes to the knowledge base are made in order to run the expert system from within CAD/CAM. When the knowledge base is invoked from CAD/CAM using STONErule, the same reasoning path is followed as if the expert system were being run from an interactive PC, TSO, or CICS session. If a value for a user sourced parameter is needed, STONErule is notified automatically and dynamically creates the required input panel directly on the CAD/CAM screen. Once the user has responded to the question, STONErule sends the response back to the knowledge base, which continues its reasoning normally.

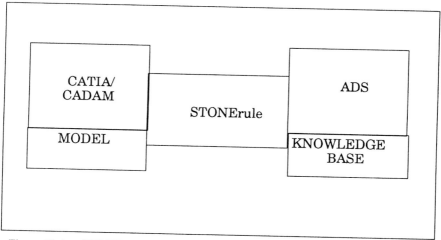

Figure 28-1   STONErule Architecture

An Example

As an example, consider the design of light aircraft landing gear. As a part of the design process, manufacturing specifications are generated. In particular, the machining tolerances for different parts are based on the type of part, and perhaps some other rules.

In this example, an ADS knowledge base has a parameter called 'part identifier', which is the name of a part that is to be reviewed for machining tolerance specification. In this example, the parameter is defined in the knowledge base as a string, and is user-sourced.

This parameter may be used in a rule of the following form:

IF   the part identifier is 'piston casing'

THEN   the machining_tolerance is 0.05
ELSE   the machining tolerance is 0.08
        end

Under normal conditions, the parameter would be resolved by asking the user the following question: What is the part identifier? (answer by entering a string value).

ADS screens, illustrating the parameter definition and rule entry format during knowledge base development, and the format of the question during an interactive consultation are shown in the following three figures.

The equivalent input screen for this knowledge base run under STONErule is shown in the following figure. here, we can see that the question appears directly on the CAD/CAM screen (in this case

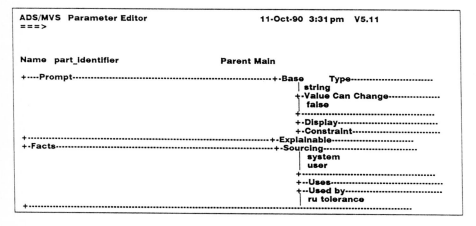

Figure 28-2   ADS Parameter Definition Screen

Figure 28-3   ADS Rule Definition Screen

```
ADS/MVS Consultation Monitor                11-Oct-90  3:31 pm   V5.11
===>

Enter a string value
-Prompt------------------------------------------------------------------------
What is part_identifier?




--------------------------------------------------------------------------------
--Answer------------------------------------------------------------------------
```

Figure 28-4   ADS Consultation Question Screen

CATIA), and the design model   is active. The user types in the
response, which is then transmitted, by   STONErule, to the
knowledge base.

In order to optimize the interactive session, the developer would
inform STONErule that the information is available from the model
itself. STONErule would then prompt the user to select the part with
the mouse, and the part identifier would be read from the model and

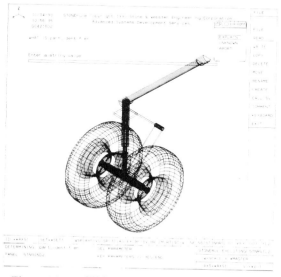

Figure 28-5   STONErule Consultation Question Screen

bound to the ADS parameter automatically. The following screen illustrates the STONErule automated input screen.

Conclusions

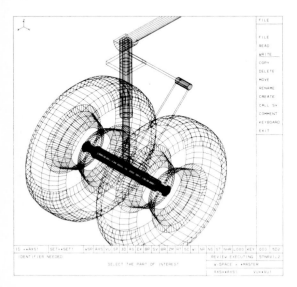

Figure 28-6   STONErule Geometry Selection Screen

In general, for any expert system application, in order to successfully implement the system in practice, it is important to (i) focus on small, practical problems, (ii) concentrate the development effort on the application, not the expert systems technology, and (iii) integrate the expert system into the working environment of the end users. For many design and manufacturing applications, it is necessary to present the expert system application to the designer or engineer as an integral part of the existing CAD/CAM system. The integration of CATIA and CADAM design systems with ADS expert systems, through STONErule, provides the requisite technical architecture, and allows developers and end users to focus almost entirely on the domain knowledge and problem-solving strategy.

By providing a transparent, seamless integration of expert systems and CAD/CAM applications, numerous benefits accrue. Most notably, the simultaneous application of a variety of knowledge bases to the design process provides true concurrent engineering. The result is a shortened design-manufacture time, improved productivity of the engineers and designers, and improved product quality through consistent application of the best design and manufacturing expertise available.

In the 1960's and 1970's artificial intelligence was a research topic. In the late 1970's and through the decade of the 80's the focus was on applications of expert systems to real-world problems. STONErule is an example of enabling technology that will extrapolate the progression of expert systems through the 1990's, bringing expert systems applications into the conventional business and technical computing environments simply as a component of integrated solutions.

Author Biographical Data

Gavin Finn is a consulting Engineer in the Advanced System Development Services Division of Stone & Webster Engineering Corporation. He is responsible for all aspects of expert system projects, including technology development, marketing, and project management. He has lead the development of expert systems applications for the manufacturing, chemical process, aerospace, and electric utility industries. Mr. Finn is the product manager for the STONErule CAD/expert systems interface system.

He was named the "Young Engineer of the Year" for 1990 by the Massachusetts Society of Professional Engineers. Mr. Finn is a member of Technical Council on Computer Practices Committee on Expert Systems of the American Society of Civil Engineers.

Mr. Finn has taught courses and seminars on Artificial Intelligence and Expert systems at Union College, the University of Maryland, MIT, and the Computational Mechanics Institute in Southampton, England. He has published more than twenty papers, articles and chapter on expert systems applications. He holds a Master of Science degree in Civil Engineering from MIT, and a bachelor's degree in Civil Engineering from Oklahoma State University.

An Expert System for Designing and Processing Autobody Parts

Mahmoud Y. Demeri
Research Staff
Ford Motor Company
Dearborn, Michigan

Introduction

This paper describes an expert system developed to aid engineers in designing and processing autobody parts. The system is rule-based and simulates the reasoning of human experts in the stamping engineering domain. Traditionally, part design, material selection and process development are based on prior experience and rules of thumb. This makes designing and processing of autobody parts a suitable candidate for expert system applications. Knowledge in the expert system consists of information and guidelines derived from experience acquired from the development and testing of production and experimental parts. Information in the knowledge base is structured as sets of conditional relationships. A major objective of the expert system is to give engineers useful recommendations regarding part design and processing. In this capacity, the expert system is used as a consultant and not as a decision maker. The expert system is implemented on a PC-based shell with a highly interactive environment.

Most knowledge about stamping autobody parts is acquired from observations made during the development and testing of production and experimental parts. Design verification tests are conducted to ensure compliance with engineering standards. The key to successful forming of autobody parts is to use the guidelines and recommendations based on prior experience in solving similar problems or on extrapolations based on such experience.[1] Traditionally, rules and guidelines are developed by experts in the field through numerous tryouts and years of experience. Some of the information gathered is formulated as guidelines and the rest is either retained by the expert or lost because of the lack of a mechanism to formalize the experience. Also, as the number of guidelines and rules increase, their access and utilization by application engineers becomes more difficult.

It is believed that expert systems can provide domain experts with the tools to: (1) capture and formalize their experience , (2) organize their guidelines and rules of thumb in a systematic manner, (3) facilitate the accurate and consistent utilization of such rules and guidelines and (4) build an apprentice system to teach new engineers the tricks of the trade. An expert system is not meant to replace a human expert but to support him in formulating and dispensing his acquired knowledge. Expert systems do not only organize, preserve and dispense knowledge efficiently, but they also allow non-experts to utilize the knowledge of experts. The following sections describe the development of an expert support system to help engineers obtain proper guidelines and recommendations for designing and processing autobody parts. The expert system acts as a consultant providing designers with essential information on how to solve specific sets of domain problems.

Expert System Shells

An expert system actually has three major components, a knowledge base, an inference engine and a user interface. The knowledge base contains data structures representing knowledge in the form of facts and rules. Facts represent declarative knowledge while rules represent procedural knowledge. The major techniques for knowledge representation are production rules, frames and semantic networks. Production rules, in the familiar form (IF... THEN rules), are simple to formulate and implement. The inference engine performs the inference procedure that solves a specific problem. This is accomplished by using either a search or a deduction strategy. The search strategy uses either forward or backward chaining methods. Searching for a solution is directed by a control mechanism which determines the best

---

1   M. Demeri, "Expert Systems in Forming Processes", in Expert System Applications in Materials Processing & Manufacturing, Edited by M. Demeri, A TMS Publication, 1990, P19.

strategy to achieve a goal. The user interface provides communication between the user and the program. This includes editors and debuggers which provide an environment for software development.

An expert system shell is a development software tool that provides a method for knowledge representation and inferencing. Expert system shells usually provide an inference engine, support for representing knowledge and user interface. In an expert system shell, the inference engine and knowledge representation are fixed while the knowledge base varies depending on the application. Knowledge can be replaced to solve different problems, while the inference engine and user interface remain the same. A number of commercial expert system shells can be used for prototyping and developing end-user applications. In choosing a software tool, a number of criteria must be established for tool selection. Such criteria include: basic features, development environment, end use, ease of learning, cost and technical support.[2]

A major consideration in choosing an expert system shell for the current application is the way knowledge is represented. Knowledge bases are structured into hierarchical rule sets whose structures are called frames. Each frame addresses a part of the overall problem. This structure permits the use of a rule set more than once during the same consultation and allows one rule set to inherit information from another rule set.[3] Other considerations in choosing an expert system shell include the tool being PC-based, easy to use and have facilities for developing and testing the knowledge base. An efficient expert system shell reduces the task of building an expert system by allowing the developer to concentrate on the knowledge base, rather than the mechanics of the system.[4]

Selection of an Expert System Development Shell

Based on the criteria considered for choosing an expert system shell, a software package from Texas Instruments called Personal Consultant Plus (PC-Plus) was selected to develop the current application. PC-Plus is a highly interactive environment for developing and testing expert systems on personal computers. It has a window-oriented interface, on-line help and a full screen editor. It belongs to the class of rule-based tools which offer frame structures that allow the

2   M. Richer, An Evaluation of Expert System Development Tools, Expert Systems, Vol. 3, No. 3, July 1986, P 166.

3   R. Vedder, PC-based Expert System Shells: some desirable & less desirable characteristics, Expert Systems, Vol. 6, No. 1, February 1989, P 28.

4   J. Durkin, Introducing Students to Expert Systems, Expert Systems, Vol. 7, No. 2, May 1990, P 70.

knowledge base to be divided into logically separate but related segments. PC-Plus, written in PC Scheme (a simple modern Lisp), offers an English-like rule-entry language called Abbreviated Rule Language (ARL). The inference engine typically uses backward chaining but forward chaining can be used if specified in the rules. PC-Plus can access DOS files, dBase databases and Lotus spreadsheets. It displays graphics and has a mechanism for handling uncertainty from the developer and the end user. A very desirable characteristic of PC-Plus is its suitability for creating prototypes rapidly and efficiently.

The minimum system requirements for development version of the PC-Plus is an 80286 or 80386 based IBM compatible computers. At least 640 Kilo bytes of RAM, 2 Megabytes of extended or expanded memory and at least 1.5 Megabytes of hard disk space are required. The delivery (application) version can run on less powerful computers and requires less available memory. The current application was developed on an IBM-PS2 model 70 computer with 640 Kilo bytes of RAM, 6 Megabytes of extended memory and 120 Megabytes of hard disk capacity.

Structuring the Knowledge Base

The first step in developing a knowledge base is to define the problem domain, then to identify the problem to solve in that domain. The domain of this knowledge base is the designing and processing of autobody parts. The problem is to provide engineers with the proper design and process guidelines for the successful stamping of autobody parts.

The second step is to acquire and structure knowledge in the domain. The knowledge base for this domain is very large. So, in order to have a workable application in a reasonable time, the prototype domain is narrowed substantially. By initially building a prototype, rather than a full scale application, knowledge structure and information organization can be tested. After testing, modification and validation, the prototype can be expanded into a full scale knowledge base. Acquiring knowledge usually requires the services of a knowledge engineer, but in this application the author relied on his expertise to develop the prototype knowledge base. Information for the knowledge base is gathered from different written sources, scattered documents and the authors's personal experience in the domain field. For the most part, knowledge acquisition for this application involves selecting and organizing information from a large collection of scattered guidelines and rules.

The PC-Plus knowledge base system consists of three basic structures that control and organize information; frames, parameters and rules. Each has its own set of properties which define its characteristics. Frames provide information about the domain, structure and operation of the knowledge base. A knowledge base contains one or more frames that are used to group parameters and rules into logically organized structures. Parameters are individual pieces of knowledge that have values. Rules in the form (IF-THEN statements) express relationships between the parameters and their values. In the PC-Plus knowledge base, at least one goal must be defined. A goal is a parameter whose value is the result of a consultation. After a goal is entered, information about the intended application is then structured into frames, parameters and rules.

Generating a knowledge base is facilitated by development of a chart showing the logical steps needed to arrive at a conclusion. Such a chart for the current application is shown in Figure 29-1. It shows the flow of logic that produces a conclusion during a consultation. At the top is the knowledge base goal, described as providing design and process guidelines. This is the parameter GUIDELINES, its values being the conclusions reached during a consultation. Goal values are determined after answering questions that identify the facts needed to arrive at a conclusion. Questions are used to develop the knowledge base parameters. These parameters are shown on the right side of the chart and include: PART, MATERIAL, APPLICATION, FORMING and ALUMINUM ALLOY. Answers identify the possible parameter values shown in boxes with broken lines. For example, the PART parameter has a list of nine values which include Fender, hood, bumper, etc. Similarly, the MATERIAL parameter has the values: Aluminum Alloy, Low-Carbon Steel, High-Strength Steel and Plastic.

It is important to realize that the chart in Figure 29-1 shows the flow of logic for a specific case. The final answer depends on the values assigned to knowledge base parameters. Each parameter value follows a unique path to reach a possible answer for that goal. Once knowledge is structured in chart form, rules can be derived by working down from the top of the chart. Figure 29-1 shows how the charting method is applied to derive rules by simply finding relationships between the knowledge base parameters and their possible values.

The (IF) part of a rule describing the logic in Figure 29-1 is written in PC-Plus' Abbreviated Rule Language (ARL) as:

IF :   PART = Hood-Outer and
                   MATERIAL = Aluminum Alloy and
                   APPLICATION = Forming and
                   PROCESS = Stretch-Drawing and
                   ALUMINUM-ALLOY = 2036-T4

Identifiers on the left side of the equality signs are knowledge base parameters. Those on the right side are the values assigned to them. Since the goal is a parameter, whose value is given at the end of the chart, the conclusion part of the rule is given in ARL by:

THEN: GUIDELINES = Some Recommendations

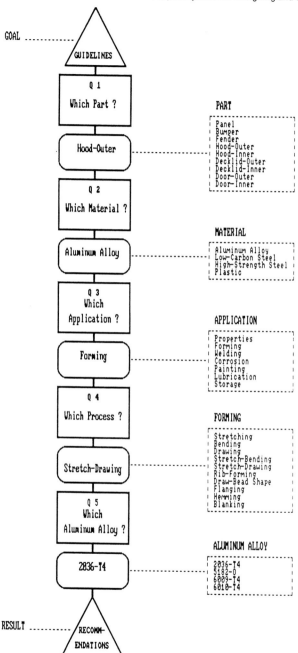

Figure 29-1 Knowledge Structure for Designing and Processing Autobody Parts

Figure 29-2 shows the complete rule in both English and ARL. The order in which parameters in a rule are arranged is important. The order determines the path taken during a consultation to reach a conclusion. Also, parameters must be defined before they can be used in a rule. A consultation is an attempt by the expert system to find a value for the goal parameter 'GUIDELINES'. When PC-Plus finds a rule that gives the goal value; it tests the rule by looking at values for other parameters in the IF statement. This is called backward chaining. If the premise is true, the expert system applies the conclusion, in the THEN part of the statement, and a value of the goal parameter is determined. Rules for other situations are derived in a similar way.

RULE001

If

1) the name of the part being considered for designing & processing is HOOD-OUTER, and
2) the type of material being considered is ALUMINUM-ALLOY, and
3)   1) the name of the aluminum alloy being considered is 2036-T4, or
       2) the name of the aluminum alloy being considered is 6009-T4, and
4)   the type of application being considered for the part is FORMING, and
5)   the type of forming operation being considered is STRETCH-DRAWING,

then it is definite (100%) that the guidelines that should be considered in designing & processing this autobody part is (*) Large flat areas should be avoided. (*) Positive crown required. (*) Hatch-type hood designs are preferred to eliminate wavy metal and undesirable surface highlights. (*) Front and sides must have downstanding flanges. . .

IF   : : PART= HOOD-OUTER AND MATERIAL=ALUMINUM-ALLOY AND ALUMINUM-ALLOY=2036T OR ALUMINUM-ALLOY=6009-T4 AND APPLICATION=FORMING AND FORMING=STRETCH-DRAWING
THEN   : : GUIDELINES= TESTVAL :LEFT 5 :RIGHT 80 :LINE 2"(*) Large flat areas should be avoided." "LINE 2'(*)Positive crown required. " :LINE 2 '(*) Hatch-type hood designs are preferred to eliminate wavy metal and undesirable surface highlights." :LINE 2"(*) Front and sides must have downstanding flanges."

Figure 29-2   A Knowledge Base Rule Written in both English and ARL

```
Activities:

CONSULT
DEVELOP




Run a consultation with this database. Press F1 for help.
```

Screen 29-3 Activities Screen

Building and Testing the Expert System

After structuring the information on designing and processing autobody parts, the task of developing the knowledge base and building the expert system begins. The major effort in building an expert system is obtaining and structuring knowledge. In PC-Plus, building and testing an expert system are facilitated by a highly interactive environment and screen editor. The expert system is developed by selecting items, entering information or answering questions presented by a sequence of well developed screens.

The first set of screens deal with defining the three basic elements of the knowledge base; declaring the domain, naming the frame and deciding on the goal parameters. The next set of screens address the knowledge base parameters, determining their types and declaring their expected values. Another set of screens facilitate rule grouping, entry, description and translation. Rules in PC-Plus are

```
Current Objective:

         AUTOBODY DESIGN & PROCESS CONSULTANT

                      M.Y. Demeri
                     Research Staff
                  FORD MOTOR COMPANY

This program helps engineers to obtain design & process guidelines for cer-
tain APPLICATIONS of AUTOBODY PARTS.
AUTOBODY PARTS: bumper, door-outer, decklid, hood, panel, etc.
APPLICATIONS: forming, welding, corrosion, properties, etc.


** End - RETURN/ENTER to continue
```

Figure 29-4 Current Objective Screen

What is the name of the part being considered for designing and processing?

PANEL
BUMPER
FENDER
HOOD-OUTER ▬
HOOD-INNER
DECKLID-OUTER
DECKLID-INNER
DOOR-OUTER
DOOR-INNER

1. Use the arrow keys or first letter of item to position the cursor.
2. Press RETURN/ENTER to continue.

Figure 29-5   Part Name Screen

What is the type of material being considered?

ALUMINUM-ALLOY ◀
LOW-CARBON-STEEL
HIGH-STRENGTH-STEEL
PLASTIC

1. Use the arrow keys or first letter of item to position the cursor.
2. Press RETURN/ENTER to continue.

Figure 29-6   Material Type Screen

consequent. This means that a rule is used when it is needed to set a value during the backward-chaining process. Rules can also be specified as antecedent. Such rules are used as soon as their premises become true and without any reference to the backward-chaining process. Additional information to help users respond

What is the type of application being considered for the part?

PROPERTIES
FORMING   ◀
WELDING
CORROSION
PAINTING
LUBRICATION
STORAGE

1. Use the arrow keys or first letter of item to position the cursor.
2. Press RETURN/ENTER to continue.

Figure 29-7   Application Type Screen

accurately to prompts is provided by the HELP facility. The HELP property can be added to any knowledge base parameter.

What is the type of forming operation being considered?

STRETCHING
BENDING
DRAWING
STRETCH-BENDING
STRETCH-DRAWING   ◀
RIB-FORMING
DRAW-BEAD-SHAPE
FLANGING
HEMMING
BLANKING
GENERAL-FORMING

1. Use the arrow keys or first letter of item to position the cursor.
2. Press RETURN/ENTER to continue.

Figure 29-8   Forming Operation Screen

What is the name of the aluminum alloy being considered?

2036-T4 ◀
5182-0
6009-T4
6010-T4
GENERAL-ALUMINUM-ALLOY

1. Use the arrow keys or first letter of item to position the cursor.
2. Press RETURN/ENTER to continue.

Figure 29-8a Aluminum Alloy Screen

Conclusions:
The guidelines that should be considered in designing & processing
this autobody part is as follows:

(*) Stretchablity is less for alummpanels compared to those of steel
due to lower elongation, less ability to distribute local strains and a
lower forming limit diagram.

(*) Drawability of shallow dran panels will be compatible to drawing quality steel in tools incorporating aluminum design guidelines.

(*) Large flat areas should be avoided.

(*) Positive crown required.

(*) Hatch-type hood designs are preferred to eliminate wavy metal
and undesirable surface highlights.

(*) Front and sides must have downstanding flanges.

End - RETURN/ENTER to continue.

Figure 29-8b Aluminum Alloy Screen

When all aspects of the interactive process of creating the knowledge base are done, the basic framework of the expert system is completed. The prototype system is now tested to see if it reaches the right conclusions for each combination of questions and answers.

The expert system is tested by running a series of consultations to find out if the prototype is working as intended. The expert system should not only reach the right conclusions, but should also have clear, correct and logical prompts. PC-Plus has a number of debugging facilities to test the program. The REVIEW command is used to test all possible combinations of prompt responses to ensure that the program reaches the right conclusion for each set of prompts and responses. The WHY command explains each prompt, while the HOW command rationalizes each conclusion. Problems in rules and their associated parameters can be identified by the TRACE function.

Consultations

An engineer may consult with the expert system by simply answering a sequence of prompts. A consultation can be long or short depending on the answers given. Since varying responses affect conclusions, the expert system can be used by engineers to explore different possibilities. The following sequence of screens show a typical application of the Autobody Design & Process Consultant. In this example, the expert system is consulted to provide information and guidelines for designing and processing a Hood-Outer.

The first screen in the consultation sequence, after selecting the appropriate knowledge base, is the activities screen (See Figure 29-3). The engineer may select one of two options, consult or develop the expert system. The response in this case is to consult. Figure 29-4 is the current objective screen which gives the title of the expert system and describes its objectives. Figure 29-5 prompts the user to choose the name of the part being considered from a list of nine parts. For this consultation, the Hood-Outer is selected. Figure 29-6 prompts the user to choose the type of material from four classes of materials. Aluminum-Alloy is picked as the candidate material. Figure 29-7 prompts the user to select the application from a list of seven applications. Forming is chosen. Figure 29-8 prompts the user to select a forming operation from eleven possible forming operations. Stretch-Drawing is selected. Figure 29-8a prompts the user to pick a specific aluminum alloy from five alloys. The aluminum alloy 2036-T4 is selected. All the prompts are answered, and the result is displayed in the last screen as shown in Figure 29-8b, which is the conclusions screen. Conclusions are a list of guidelines that should be considered when designing and processing Hood-Outers.

At the end of the consultation, the user may print the recommendations, save a record of the consultation, continue with another consultation or quit. Other options such as invoking the HOW command to explain the reasoning that led to the conclusion is also available.

Expanding the Prototype Expert System

After testing and all necessary modifications are completed, the prototype expert system can be expanded to include most aspects of the knowledge base domain. Expansion is carried out one step at a time, by adding new parameters and new rules. After each step, testing and modification are performed before going on to the next step. When the knowledge base is complete, the systems will be ready for more elaborate consultations. Of course, the expert system may not contain all the knowledge it needs for all possible situations. The expert system may occasionally indicate the inability to reach conclusion about a consultation. This happens when the expert system cannot find a value for the goal parameter. This problem can be solved by adding more rules to the knowledge base. As more rules are added, the expert system becomes more useful. In this type of application, however, an expert system is never complete. The expansion and growth potential of an expert system depends largely on the design of its knowledge structure. A good design permits the addition of new knowledge base parameters and rules as often as needed.

Although the current knowledge base contains about fifty rules, the amount of information involved is substantial. The rules and guidelines were extracted from a fifty page design manual and a number of information sheets. The majority of the rules deal with designing and processing Aluminum Alloys. The domain of this material is much narrower than that of other materials. Therefore, information on Aluminum Alloys were chosen to populate the prototype knowledge base. Rules for other materials have been added to verify the knowledge structure. Knowledge on other materials will be added in the future. It is estimated that a complete production version of this system will have over five hundred rules. The knowledge structure is comprehensive, so a small number of new parameters will be added to the parameter list. Future expansion of the expert system will, more likely, be accomplished by adding more rules than parameters.

Conclusion

The expert system was shown to engineers who may be potential users. Initial reactions were generally positive and enthusiastic. The usefulness of the system will be more apparent when the knowledge base is expanded. After expansion, the expert system will be made available for engineers who design and process autobody parts. This includes car designers, die & tool makers, body, process, CAD/CAM, development and stamping engineers. The expert system should provide essential, consistent and fast information to help expert and non-expert users in designing and processing autobody parts.

Author Biographical Data

The author holds a B.S. in Physics, and M.S. in Solid State Physics, an M.S. ion Electronics and Computer Controlled systems and a Ph.D. in Metallurgical Engineering. He is a senior scientist at the Scientific Research Labs, Ford Motor company. His main interest is in materials processing and manufacturing. He is involved in the application of databases and expert systems to this area. He has published a number of papers and recently edited a book entitled *Expert System Applications in Materials Processing & Manufacturing.*

Acknowledgement

The author wishes to thank Mr. Perry MacNeille for proof-reading the manuscript.

Quality and Safety

30

Lubrizol Material
Safety Data Sheet System

(216) 943-4200

Ms. Rosalie Troha
Lubrizol Corporation,
Wickliffe, Ohio

Introduction

The Lubrizol Corporation is a specialty chemical company located near Cleveland, Ohio. Lubrizol was founded over 60 years ago and has enjoyed a reputation for being a leader in the innovative development of specialty chemicals for use in industry and agriculture. A small manufacturing facility is located on the corporate headquarters site. Two large manufacturing facilities complete the domestic operations, one located in Painesville, Ohio and the other spanning two sites in Deer Park, Texas and Bayport, Texas.

Because it is a worldwide manufacturer and supplier, Lubrizol has a duty to provide information about the makeup and handling of its specialty chemicals to various organizations including customers, carriers, communities and plant personnel. This information is contained in a document known as a Material Safety Data Sheet, or MSDS.

As a company selling products containing chemicals which may be hazardous if not handled properly, we are required to have MSDS for all products on site. We are also required to supply MSDS to our customers with the shipment of a product they are purchasing for the first time, with the shipment of a product whose MSDS has

448

changed since their last order, and at least once a year. The generation of this document is no small matter as the chemical makeup of each of its components must be considered to determine potential health and safety effects.

At Lubrizol, material safety data sheets are the responsibility of our Corporate Resources Assurance Division (CRAD).

Prior to 1989, the writing of these documents was a manual function partially automated by the use of a word processing package. Because of the volume of MSDS needed to support 8000 products, intermediates, raw materials and research materials, several people were designated to write these documents. Since understanding of complex government regulations applicable to the content of an MSDS is subject to interpretation, a sheet written by one individual could contain different verbiage than a sheet written by another on the same product. Although both would be in compliance and within the limitations imposed by the government, they might send different messages to the reader. Due to the amount of time and resources necessary to keep our MSDS file updated, it was determined that the MSDS process should be automated as much as possible. Computerizing the process would also bring consistency and standardization to the content of the documents.

A project was begun in December 1987 to provide a system which would be able to store data about our chemicals, utilize data to determine the content of the MSDS for any material, and determine when an MSDS had to be sent to a customer. After the preliminary systems design was complete, it was decided that the system would be installed in two phases. The first phase would allow the data collection to take place. It provided for a series of online transactions to be used by the people who formerly wrote the MSDS, to capture raw data about our chemicals in a relational database. The second phase would encompass the generation and printing of the document, either on request or because of a customer requirement.

Building the Material Safety Data Sheet System

When we first began the project, Lubrizol had no artificial intelligence package in place, nor did we plan on purchasing one. In concurrence with the development of the first phase of the project, the potential of using an expert system was raised. At that time, we had little expertise in the technology nor experience with the vendors who supplied it. An evaluation of five of the more popular artificial intelligence packages was begun by members of the MIS team at the time. We began the evaluation by documenting the requirements which the expert system would have to fulfill. These included technical requirements as well as cost, ease of use, features provided, etc. The evaluation took three months and consisted of reviewing literature, as well as visual demonstrations of each product. When it was complete, we decided that none of the packages met our needs. However, we also recognized that the technology was still maturing, and that we could anticipate improvements in the artificial intelligence market.

After the first phase of the project was installed into production, development began on the second phase. As the requirements of the system were refined, it became obvious that we were facing a few challenges. First, the number of government regulations were increasing, as were the changes to the existing regulations. The federal government was not the only body demanding compliance. The states were also enacting 'right to know' laws. This dictated the need for a system which could be quickly and easily maintained. Our second challenge was the need to "cascade" the effect of a change made to one material onto others products containing that material. Not only would the MSDS need to be regenerated for the changed material, but also for every other material which contained the same chemical. Which brings us to our third challenge. What if enough commonly used materials were changed on one day to create a generation request for every material on our file? In a worst-case scenario, we could have thousands of requests for MSDS to be generated in one evening. What kind of system could handle that volume? (Actual capabilities for throughput are discussed later.) It did not take very long for us to realize that we were not dealing with a traditional COBOL system.

The idea of using an expert system was resurrected. Although we had not chosen an expert systems package after the evaluation, we had a definite opinion on the front-runner. A quick study was done to see how the package had matured. We felt that the software had progressed to a point which better fit our original requirements. Recognizing we were choosing a risky alternative to traditional development, we decided to purchase ADS from Aion Corporation as our expert system. Our plan was to develop a prototype of the logic needed to generate an MSDS on an IBM PS2 Model 70, then port it up to the IBM 3090 mainframe for volume testing and eventual implementation. For the rest of the system, the COBOL-generating lower-case tool TELON would be used.

Our next step was to have the users document their business rules for writing a material safety data sheet in a format that would be easily convertible into expert system rules. For this we developed a variation of a decision table (Table 30-1). This format had more advantages than we originally anticipated. Not only did it help the MIS staff turn business rules into computer logic, it helped the users organize their thoughts to better communicate their needs to MIS. It also helped us document test conditions and expected results when we had reached the testing stage.

Over one hundred decision tables were documented by the users. These were turned into states, objects and rules in the expert system. An example of a rule determining how to set a flag indicating

Test 2. Set Corrosion Flag

PH <= 2?	PH >=12.5?	Strong Acid Ind = Y ?	Strong Base Ind = Y ?	Corrosion Flag
Yes	-	-	-	Yes
No	Yes	-	-	Yes
	No	Yes	-	Yes
		No	Yes	Yes
			No	No

RELEVANT FIELDS
                pH: BDCH-PH
    Strong Acid Ind: BDCH-STR-ACID
    Strong Base Ind: BDCH-STR-BASE

Table 30-1   Example of a decision table

the corrosive properties of a material is shown in Table 30-2. Testing was performed on each state individually, and then on groups of states together. Results of testing were compared against expected results previously prepared on paper by the users. While programming the modules that would handle the decoding, it became apparent that we had another opportunity to use the expert system. The portion of the system which reads the coded document from a DB2 table and creates a print image of the final document was also developed in ADS. Using the expert system for this feature allowed us to more easily control spacing, pagination, and other hard-to-code forms control techniques.

```
$Rule
    set_corrosion_flag_rule
$Comment
        set_corrosion_flag_rule
        =========================
        Sets the corrosion flag based on indicators
        on base data
$Rule Definition
    /* See comment  */

    IF bdch_pa_4.bdch_ph <= 2              OR
       bdch_pa_4.bdch_ph >= 12.5           OR
       bdch_pa_4.bdch_str_acid = 'Y'       OR
       bdch_pa_4.bdch_str_base = 'Y'       OR
       (single_comp_bool_pa = TRUE AND
        (single_comp_phyprop.bdch_str_acid = 'Y'   OR
         single_comp_phyprop.bdch_str_base = 'Y'   OR
         single_comp_phyprop.bdch_steel_corr = 'Y'))
    THEN
        corrosion_flag_pa is TRUE
    ELSE
        corrosion_flag_pa is FALSE
    END
$Rule Type
$Priority
    0
```

Table 30-2   Example of an expert system rule

Our system does not make use of object-oriented design techniques, although our expert system is capable of handling them.

Data is retrieved from DB2 tables via the expert system, then manipulated through both forward and backward chaining capabilities of the inference engine. The results of that exercise are a coded form of the MSDS. A series of tables maintained by the users contain the verbiage which is later used to decode the document for final printing. Table 30-3 contains page 2 of a 6-page MSDS containing the decoded text.

Although not without its anxious moments, the development of the system was an impressive achievement. The most exciting moment for the team was not the day the system went into full production, but in the testing phase months before when the first MSDS was generated using the expert system logic. It contained many data and logic errors, but was a sight to behold for the accomplishment it represented.

The system has been in full production for nearly a year. Although only the CRAD users have access to update data on the DB2 tables, users from all of the domestic locations have the online capability to view the document as well as request that a hard copy be produced.

```
SECTION 3 - HEALTH HAZARD DATA
```

```
                    -- ACUTE EXPOSURE --

ORAL TOXICITY:     The LD50 in rats is > 5000 mg/Kg.  Based on actual data.
EYE IRRITATION:    Not expected to cause eye irritation.  Based on actual data.
SKIN IRRITATION:   Not expected to be a primary skin irritant.  Based on actual
         data.    Prolonged  or repeated skin contact as from clothing wet with
         material may cause dermatitis.  Symptoms may include redness,  edema,
         drying, defatting and cracking of the skin.
DERMAL TOXICITY:   The  LD50  in  rabbits  is  >  2000  mg/Kg.  Based on similar
         materials.
INHALATION TOXICITY:     High concentrations may cause headaches,  dizziness,
         nausea, behavioral changes, weakness, drowsiness and stupor.
RESPIRATORY IRRITATION:     If  material  is  misted  or if vapors are generated
         from heating, exposure may cause irritation of mucous  membranes  and
         the  upper  respiratory  tract.  Based  on  data from components and
         similar materials.
DERMAL SENSITIZATION:     May cause skin sensitization.  Based  on  data  from
         components or similar materials.
INHALATION SENSITIZATION:  No  data available to indicate product or components
         may be respiratory sensitizers.

                    -- CHRONIC EXPOSURE --

CHRONIC TOXICITY:      No data available to  indicate  product  or  components
                       present at greater than 1% are chronic health hazards.
CARCINOGENICITY:       No data available to indicate any components present at
                       greater than 0.1% may present a carcinogenic hazard.
MUTAGENICITY:          No data available to indicate product or any components
                       present   at   greater   than   0.1%   are   mutagenic  or
                       genotoxic.
REPRODUCTIVE TOXICITY: No  data  available  to  indicate  either  product  or
                       components  present at greater than 0.1% that may cause
                       reproductive toxicity.
TERATOGENICITY:        No data available to indicate product or any components
                       contained at greater than 0.1% may cause birth defects.

                    -- ADDITIONAL INFORMATION --

OTHER:                 No other health hazards known.
EXPOSURE LIMITS:       Contains  mineral  oil.   Under  conditions  which  may
                       generate  mists, observe the OSHA PEL of 5 mg per cubic
                       meter, ACGIH STEL of 10 mg per cubic meter.
```

Table 30-3

MSDS data is transmitted daily from corporate headquarters to Lubrizol in Bromborough, England to ensure that their MSDS system uses the same basic chemical information from our databases. A system which creates Canadian English MSDS by running a near-clone of the knowledgebase on a PC, retrieving its data from the mainframe, is scheduled for production soon. More enhancements are planned for the future, including the ability to generate the document online by running the expert system in real time and the ability to send the print image of the document to a FAX machine rather than a printer. We feel that we can technically enhance the system even further. Today, it takes five hours to run the batch process on an IBM 3090-200 series mainframe. Up to three of those hours are consumed by the expert system portion that generates the MSDS. Because we cannot predict how many requests for MSDS generation will reside on the input file on any given day, we have established a time-out of 180 minutes on that job. In three hours, we can produce 500 -600 MSDS. In order to process urgent requests first, an enhancement was made enabling requests to be prioritized by the system and generated accordingly. We also believe we can improve throughput by changing the data retrieval from being part of the expert system to a COBOL program which will pass the data to the knowledgebase.

As the system stands today, there are two knowledgebase modules, one to generate the coded MSDS, and the other to decode the MSDS into printed format. In total, the knowledgebase contains 287 parameters, 154 rules and 22 states. There are 17 DB2 tables containing 2.9 million rows of data. We have over 8700 base materials comprising over 7300 products. We generate MSDS for all raw materials, intermediates, finished products and research materials.

Conclusion

The culmination of achievement for our MSDS system came on June 25, 1990 in Washington, D.C. when we were named winners of the 1990 Computerworld Smithsonian award over four other finalists in the Manufacturing division. The award recognizes technologically advanced applications contributing to the benefit of society. We were nominated for the award by Andersen Consulting, who also participated in the development of our system. The winning of the award was thrilling and gratifying, and we at Lubrizol are proud to have been part of such a prestigious occasion.

Author Biographical Data

The Lubrizol Corporation, headquartered in Wickliffe, Ohio, is a specialty chemical company engaged in the chemical, mechanical and biological technologies serving world markets in transportation, industry, and agriculture. Lubrisol employs over 5000 employees located at the corporate headquarters, sales offices, laboratories, and manufacturing plants worldwide.

Rosalie M. Troha has been an employee of Lubrizol for ten years. Rosalie's entire career at Lubrizol has been in the Management Information Systems Division. Her current position is Manager of Applications Development supporting the Logistics Division, the Corporate Resources Assurance Division and the Manufacturing Division.

An Expert System for Quality Control in Manufacturing

James R. Brink and Sriram Mahalingam
E-KE,
Dublin Ohio

Battelle Memorial Institute,
Columbus, Ohio

Introduction

The subject of this article is an expert systems application in the packaged food industry. This industry involves the acquisition of raw materials such as wheat, corn and sugar that meet specific criteria and processing the materials into a finished product such as cookies, crackers and canned goods. It also involves shipping the products to retail outlets (and ensuring the availability of shelf space), mass marketing to the consumer, and the planning and control of inventory. Although, expert systems technology applications are possible in all of these areas, the expert system described here applies to the processing portion, specifically to quality control aspects of processing raw materials into finished product.

Baking is an ancient art that is a standard staple of most households. Producing baked goods on a commercial scale is similar, except on a much larger scale. Essentially, the raw materials (e.g., wheat and sugar) are mixed according to a predefined recipe, they are formed into the desired shape, baked and packaged. During this process, elapsed time may also be part of the recipe to allow the dough to rise. Throughout this process, there are many potential opportunities for expert systems application: preventative maintenance of the equipment (especially, the packaging equipment), scheduling of personnel & production lines, and quality control/review of the process from raw material to baked good.

RJR Nabisco is a Fortune 500 company with a global presence in this market place. Its brand names are known throughout the world for high-quality baked goods. The company's products are manufactured, that is baked and packaged, at many plants (bakeries) throughout the world. Each product is manufactured at several different bakeries, but the quality of the product is consistent across the bakeries. This is no easy task given the variability in raw materials and operating conditions.

Within each bakery, several products are produced. Some production lines are dedicated to a single product (e.g., Oreo cookies) while other lines are set up for multiple products. Nabisco has built quality into its process based on years of research and innovation.

Project History

By the mid-1980s, the media generated interest in the Artificial Intelligence/Expert Systems (AI/ES) technologies reached the packaged food industry. In particular, the Texas Instruments' development of a diagnostic system for cookers at Campbell's Soup demonstrated that the technology could be successfully applied to the packaged food industry. Based on an initial presentation of AI/ES potential (early 1986), Nabisco corporate R&D decided to contract with us to give advice and provide knowledge engineering expertise.

Because Nabisco had not developed an expert system before, the development plan was constructed with several "go/no go" milestones so that results could be measured, unforeseen problems resolved, and proper decisions made. Major milestones included final problem selection, prototype development, field evaluation, production system implementation, and final deployment.

General AI Strategy

The normal software engineering approach is to use the AI/ES technology if considered appropriate as a result of the analysis of a task. However, in this case, the team intended to select a task based on its potential to demonstrate the value (usefulness, user acceptance, etc.) and requirements (staff resources, calendar time, retraining of employees, etc.) of the AI/ES technology. The customary AI/ES characteristics sought in the chosen task included:

1. The task required experience/expertise for solution and the expert was available to the knowledge engineering team.

2. A junior person could complete the task via a telephone conversation with an experienced person in a reasonable amount of time.

3. The task primarily involved classificatory diagnosis, a problem solving strategy well-known to the AI community.

4. Problem solving in the task did not appear to be amenable to conventional software approaches.

5. The problem solving knowledge was not changing rapidly.

Not only were these AI/ES criteria used, but company specific criteria were added:

1. The task must be important to the company and demonstrate clear value for using AI/ES.

2. Even if the system were not to prove useful on the production line, it should at least be useful as part of a training program for line operators. This would account for the possibility that unforeseen operational/integration problems would preclude its use rather than the AI/ES technology itself.

3. Development of the prototype should not have an impact on production.

4. Specific hardware/software choices should be dictated by the needs of the prototype and its success, not the ultimately deployed system.

5. Technology transfer to Nabisco must occur so that Nabisco could maintain/enhance this system and develop others.

Prototype Selection

The resulting Ritz Line Expert System met all the criteria. The Ritz cracker line had been in production for decades (i.e., the knowledge was not changing rapidly) and a recently retired baker was available to provide the expertise. Because of the importance of Ritz, a successful system would generate interest in AI/ES throughout the company. In fact, because of the media attention to the technology, and Nabisco's commitment to its products, this project drew the attention of senior corporate officials. This was unexpected, but project success would ensure confidence in the technology at the highest corporate levels.

The Ritz Line Expert System is a quality control application that offered potential for a real impact on costs. Nabisco insists on a maintaining a high quality product, and if the quality deteriorates slightly, a large amount of product can be lost. If operators manning the line are inexperienced, solving the problem can take a lot of time and/or increase the amount of product lost. Thus, a system was conceived which could help the supervisors and operators who run the production line maintain the quality of the product.

Thus, the goal of a quality control system is not to increase quality, but to maintain the same high quality more efficiently. These efficiencies can occur through reducing lost product, increasing up-time of equipment, allowing operation of the line by more junior staff (especially important during night shifts), reducing training costs, increasing the percentage of time that top quality product is produced, and facilitating the consistency of quality across bakeries.

Prototype Development

Prototype development of the Ritz Line Expert System began early in 1987 using the "generic tasks approach" (although briefly described below, more detailed information can be found in [1] and [2]) on a Xerox 1186 lisp machine running the LOOPS/Interlisp object oriented programming environment. At that time, lisp machines were still the choice for development of expert systems. Most other tools were weak in terms of flexibility, which could jeopardize a project of this nature. In addition, we had performed several projects where the prototyping was done using CSRL (one of the generic tasks described more fully in [3]), and the production system was written in a rule-based language (we often viewed this as a "working" design followed by implementation). We found using the structured approach of CSRL in addition to the power of an object oriented system to be expeditious, given the capabilities of commercially available rule based systems.

The prototype was completed in approximately 6 calendar months using several interview/develop/review cycles for acquiring the knowledge. Because the problem solving behavior of the expert closely matched the CSRL structure, the expert was soon able to view (graphically) the representation of the knowledge and contribute to refinements. After extensive reviews of the prototype, the hardware/software system was installed in the bakery for further evaluation and testing.

The prototype system took advantage of the extensive graphics capabilities of the Lisp machine, including active images. Operation of the prototype was a straightforward "point-and-click" interaction with the graphical interface. If something started to go wrong (e.g., the crackers are a bit misshapen), the operator interacted with the system which provided advice on how and where to correct the problem. Sometimes the system could not precisely pinpoint the problem with the data provided, so it would recommend a course of action and ask the operator to check back after the action was taken. Based on the results, the system further analyzed the situation and made new recommendations if required.

After the operators gained some experience with the system, and after the system was slightly modified to reflect current conditions, a testing plan was developed to validate the system. The testing involved collecting data about actual events on the Ritz line and comparing these (independently) to the expert system's performance. The system's recommendations proved to be useful and operator response to interacting with the system was extremely positive (in fact, they did not want to go back to operating without the expert system after the test).

Production

As a result, Nabisco decided to move the system to a production environment. Although excellent for prototyping (at the time), the lisp machine was not appropriate for the production environment. In any event, it was clear that the problem solving for this application could be completely modeled with a backward chaining inference engine. Because the CSRL structure could be represented explicitly in the rules, a knowledge base of a few thousand rules could easily be supported during the long term maintenance of the system. After careful review of the extent of the knowledge and a review of environmental concerns (bakeries are hot and humid), Level 5 from Information Builders running on a personal computer was selected for the production hardware/software platform.

The project staff from Nabisco assumed responsibility to build a PC-based version of the Ritz Line expert system. Because the CSRL approach embodies sound knowledge engineering principles, the conversion to a Level5 system was straightforward. The PC-based system was enhanced by Nabisco into a full production system with more complete knowledge of the Ritz line.

The user interface to the system was also redesigned based on input from the line operators and conditions on the line. Connecting the expert system directly the baking equipment was not necessary because sensory information (e.g., shape, color, etc.) was easily obtained from the operators via menu selection, and executing actions (e.g., change the temperature) on the equipment was straightforward. Thus, the system was placed in a specially constructed box on wheels so that it could be rolled with the operator and be protected from environmental hazards (e.g., flour).

Reports from Nabisco indicate that the operator/user response was overwhelmingly positive and that the project was a success worth emulating in other bakeries and other product lines. It is worth noting that the full involvement of line supervisors and operators during the test, enhancement, and user training phases contributed to the project's success.

The Software Development Approach

The Ritz Line expert system was developed using the Generic Tasks approach to the development of expert systems pioneered by the Ohio State University Laboratory for AI Research ([1], [2]). This approach is based on the premise that experts not only possess vast knowledge but know how to use and organize that knowledge to efficiently solve a problem.

The generic tasks approach embodies the notion of a group of specialists cooperatively solving a problem. Each specialist is a mini-expert-system that may comprise a rule based system. Controlling the problem solving process and the decision of who communicates with whom and when (in terms of the object oriented environment) is the primary purpose of defining generic tasks. This control is determined primarily by a pre-specified organization of the specialists such as a hierarchical structure where the top specialists access lower level specialists. Finally, each specialist has its own local control mechanism with enough knowledge to independently determine the next specialist to be accessed.

From a software engineering perspective, this approach has several distinct advantages. First, the definition of specialists puts related pieces of knowledge in one place, thus limiting the search of the knowledge base. Second, the structured architecture of these systems facilitates program maintenance and provides a concise way of storing knowledge and inference procedures used by more than one specialist. Third, uncertainty can be handled with much greater sophistication and with a variety of uncertainty methods. Fourth, a variety of knowledge representation schemes can be used to accommodate handling unforseen failures, thus reducing the "brittleness" so often ascribed to expert systems. This brittleness is really an artifact of the pure rule based approach to capturing and using knowledge.

In particular, we used the CSRL generic task which is designed for classificatory diagnostic systems. CSRL is an approach/methodology to developing such systems, but for a time, a lisp-based shell also embodied the fundamental notions. In the CSRL approach, specialists are defined which interact in an "establish-refine" paradigm. Each specialist (which appears as a node in a hierarchy) is provided knowledge for determining its degree of relevance to the current hypothesis in the diagnosis (i.e., can establish itself). Each specialist is also provided with knowledge about whether or not and how its sub-specialists should further "refine" the hypotheses and conclusions (this is the "control" of the problem solving). One can view each specialist as set a of rules which are specific to the specialist's being able substantiate its relevance to the problem at hand and to potential diagnostic conclusions. Figure 31-1 shows a portion of this hierarchy of specialists from the prototype system.

This paradigm is similar to establishing subgoals in a backward chaining environment, however, the selection of which subgoals to pursue is explicit and knowledge based. In a pure rule-based system, the same rule structure must be used for subgoal selection and the different types of knowledge decisions ("established" and "refined" call for different types of knowledge with different levels of acceptable uncertainty).

Thus, the goals/subgoals for the backward chaining inference engine were defined based on the CSRL specialists and their hierarchy. A portion of the mapping of specialists in the hierarchy to the goal/subgoal structure in Level 5 is shown in Figure 31-2.

The knowledge within each specialist was translated fairly directly into a set of rules in the Level 5 syntax that determine the degree of relevance of the corresponding goal. For example, the specialist Gauge Rolls determines the likelihood of the problem involving the dough rolling mechanisms on the line. Based on its position in the hierarchy, it will assume that the problem has to do with cracker "thickness" and will use the data and conclusions reached by the specialists Thickness and DoughWeight.

Let us suppose the operator initiates the system to find a remedy for thin crackers. The expert system will first determine if the wet

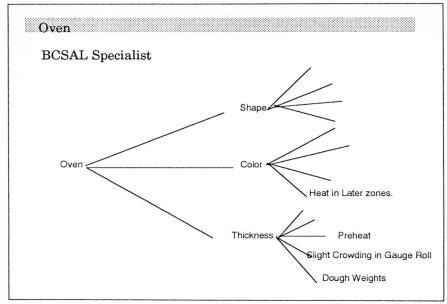

Figure 31-1   Hierarchy of Specialists

dough weight is within tolerance. Next it will check to see if "thin crackers" are due to dough crowding in the gauge roll. Figure 31-3 shows a CSRL-rule within the specialist GaugeRolls which rates the GaugeRolls hypothesis on a scale of -3 to 3. This rule in turn uses other CSRL-rules which check if there are faint dark streaks on the cracker and if there are dark spots on the cracker. Based on the presence or absence of these findings, the GaugeRolls hypothesis is evaluated. Figure 31-4 shows the equivalent set of Level 5 rules. The rating scale used in Level5 is from 0 to 100.

Thus, the final system was a structured, rule based system with backward chaining as the principle controlling mechanism. The beauty of the CSRL approach was that knowledge was represented at the right level of abstraction for accomplishing the problem solving task. As such, not many specialists were required to represent the majority of the problem space. Although the number of rules for each specialist varied depending on the complexity of the specialist's task, the localization of effect made it a manageable number. The final system, as present on the factory floor, had approximately 1300 Level5 type rules.

1. There is a problem
   1.1 Oven
       1.1.1 Thickness
               1.1.1.1 DoughWeight
               1.1.1.2 GaugeRolls
               1.1.1.3 Preheat
               ....
               ....
       1.1.2 Color
               1.1.2.1 Heat in Later Zones
               ....
               ....
       ....

Figure 31-2   Mapping Hierarchy to Rules

```
[ GaugeRolls
   (kgs summary Table
        (match ThinCrackers FaintCrackers DarkSpots
        with
                (if  (GE 2 ) (GE 2) (GE 20) then 3 elseif
                     (GE 2) (GE 2)  ?  then 2 elseif
                     (GE 2)  ?  (GE2) then 2 elseif
                     (GE  2)  ?   ?  then 0 else
                                                    -3
        ))) ]
```

Figure 31-3   CSRL Specialist

RULE	To see if there is crowding in gauge roll
IF	the crackers ARE thin
AND	there are faint dark streaks on the crackers
AND	there are dark spots on the crackers
THEN	Gauge Roll CF 100
AND	DISPLAY gauge roll conclusion
RULE	To see if there is crowding gauge roll
IF	the crackers ARE thin
AND	there are faint dark streaks on the crackers
OR	there are dark spots on the crackers
THEN	Gauge Roll CF 70
AND	DISPLAY gauge roll conclusion
RULE	To see if there is crowding in the gauge roll
IF	the crackers ARE thin
THEN	Gauge Roll CF 30
AND	DISPLAY gauge roll conclusion

Figure 31-4   Level5 Rules

Testing and Evaluation Protocol

In expert systems, several types of evaluation and testing must be performed. The typical evaluation goals are:

- Usability: will the system be used by and acceptable to its intended users?

- Consistency: are there any contradictions in the system?

- Completeness: does the system handle all the problems in the domain?

- Soundness: does the system make only correct conclusions?

- Precision: is the strength of conclusions by the system justified by its data and knowledge?

Although achieving these goals is inherently difficult for expert systems, measuring the success of a developed system against these goals is necessary.

The formal system validation was performed on the prototype in early 1988. A typical performance testing process was used consisting of presenting several test cases or problem scenarios to both the expert system (together with its intended user) and to human experts. For this project, the test cases consisted of situations that arose during the few weeks prior to the installation of the prototype hardware/software platform. The results were then analyzed to provide a measure of performance.

Outcome and Conclusions

The users have reacted very well to using the system. Generally, the operators have almost no familiarity with computers. With the training introduced via the arcade game PacMan, operators have no fear of the system. Because the interface to the expert system requires just a few keys, the users quickly learn to solve the problems that occur with the help of the system. One operator said that it is like having 40 years' experience standing by your side on the line. Because of this experience, other Nabisco bakeries are asking that similar systems be developed for their lines.

From the Nabisco perspective, the expert system was clearly successful. Now in production, they are finding benefits that were not contemplated before the system was developed (e.g., more consistent trouble-shooting of problems). Furthermore, the project confirmed the potential of AI/ES approaches and Nabisco decided to develop their own AI/ES capability.

From the project perspective, the expert system was extremely successful. Although the total process took over three years, the original plan from problem selection, to prototyping, to evaluation, to production was executed without problems (actually, the KKR buyout of RJR Nabisco hurried the use of Nabisco staff for the production system in our place). Our Generally Accepted AI Principles (GAAIP) as described in the approach section played a major role in ensuring the success.

References

[1]   Chandrasekaran, B., "Towards a Taxonomy of Problem-Solving Types", AI Magazine, 4(1):9-17, Winter/Spring 1983.

[2]   Chandrasekaran, B., "Generic Tasks in Knowledge-Based Reasoning: Characterizing and Designing Expert Systems at the 'Right' Level of Abstraction", Proceedings of the Second Conference on Artificial Intelligence Applications, IEEE, Miami Beach, Florida, December 1985.

[3]   Bylander, T.C., Mittal, S., "CSRL: A Language for Classificatory Problem Solving and Uncertainty Handling", AI Magazine, (3):66-77, Summer 1986.

Author Biographical Data

James R. Brink, Ph.D. is Chief Knowledge Engineer for Expert Systems Knowledge Engineers, Inc. (E-KE) and Research Leader for Battelle Memorial Institute. Dr. Brink has more than 14 years of computer science experience (12 years at Battelle Memorial Institute), including more than 10 years in the area of Artificial Intelligence/Expert Systems (AI/ES). His primary areas of expertise include expert systems, object oriented database systems, relational database management systems (R-DBMS), intelligent user interfaces, telecommunications and neural networks. Having established the AI/ES technology thrust at Battelle, he often assists clients in identifying strategic uses of AI/ES and R-DBMS, acquiring appropriate hardware and software facilities, and implementing and integrating solutions.

He has led and developed a variety of expert systems, including systems which interpret faults in welds, diagnose problems with rotating machinery, maintain the quality of a manufactured item on the factory floor, diagnose hardware and operator faults in a real-time computer network, evaluate potential retail sites and facilitate access to existing databases. These systems have been developed on behalf of clients from the military, manufacturing, retail, and banking.

Sriram Mahalingam, Ph.D. Candidate, is a Senior Knowledge Engineer for Expert Systems Knowledge Engineers, Inc. (E-KE) and Principal Research Scientist for Battelle Memorial Institute. Mr. Mahalingam has designed and developed expert systems since 1985 for a variety of applications. He has a background in artificial intelligence, computer science, systems analysis, and electrical engineering. His work on expert systems is concentrated on classification and diagnostic tasks. Mr. Mahalingam has extensive programming skills including Lisp, C, and Pascal. He is very familiar with a variety of knowledge engineering tools in multi-platform environments using several operating systems. As a senior knowledge engineer he is primarily concerned with conception of projects, designing innovative solutions to technical problems and delivering working systems. He routinely interacts with the scientific community and keeps abreast of the state of art of AI technology.

Mr. Mahalingam has undertaken expert systems development on behalf of a variety of industrial and government clients worldwide. Highlights include a production expert system which diagnoses product problems on the production line and a study used by over 30 clients worldwide to help determine their direction for taking advantage of this technology. In addition, he has authored a number of related articles.

Dr. Brink and Mr. Mahalingam can be reached through Battelle or E-KE, Inc. which undertakes missions worldwide in affiliation with Battelle Memorial Institute in Columbus, Ohio. These firms together perform a wide range of advanced computer technology services for clients including advice, education, software development, and integration of new technologies in existing computer environments.

32

Software Architecture to Support Total-Quality Companies
Using a Knowledge-Based Process Modeler

Richard Tabor Greene
Manager, Knowledge Based Systems Circles Program
Manager, Knowledge Based Applications to Total Quality
Knowledge Based Systems Competency Center
Xerox Corporation
Fairport, New York

Introduction

This chapter describes the complete transformation of all aspects of a corporation created by commitment to total quality, and how ordinary software systems and MIS departments are left supporting an older image of doing business, not consistent with new total quality assumptions. A new software architecture capable of supporting a new total quality image of doing business is presented, based on a fundamental knowledge/data structure that models business processes. Automatic software support for Total Quality Control, Quality Function Deployment, Policy Deployment, Organizational Taguchi Analysis, Kansei Engineering, Customer Driven Companies and other leading quality techniques is achieved, not as isolated "quality" functions but as the guts of the way all ordinary functions of the business are conducted. Group process as well as individual process support is achieved. A multi-dimensional nature to that fundamental

467

knowledge/data structure is described, so that what the organizational sciences, the cognitive sciences, the humanities, the social sciences, the quality sciences and others know about organizational processes can be brought to bear automatically. Cross-domain concept hierarchy techniques as a new knowledge representation approach to achieve this multi-dimensional modeling are presented.

The Problem:

Most Fortune 500 companies and most of their millions of supplier companies have committed to total quality programs. The majority of them have had such programs more than five years. Indeed, major companies are beginning to win prizes for quality—Florida Power and Light just won the Deming Prize in Japan, and Xerox, Motorola, and others have won the Malcolm Baldrige Prize recently.

The people doing total quality in those companies and the people doing high technology are different and culturally separate. The result is the quality people are introducing dozens of new software applications, each from a different vendor and incompatible with the others, that they propose tens of thousands of employees use regularly if not daily. The computer people in each corporation, simultaneously, are working on corporate-wide architecture and CASE programs that have only incidental relation, if any, to total quality images of the business.

Indeed, at the heart of the world's key quality methods and best quality companies is a principle of direct challenge to computing systems—the principle of data minimization—only data that is actually justified by impact on customer wants and actionable should be left in a corporation after major quality systems of Quality Function Deployment and Policy Deployment are implemented. Such quality systems should be implemented by the replacement principle—only if older data and behavior systems that each part of these quality methods replace are spotted and eliminated in favor of quality versions of them, do these quality methods succeed and benefit the corporation. To tolerate plural, quality and non- or pre-quality tracking and planning systems, is to allow surfeits of data. This allows the de-coupling of managers from the corporate agenda in favor of individual agendas, justified by different subsets of the flood of excess data in the firm. This is the autonomy instead of corporate action by subsetting floods of data principle. Finally, there is a quality principle of designing systems to manipulate and exchange commitment among people not concepts. Most computer systems are exclusively imagined, produced, sold, and used for concept exchange, resulting in great ideas poorly implemented or not implemented at all. I summarize below:

The Data Minimization Principle:

Only data justified by impact on particular customer wants and actionable is to remain in corporations after major quality approaches are implemented.

The Replacement Principle:

Quality approaches can be successfully implemented only if what each of their parts replaces in the firm is spotted and replaced.

The Corporate Action Principle:

Corporate improvement through action is possible only when individual autonomy is reduced by eliminating excess data that can be subsetted differently by different managers, justifying de-coupling manager agenda from the corporate action agenda.

The Commitment-Not-Concept Exchange Principle:

Design systems for supporting exchange of commitment and building relationships among people, not for concept manipulation and exchange alone.

All four of these principles challenge the size, complexity, role, and future import of computer systems and the people and departments whose careers are based on emphasizing computer system importance. The unpleasant fact of life is, in spite of the many ways quality can improve computer systems, and the many ways computer systems can improve quality implementation, there is some direct antagonism at the core of the two principles—quality and computation.

This has not become evident in companies thus far because by and large their size allows the quality people and programs to bypass without contact the computer people and programs and vice versa.

This lack of open conflict, however, prolongs and increases two problems. First is the deployment to tens of thousands of employees quality software tools without a corporate integrated architecture and supplier for them. The second problem is the deployment of major computer systems envisioned in pre-total-quality days.

A Way of Finding A Solution:

As quality approaches are implemented in companies, it becomes clear that a few major things are missing. These need to be there if even the most basic quality tools are to be actually used by whole workforces and benefit the company.

Artificial intelligence appliers find as they go around corporations defining possible AI applications that many of the systems they specify constitute what the quality people discover is the missing prerequisite.

The missing prerequisite is: a model of the organizational process used by each part of the business. The missing prerequisite is not as simple as it seems. Modeling business processes is a fractal phenomenon—it has to be done on all size scales. Major quality principles here come into play as well:

The New Unit of Competition Principle:

In the total quality view of doing business it is not corporations that compete with corporations but Cross-Units that compete with Cross-Units. Cross-Units are cross functional combinations of: cross-industry groups, cross-competitor platform development institutions, cross-company Silicon Valleys around major corporations, pioneer customer groups, customer organization groups, and other components. It is these units that must implement all major quality approaches to achieve competitiveness. It is these units that must implement integrated computer systems.

The Fractal Social Process Principle:

Total quality sees a vertical cascade of business processes across successively bigger and smaller size scales from the Cross-Unit to the workgroup/workweek, and a horizontal cascade of business processes across the components of each Cross-Unit. Social fractality involves a copying dynamic between vertical layers and a copying dynamic between horizontal components.

Another dimension of finding a software architecture to support total quality involves acknowledging the implications of systems for manipulation and exchange of commitment, not concepts alone. Much of the actual work of quality approaches is groupwork. Most software systems are based on supporting individuals persons working with individual applications on individual workstations that are incidentally network connected. So there is a tendency to assimilate software support for total quality approaches to old database images of the role of computation, as the skills and software tools for developing software systems for supporting group consideration processes are weak or lacking in corporations.

An important distinction here is the one between Computer Forms of Cooperative Work and Computer Augmentations of Cooperative Work. The new field of Computer Supported Cooperative Work (groupware) divides into these two sub-domains. Total quality companies have a principle that pertains here as well:

The Locus of Work is In the People not In the Systems Principle:

Successful implementation of any total quality approach or technique requires that from beginning of use of the approach to final mastery of it, the primary work done by the approach be one of educating human perceptions and commitments in the workforce. It is not marking symbols in systems to support the workforce. Such symbol marking support as is used should be minimized to not draw the work out of the minds of the workforce but rather to optimize the operation of the approach or technique in the minds of the workforce.

A proviso to this need to augment human cooperative work with systems rather than switch to systems as automatic performers of work (use Computer Augmentations of Cooperative Work, not Computer Forms of Cooperative Work) is this—humans in the West refuse, over time, to do work in meetings. Over time, in all quality approaches tried thus far, what starts out as meetings wherein work gets done in a group setting, becomes in a few weeks, coordination and status reporting events, with the work done off-line out of sight by individuals. In part this is because managers do not work in the West—they coordinate instead and lose technical capability to get involved in details and technical aspects of things. Managers, thus uncomfortable with their own ability in a group setting to be seen as equal contributors to the technical guts of problem solving, reduce meetings, subtly, imperceptibly, to status reports and coordination events. This raises the following principle for total quality organizations:

The Evolution of Meetings from Places-to-do-Work to Places-to-report-Status Principle:

Every quality approach and technique must from the beginning include specific tactics for fostering, measuring, and incenting work conducted in group settings and tactics for preventing, measuring, and dis-incenting transfer of such work to individual settings and the etiolating of meetings themselves.

Eight principles have been introduced thus far in this chapter. They rule out most of the software systems currently on the market to support quality and they rule out most of the very little work in ordinary MIS shops for support of quality. They constitute a kind of definition by inverse of what adequate software support for a total quality corporation is. The section that follows presents a more constuctivist vision of adequate support for a total quality corporation.

The Scientific Requisites of Finding a Solution—A Model of All Quality Approaches:

The most elemental beginning point possible for thinking about software support for total quality is a really good model of what total quality is. Long gone are the days when quality was "free", quality was an "attitude", quality was "easy". In their place are present days in which quality is pluriform. There are dozens of approaches and methods involved in quality worldwide. What is more there are, in most corporations, half a dozen or more approaches to total quality being tried at once. In addition, each approach "takes" differently to different parts of the organization. That leads, in turn, to each part of the organization having a different quality background, history, and capability. That in turn leads to present quality approaches "taking" differently in the different parts of the corporation. This is a very complex domain for implementation and this variety alone is enough to reduce most Western corporation quality programs to pale imitations of their Japanese counterparts.

I summarize this variety below:

1. twenty different approaches to total quality worldwide to choose from

2. half a dozen different approaches being tried in any one corporation

3. each approach takes better or worse in different parts of the corporation

4. each part of the corporation has a different quality history

5. each part of the corporation has a different need and readiness for any one new quality approach being tried

6. each new approach takes better or worse in each part of the corporation.

Add to this a plurality of gurus, consultants, vendors, and internal factions fighting to control quality and you have a situation guaranteed to achieve lots of new launches and programs with none of them implemented well and no good "finishes" of programs in the corporation to match the flashy career-building "launches".

This brings us to another principle of total quality implementation:

The Compensation-for-National-Neuroses Principle:

No quality approach can be implemented well or successfully without particular tactics throughout its implementation to counter-act the neurotic tendencies of the national culture it is being implemented in.

For the United States these neuroses include:

1. a bias toward launching things over finishing things

2. a bias for flashy visible doing of shallow things over quiet unobtrusive doing of deep things

3. a bias for great concepts as a product over great loyalty and commitment as a product and dozens of others presented in my book, being published in 1991, titled: A Psychotherapy for Curing America's Productivity Problem.

The Total Quality Paradigm of Doing Business—A Matrix of All Quality Approaches:

I conducted a phone survey of over forty quality managers in Japan, Europe, and the US, asking them a number of questions. A subset of these questions resulted in those surveyed naming all the major approaches to achieve total quality that they knew of and naming principles shared by all of these approaches. There is not space in this chapter to present the entire model, but the two dimensions of the matrix that defines the new Total Quality Paradigm of Doing Business can be presented briefly below.

I do not use the word Paradigm lightly. I have read carefully the history of science research that largely refutes Thomas Kuhn's theory of paradigm shifts in science. Therefore, by "paradigm shift" I mean not a period of science-as-usual reaching a crisis that, when resolved, ushers in a new "paradigm" (Kuhn's model). Rather, I mean a second model of what doing business to the model that all the rest of the world shares.

I believe that Western corporations by copying particular approaches, of the twenty listed below, have installed uncompetitive forms of total quality to the forms present in Japan, in spite of much progress. The corrective is implementation, in each of the twenty approaches, of the twenty-nine paradigm points listed below.

All Approaches to Total Quality in Function and Rough Chronological Sequence:

- Debuffering Approaches:
    1. Just-in-Time
    2. Statistical Process Control
    3. Quality Circles
    4. Total Quality Control

- Scientific Styling Approaches:
    5. Total Preventative Maintenance
    6. Taguchi Analysis
    7. New & Old Tools
    8. High Technology Circles

- Whole Workforce Deployment Approaches:
    9. Automation Deployment
    10. New Technology Deployment
    11. Quality Function Deployment
    12. Policy Deployment

- Organizational Transparency Approaches:
    13. Customer Aided Design
    14. Customer Managed Corporation
    15. Kansei Engineering
    16. Middle-Up-Down Management

- Cognitive Competitiveness Approaches:
    17. Cognitive Quality of Worklife
    18. Meta-Cognitive Organization
    19. Social Democratic Quality
    20. Democratic Scientific Management.

The Principles Shared by All Quality Approaches

Ordered From Materially to Spiritually Demanding

- Compensations for Human Frailty:
  1. Compensate for National Neuroses
  2. Work Through Mediate Structures
  3. Implement All Things Through a Replacement Strategy
  4. Achieve Organizational Learning Capacity from Incidents

- The New Bases of Competing:
  5. The Scientific Method as Basic Workgroup Process
  6. Whole Workforce Deployment of Business functions
  7. Sincere Relations to Things Within & Beyond the Lifespan
  8. The New Unit of Competing—the Cross-Unit

- The Links:
  9. Customer and Competitor Data Drive All Decisions
  10. Invisible Management of Self-Managing Workforces
  11. Competitive High-quality Copying System—Fractal Sociality
  12. Unifying Saying with Doing

- The Doers:
  13. The Whole Organization as Doer
  14. The Whole Person as Doer
  15. The Customer as Doer
  16. The Meta-Dimension to Doing

- The Platforms:
  17. The Performance Platform and Victory
  18. The Technology Platform as Surpassing Catch up
  19. The Service Platform and Customer Enchantment
  20. The Cognition Platform and Data Drivenness

- The Values:
  21. Monastic Management
  22. Issueless Management
  23. De-professionalization of Managers and Staffs
  24. Competitive Mundanities

- Competitive Cognition:
  25. Percept Versus Concept Companies
  26. Competitive Processes Versus Competitive Products
  27. Invisible Asset Accounting
  28. Business as Translation
  29. Affect Based Processes.

This matrix of twenty column titles—the quality approaches—and twenty nine row titles—the paradigm principles shared by each of the twenty approaches—specifies a total of 580 quality capabilities that must be supported in total quality companies. It is this entire quality architecture, not any one approach being pursued, that must be supported by a company information systems architecture. Clever AI applications supporting fragments of this matrix will, over the long term, serve more and more as hindrances to total quality competitiveness. The items of this matrix constitute the Japanese Contribution to a quality architecture for companies.

The converse is not true—one does not serve a company or oneself by defining some grandiose giant software application that, built in the next year or two, supports all the 580 quality capabilities. Rather, the 580 capabilities serves as the specification of the customer requirements of an overall architecture only particular parts of which get implemented as corporate need dictates. Nevertheless, this is quite different than the current efforts to, without having a vision of the whole quality arena, build clever software to support particular quality approaches like Quality Function Deployment.

The Solution:

Imai, in his book Kaizen, demonstrated that no approach to achieving total quality could be successfully achieved without a theory of organizational change. The early quality gurus—Deming, Juran, Crosby, Conway—had a tendency, much reduced in recent years, to talk as if organizations were either not there or essentially simple. The result was shallow training of supervisors in SPC that never filtered down to line workers, and total resistance by professional workers to all quality training and methods in the name of "we don't need to be told how to do our work". In turn, the result of that was sub-Japanese-benchmark performing of quality approaches and methods, accepted happily since it was a vast improvement on previous performance even if not enough to insure industrial survival over the long term.

Those companies who bothered to develop organizational change perspective and wisdom to guide quality implementation did well—Motorola and Xerox among them. The Organizational Development community itself went through stages:

1. ignoring total quality as parvenu—till last year

2. denigrating it as rehash of earlier OD insights presented abominably by ex-statisticians

3. admiration and capitulation before the quality movement's ability to achieve on a wide scale several of the OD community's profoundest values—participation, decentralization, self-management

4. serious development of OD models of the key quality constructs.

One major result of applying the greater body of organization change knowledge to the implementation of particular quality approaches was the need to:

The Organizational Development Contribution:

1. detach organizations from total organization around aristocratic detachment of professional staffs for leap innovation by individuals

2. attach organizations to organizing around democratic engagement of whole workforces in incremental innovation

3. enable organizations to develop a new democratic group form of leap innovation.

To achieve these three, it was necessary to combine Swedish organization development approaches with Japanese total quality approaches.

The Swedish and European Contribution:

1. socio-technical systems theory

2. autonomous workteam theory

3. action research networks

4. cognitive health measures

5. learning workplace theory

6. organizational cognition enhancement.

The final player came not from the organization development community but from technology. I call it the Cognitive Economy Theory:

The American Contribution:

1. the cognitive sciences

2. artificial intelligence technologies

3. groupware technology

4. software agent and tool technology

5. self tutoring and self managing software systems technology

6. cross-domain concept hierarchy—knowledge representation schemes

7. mail-able capabilities—agents and tools—technology

8. virtual realities

9. holographic life and work spaces

10. virtual sharing of rooms, houses, tasks, times.

So, unfortunately for those who wish to build some simple software system to support some simple quality technique such as Quality Function Deployment:

1. 580 quality capabilities need supporting

2. great variety of readiness and susceptibility for quality needs to be handled

3. three forms of innovation need supporting

4. six Swedish and European approaches to achieving any organizational change need supporting

5. ten American forms of technology extending how systems can augment and automate social functions among people need inclusion.

Take into account that over 20 Cognitive Technologies (see Implementing Japanese AI Techniques, McGraw-Hill, 1990, by the author for a detailed list) and over 50 Artificial Intelligence Technologies exist (see the author's forthcoming book Knowledge Modeling—How to Create Expert People, Organizations, and Systems), and the reader can readily appreciate that the software that is to support total quality is very complex in itself and the quality it is to support is very complex too.

Making a Start on This Daunting Task:

It is clear, by now, that the long term customer requirement for software support of total quality ways of doing business is challenging in the extreme. This is mitigated by short term customer requirements to help companies with the quality approach of the moment. Too much compromise with this short term goal produces individual database-like software that short circuits group consideration and education processes that are the heart of the productivity effects of modern quality approaches. Such systems not only do not support quality well, they prevent other work from achieving good support.

Yet facing the long term customer requirement full face makes one shrink before the complexity and magnitude of the task. The saving grace is this: the model of the processes by which all work gets done in a business is the core data/knowledge structure for all the quality approaches that currently exist.

Using this as a starting point, the beginnings of an architecture to support all major quality approaches can be sketched out as follows.

First, most workgroups currently in any business have no model of how they do their work. In fact, the same task is done very differently by different people and differently by any one person at different times or occasions. This, in itself, is not as bad from a quality perspective as the lack of models for these various processes. Only then does the fact that each individual lacking a model of his/her own process, fails to make and remember improvements in his/her own processes and fails to learn from process improvements by others. It is the missing of this organizational learning that quality approaches point out.

Second, in workgroups, which it is discovered that current processes in the business are based on personal opinion, incidental habit, unscientific and erroneous conclusions from subjective impression data. It is the supplanting of this basis of process design by truly scientific statistically valid data collecting, hypothesis formation, and experimental design (starting as rudimentary control charting and extending to multi-variate Taguchi orthogonal array designs) that total quality approaches bring about.

Third, a weave architecture to both social and technical systems arises from total quality approaches, with the Voice of the Customer driving the company (preferably the Cross-Unit) cross-functionally toward short-term customer wants. The Voice of the President drives the company cross-layer (top down, bottom up) toward long-term customer wants. Both of these join at every function (component) and layer to specify how current business processes need to be tweaked, sloughed, or maintained to respond to customers.

Fourth, the primary actor supported by all quality approaches is the cross-functional, cross-unit, and cross-layer team, doing work in group settings, not reporting status in group settings and doing work individually off-line.

Fifth, professional employees performing knowledge intensive, highly integrative and synthetic, "creative" work see both the definition of existing processes and the possible imposition of common processes (including even the principle of learning from the way others do similar or related work) as direct impediments to the quality of their work life. Quality practitioners, generated from manufacturing shopfloor backgrounds, indeed, have either been daunted by the seeming formlessness and infinite variety to whatever processes such professional staff use, or have actually rigidly in quite inappropriate ways tried to impose forms of process definition suited to machine tool Statistical Process Control onto stunted and inappropriate models of professional processes. In general, it takes a brilliant technical professional to model another technical professional's processes or enable that professional to model his/her own processes. So it is not that creative professionals cannot model their own processes or that they cannot joyfully and beneficially, in their own terms, borrow from and learn from processes of their confreres and consoeurs. Rather, it is that the processes they must model have to be modeled as sophisticatedly as the professionals themselves conduct their work, not an easy undertaking but an entirely feasible one, given effort and talent for it.

Until the advent of expert systems, from artificial intelligence, there were no techniques for modeling highly knowledge intensive processes. AI has, as a side-effect, created a body of people in industry, capable of working with professionals to model their processes in ways that reflect the true sophistication of those processes. That does not mean that modeling their processes is something the professionals will welcome, find non-threatening, find worth their time, or even tolerate. But attitude aside, it is feasible, worth doing, and essential to creating total quality companies with robust organizational learning properties and flows of creative new standards of performance discovered and propagated across all areas of activity of the company (Cross-Unit).

Sixth, both the impact of particular quality methods on the customers of the business and the degree of conformance of those doing the methods to robust performance criteria of those methods should be measured. If impact is measured without conformance, then second rate initial try-outs of complex methods are found, usually by highly educated technical professionals or career-dominated managers, to not work, justifying dropping interest in quality in general.

Seventh, measures of the quality with which particular approaches to quality are implemented get developed by total quality companies, so that the successive implementations of successive different quality approaches are measured, compared, and managed into a rate of gradual improvement. This measurement process makes visible sources of misunderstanding, mistreatment, or misdirection of quality methods.

Eighth, several of the major quality approaches, in particular Quality Function Deployment and Policy Deployment, require that actions in the company, all of them, be tracked and accounted for in terms of their impact on particular short-term customer wants(QFD) and particular long-term customer wants(PD). If the replacement principle, introduced earlier in this paper is followed, they supplant other forms of tracking activities in the Cross-Unit (company).

There are another 50 practical implications of the principles already introduced in this paper that were used to specify the Software Architecture for Support of Total Quality Ways of Doing Business shown below. Space does not permit presenting them here. The reader can, however, extrapolate from what has already been presented, most of them.

An Initial Software Architecture for Support of Total Quality Ways of Doing Business:

Support for Group Process Input to the Modeling Tool	Support for Individual Input to the Tool	Support for Automatic Sample InputAcross a Whole Workforce or Its Parts

Horizontal Quality Support Tracking Process Changes Needed to Meet Short Term Customer Wants(QFD)		Vertical Quality Support: Tracking Process Changes Needed to Meet Long-Term Customer Wants(PD)

Tool Allowing All Workgroups
to Participatorily Model Their
Own Business Processes

The Automatic Analysis of Process Support Level:

Human Empowerment Analysis Engine	Task Analysis Analysis Engine	Knowledge Flow Analysis Engine	Other Automatic Analysis Engines

The Statistical Analysis and Display Support Level:

Statistical Process Control	Process Capability Real-time Display	Within Process Experimental Design Advisor	Between Process Experimental Design Advisor	Process Component Capability Developent Experimenal Design Advisor	Process Quality of Implementng Quality Approaches Display

The Cognitive Tools Support Level:

Group Use of Seven Old Tools Support	Individual Use of Seven Old Tools Support	Group Use of Seven New Tools Support	Individual Use of Seven New Tools Support

Cross-Unit Support Level:

Customer Want Qualitative Understanding Engine	Customer Want Change Tracking Engine	Competitor Attainment Qualitative Understanding Engine	Competitor Attainment Change Tracking
Determinants of Customer Wants Modeler	Industrial Combinatorics Product Invention Tool	Determinants of Competitor Attainment Modeler	
Pioneer Customer Leadership of Regular Customer Support Tool	Customer Organization Leadership of Regular Customers Support Tool	Customer Use Detecting Products Tool	

Quality Implementation Tracking Tool:

Support Tool of Group Process to Mesh Quality Approaches with Non-Quality Goings on in the Company (Cross-Unit)

Support Tool of Group Processes to Mesh Quality Approaches with Each Other

Implementation Issue and Schedule Predictor Tool for Parts of the Business Having Different Quality Histories and Readinesses.

The Work Already Done on Such an Architecture:

One of the reader's biggest questions about this whole enterprise is likely to be doubt that anything but obvious task analysis can be supported by software. The reason for this doubt is the richness of characterization of human elements in business processes tends to drown people in excess data, factors, variables. Any attempt to model enterprises or organizational processes that includes human factors and motivations seriously enough to correctly calculate their effects on things, is likely to require inputting everything known about organizations. There is plenty of solid evidence that everything known is insufficient to calculate even the most basic sequences of actions in real organizations. This skepticism-grounded-in-excess-data-required is a problem when grandiose functioning of organization modelers is attempted. The Japanese are famous for making systems succeed by a million tiny bites, using approaches mutually theoretically incompatible but practically useful from an engineering perspective. That is the approach we are using. The result is the following principle of system construction:

The Minimal Input Principle:

Devise the minimal inputs in terms of data/knowledge characterizing a process required to support some very particular and limited kind of automated analysis of the process.

If this principle is adhered to, then hundreds of robust principles of analyzing processes, latent in the Japanese Contribution, the Organizational Development Contribution, the Swedish and European Contribution, and the American Contribution mentioned above can be supported by software and conducted automatically.

For example, though the details of the system are proprietary at this point in time, we have achieved several kinds of powerful automatic design of processes around human empowerment principles using a data structure called "human competencies". We have also achieved automatic overlay of this analysis on top of a task analysis using a task-variation data structure so that we can spot where excess competency in relation to tasks and excess task in relation to competency match process steps where variations are caused. We can also automatically map particular customer wants that are likely (and often actually found) to be over-ridden by strong competencies in people doing work tasks. Work competency often is a source of skipping or overlooking customer wants. Work is currently proceeding on automatic mapping of customer want data from QFD onto processes throughout an organization and automatic long term customer want data in the form of PD's Voice of the President onto those same processes. This is being done so that any internal action can be tracked in terms of which short and long term customer wants it impacts. This constitutes a powerful new type of accounting system with potential to, over time, outweigh the harm done by usual cost accounting systems to productivity (as put forth by Robert Kaplan at Harvard, Bill Moscone at Coopers & Lybrand, and others).

The Knowledge Architecture Needed to Support These Achieved Functions:

To characterize organizational processes in ways rich enough to calculate over human competencies, human motivations, informal human relations built at work, dysfunctional personality effects on processes, and the like, requires a richness of concepts far beyond relational databases and object oriented databases. The issue is one of a whole ontology of concepts, usable across many very different domains.

Many people in industry have already concluded that current expert systems and AI knowledge modeling tools are woefully inadequate. We are actually in the midst of thousands of expert systems being built and deployed, each of which has a "resource" concept with the basic ideas of demand and supply for the resource, possible loss of supply or monopoly supply, possible substitute resources for any one resource, and the like. Though this basic concept occurs in many expert systems today, it gets represented in different ways in each one, and put into different code in each computer. The result is ten thousand mutually incompatible and redundant ways to represent the same basic concept.

These two motivations combine to form one requirement—a cross-domain concept hierarchy must be implemented so that concepts are entered into it only once, and then can be re-used across wildly different domains.

The Software Architecture for Support of Total Quality Ways of Doing Business presented above, requires cross-domain capability. Hence, I am building a system called SURU that achieves it.

Entered Concepts At the Point of Entry are Characterized as Follows:

1. Definitional Characteristics of the Concept

2. Typical Characteristics of the Concept

3. Prototype Example of the Concept

4. Episodes Typical for the Concept with roles having Shankian Cases

5. Is-cause-of links entered for the concept

6. Is-like-a links automatically calculated upon entry

7. Is-subtype-of parent abstraction nodes automatically calculated upon entry

8. Additional is-like-a links entered upon entry of the concept beyond those automatically calculated

9. Is-like-a episodes automatically calculated by the system upon concept entry

10. Is-like-a links between episodes beyond those automatically calculated entered by the user upon concept entry.

The result of these is an average of four hours to enter any one concept into the system. The system is written in C++ on a PC and is being ported to Objective C on the Next computer at present.

Our test of the re-use of the same concepts in different domains is a Taguchi module that uses concepts for automatic software design of experiments; a Quality Function Deployment module that uses concepts for determining what business processes need to be modified when customer wants for particular aspects of products change; and a human empowerment module that maps changes in allocations of human competencies or changes in development of them implied by QFD changes in customer wants, or Taguchi determinations of factors determining organizational process outcome (this is an extension of Taguchi principles to organizational behavior domains). Due to the complexity of this knowledge representation scheme, the solver engines that operate on it can be quite simple, typically written in CLIPS within a day or less.

Author Biographical Data

After graduating from MIT in 1971 with a joint degree in Artificial Intelligence and Organization Design, Richard Greene spent six years doing participatory management interventions in businesses and government agencies in both the US and Japan. That included designing the workshop procedures for the Royal Bank of Canada's Community Branch experiment, setting up participatory town meetings in 42 Japanese towns, cities, businesses, and social clubs, and developing a metaphorical thinking curriculum for the Weston Public Schools. He followed up venture businesses in Japan, setting up an AI group at Matsushita Electric, and participating as consultant and circle member in Sekisui Chemical Co.'s Deming Award winning quality campaign in 1979.

He returned to the U.S. to get a Master's in Japanese Management and a Master's in Organizational Learning at the University of Michigan. He is now a candidate for a Ph. D. in Knowledge Modeling of Business Processes there. While studying he conducted the consults that launched Procter and Gamble on its recent highly publicized reorganization, by showing P & G what it looked like from the point of view of its best Japanese competitors. he designed workshops for GM's Saturn project, based on Swedish autonomous workteams and educative workplace theory.

After studying at the University of Michigan he created unique systems integration forms of expert systems while creating an AI applications group at EDS. His experiences there are published as *Implementing Japanese AI Techniques* by McGraw-Hill, 1990. He then consulted for Coopers & Lybrand briefly, getting proposals accepted by Thompson, Johnson & Johnson, and N.V. Phillips. He then set up the AI Circles program at Xerox Corporation, knowledge engineered Genichi Taguchi himself for a Taguchi quality design advisor, assisted Dick Leo in setup of Quality Function Deployment at Xerox, helped found the Organization Systems Design Group at Xerox, and developed a new knowledge modeling technique being published as *Knowledge Modeling* in 1991.

John Wlley and Sons is publishing his *Knowledge Quality* book in 1991, two publishers are currently bidding for publishing his *52 Ways to Improve Any Business Process in Minutes*. He is currently designing a new Cross-Domain Concept Hierarchy knowledge system and applying it to model group consideration and workshop processes that are parts of major quality techniques.

Pricing, Packaging, and Customizing

The Quotation Assistant at PMI Food Equipment Group

J. Roger Kearney
PMI Food Equipment Corp
Troy, Ohio

Introduction

If you have ever visited a commercial kitchen, you probably saw a product of PMI Food Equipment Group. The slicer you see cutting roast beef at the deli counter was also made by PMI Food Equipment Group. We manufacture a broad line of commercial food processing equipment. Our products are sold under the brand names Hobart, Vulcan-Hart, Wolf, and Stero.

They are used to prepare food (slice it, grind it, chop it, mix it) and to cook food (grill it, fry it, broil it, bake it). We also make commercial refrigerators, dishwashers, food disposers, scales, and shrink-wrap equipment.

Our customers choose models, features, and accessories from a wide range of options. Our dealers and representatives configure the machines taking into account which features are required and which are mutually exclusive. The Quotation Assistant system helps our representatives configure and price our products.

The Quotation Assistant

PMI Food Equipment Group is a division of 2.6-billion dollar Premark International, Inc. which also includes Tupperware, West Bend, and Wilsonart. Food Equipment Group products are manufactured in multiple locations around the country. Computer systems are developed centrally at PMI Food Equipment Group headquarters in Troy, Ohio. Computer system development is mostly done in COBOL using IBM's IMS and DB2 database managers on an IBM 3090 mainframe computer.

In late-1987 we bought the PC version of the Aion Development System (ADS) from Aion Corporation, Palo Alto, California. We had no specific application in mind. We wanted to learn about expert systems to determine whether and how the technology might help our application development effort. Later, we purchased the IBM-mainframe version of ADS and now operate the system on both platforms.

In late-1988, as an experiment we built a small prototype system that helped configure one of our dishwasher products. The system asked the user a series of questions about the desired machine, considering the various rules about which feature to combine with which other feature, etc. We learned the rules and prices from the product catalog. After configuring the dishwasher the system printed a description of the dishwasher, listing all features and prices. When the vice president of MIS saw the prototype, he immediately called the vice president of sales. When he saw the prototype, the vice president of sales wanted to give a copy of the system to the Quotation Department as soon as possible. At that time, we in the MIS department did not even know there was a Quotation Department.

We learned that the Quotation Department consisted of three quotation specialists supported by a clerk-typist. A quotation is necessary because of the complexity of our products and our prices. A request for a quotation (RFQ) may come from a formal document, or from a simple hand- written request or from a telephone call. An RFQ may contain specifications for one machine or one hundred machines. Many quotation requests come from our dealers who bid on specific projects such as equipping a restaurant in a new hotel. PMI Foodservice Representatives work with dealers in many ways including quotation support. The Foodservice Rep prepares simple quotations in the field using a catalog, a calculator, and either a pen or a typewriter. Complex quotation requests went to the Quotation Department in Troy for processing. The Quotation Department mailed or faxed the quotation to the dealer.

Mistakes are extremely easy to make while preparing a quotation. Required features are sometimes omitted; two mutually exclusive features are included; the wrong price is used; or the user makes a typographic error or a mistake in arithmetic, etc., etc. Extensive proofreading caught most errors. Errors on quotations are both embarrassing and expensive. A mistake may anger a customer and result in a lost order.

In early 1989, we decided to develop a production expert system, based upon our prototype, to help the Quotation Department. The system would operate on an existing IBM PC/AT shared by the three quotation specialists.

Since there was no staff available to develop expert systems, two of us who believed in the approach decided to develop the system in our spare time. After several months of very slow progress, we decided that it was imperative that we have a group dedicated to expert systems if we were ever going to make significant progress. With management's blessing, in May 1989, we formed an expert systems group with three persons working full time on the quotation system. In July 1989, we delivered a working version of the system to the Quotation Department for testing. Within a few weeks, before we had finished correcting minor mistakes, the quotation specialists began using the system to generate real quotations. The three members of the Quotation Department enthusiastically accepted the system but had many suggestions for extensions and enhancements. Over the next few months we accommodated as many of the ideas (including some new ideas of our own) as possible. As word of the system spread to field personnel, we began to get requests for additional copies of the system. By the end of 1989, we decided to conduct a trial of the system in the field.

Several years earlier, an attempt was made to automate the sales force by providing laptop PCs equipped with a spreadsheet program, a word processor, and a calendar system. We stopped the project after a very short trial. The computers, with the software provided, were hardly used at all. The marketing representatives did not see the computers as relevant to their job. Based upon this trial some people believed that marketing reps did not need or want computers.

For the new field trial we selected four marketing representatives, one each from Los Angeles, St. Louis, Newark and Birmingham. The four reps chosen were excellent salesmen but had very little experience with computers. They came to Troy for a one-day training session in mid- January 1990. A small printer and a Compaq SLT/286 laptop computer with three megabytes of memory, 40 megabytes of hard disk space and DOS 4.01 installed were given to each person. Each computer had a menu and batch files installed to make the start-up process as painless as possible. We pre-loaded the Quotation Assistant system and ACT! software on the Compaq SLTs. ACT!, is contact software from Contact Software International, Inc., in Carrollton, Texas. ACT! maintains customer profiles, a history of contacts, and a to-do list. It includes calendaring support and a simple word processing program. At the end of the day the users were very enthusiastic about the project. They were anxious to get back to their territories so they could use their new systems.

One month later the four marketing representatives came to Troy to evaluate the project. They were unanimous in their excitement and enthusiasm about the project. They said that they were able to get quotations to their dealers faster than they could before. The quotations were professional- looking. The marketing reps were certain there were fewer errors because of the new system. All agreed that the program should be expanded to more marketing representatives. Executive management decided to provide the system to an additional sixteen marketing representatives. We also installed the Quotation Assistant on an additional fifteen desk top computers. By July 1990 the system was in use on 35 different PCs. Over all, the Quotation Assistant and ACT! have been extremely well- received by users. Even though most users had little or no experience with PCs, only two of our users had severe problems with the systems. At the users' request the computers were given to other marketing representatives. In both cases the new users had no problem learning to use the systems.

We are now adapting the Quotation Assistant system to work on the mainframe computer using DB2 databases. The system will check at run time to see what environment it is operating in. On the mainframe the system will use DB2 databases but will continue to use flat files on the PC. Customer service representatives who already have mainframe terminals and who need quotation information will use the system. We are now preparing to conduct a written survey of all system users to evaluate the entire project. Based upon the results of the survey, other marketing representatives may be trained and equipped with computers.

Quotation Assistant consists of six separate knowledge bases linked together using Aion's chain command. One of the knowledge bases is the backbone of the system. It is the entry point of the system and presents a menu of function choices to the user including the following:

- Begin a quotation
- Add an item to a quotation
- Display a quotation
- Print a quotation
- Put a note at the end of a quotation
- Copy an item on the same quotation
- Copy an item from a different quotation
- Go to utility menu
- End session

We tried to keep the system as simple as possible. We did not stress that it was an "expert system" or that it used artificial intelligence.

The goal was to have Quotation Assistant be just a very friendly computer system.

When starting a new quotation the user selects "Begin a quotation" from the main menu. The system responds by asking for heading information; to whom is the quotation addressed? What is the address? If the equipment on the quotation is for a special project, what is the name of the project? What quotation number is to appear on the quotation? The above information is all requested on one screen to which the system posts the current date. The heading information is stored in a disk file for later use.

After the heading is complete, the user is asked to describe the first piece of equipment to appear on the quotation. First the user selects one of five major product groups from a menu. The system then loads the appropriate knowledge base for the chosen product line. The system asks if the user knows the catalog number for the product. The catalog number may partially or completely define the product. If the catalog number is unknown or does not completely describe the equipment, the system asks a series of questions to complete the description of the product. The user selects the voltage, stainless steel or painted finish, whether the machine feeds from the right or the left, etc. The expert system, of course, contains the rules which specify which options are required and which are mutually exclusive. The system uses the rules to guide the user through the configuration of the piece of equipment. When the product has been completely defined and any accessories selected, the system adds prices and catalog numbers for individual features and calculates a total for the product. The user reviews the list of features and prices and accepts it, changes it, or deletes it and starts over. Quotation Assistant stores the configuration in the quotation file and then asks if another product is to be included on the quotation. If so, then the appropriate product knowledge base is loaded into memory (if the one needed is different from the one already in memory). The user is then guided through the description of the next product. The process is repeated until all pieces of equipment have been defined and priced.

The user can interrupt the process at any time. When the user is ready to continue, he selects "Add an item to a quotation" from the main menu and specifies the quotation number. If he cannot remember the quotation number, he requests a screen which lists all quotations on file with a short description and the date each was started.

The quotations are stored on the computer's hard disk. The user can add to, change, display, or print a quotation at any time. The user has the ability to place notes or comments after an item or at the very end of a quotation. The user creates a note for one specific use or stores the note in a library and pulls it into a quotation when needed. Sometimes items appear more than once on a quotation. The first item may be the same as the seventh and eighth items with only the item numbers differing. Quotation Assistant makes it easy to copy one item to another. It is also easy to copy an item from one quotation to another.

"Go to utility menu" is another choice from the main menu. The utility menu contains choices of lesser used functions. A user can change a quotation heading or can sort a quotation by item number. A user can delete an item from a quotation or can erase the entire quotation. Quotations can also be moved to or retrieved from a floppy disk.

The Quotation System consists of six knowledge bases with a total of 2,000 rules. We compiled the knowledge bases using Aion's high performance option to optimize response times. We did not use the ADS object oriented programming capabilities which became available during the project. If we were starting the project now, we would use object oriented programming techniques. Our Expert Systems group will continue to maintain Quotation Assistant as we introduce new products or change features or prices. Eventually, we will change the system so that users will update prices without the involvement of a knowledge engineer.

Conclusion

The Quotation Assistant project is very interesting and rewarding. We have used new technology to address areas of the business that were previously ignored. Our users have responded with more appreciation and enthusiasm than I have ever seen on any other information system project. Recently, Premark International, Inc., our parent company, presented us "The Most Innovative New System" award for our work on the Quotation Assistant system. The award is given annually to one of the Premark companies.

Author Biographical Data

Roger Kearney is Director of Expert Systems for PMI Food Equipment Group, a division of Premark International, Inc. His responsibilities include the development and maintenance of expert systems. He is also responsible for PMI's Information Center which provides training and other assistance for personal computer users and for mainframe end-user computing tools. Roger has been with PMI for sixteen years. He was formerly a manufacturing industry specialist for IBM. Roger is a Certified Fellow in Production and Inventory Management (CFPIM) of the American Production and Inventory Control Society (APICS). He is a member of the Aion Midwest User Group (AMUG). He has served as a speaker at APICS,

AMUG and Data Processing Management Association conferences. Roger is a graduate of Wittenberg University with a degree in secondary education. He resides with his wife and two sons in Troy, Ohio.

Product Lot Selection Advisor

Steven W. Oxman
The OXKO Corporation
Annapolis, Maryland

Introduction

For certain manufacturing companies, e.g., chemical producers, the products they make are monitored by quality control procedures which come under various titles such as Total Quality Management (TQM) and Statistical Process Control (SPC). Through these programs, manufacturing concerns attempt to make products that are consistent from the standpoint of the output characteristics.

Products that result from these programs do tend to be more consistent, but some variability is still inevitable. The products that are produced under these quality programs tend to have more quality data recorded about them. This data can be used beneficially in the selection of product lots.

Manufacturing companies are often suppliers to other manufacturing companies. Client manufacturing companies are often requiring that quality test information be provided to them at the time product lots are shipped to them. They compare the quality data of the shipped lots against the quality parameters that they have requested from the supplier.

If the quality attributes do not meet the client's quality requirements, these clients can return materials to the supplier unless there is an absolute need to continue manufacturing and there are no other raw materials to replace these out of specification raw materials.

For this reason, it is important that the supplying company understands the quality attribute requirements of its clients and ships raw material products accordingly.

One supplying manufacturing company that OXKO supports decided to develop an expert system to assist in the decision making of what product lots are to be shipped to its clients from its warehouses.

The remainder of this paper will present information about this expert system application known as the Product Lot Selection Advisor.

General Development Methodology

The methodology utilized to develop this expert system included rapid prototyping and multiple iterations development.

Rapid prototyping meant that from the time we gathered information from our domain expert, we put together a prototype of what we learned and got the prototype back to our domain expert in less than 30 days time. When possible, we turned around prototypes in two weeks.

Iterative development meant that for each iteration of interviewing the domain expert, we developed the next version of the expert system application. It included the previous version plus the information and knowledge gleaned from the most recent interview. It would be the version that we would next ship to the expert for comment and critique.

Each iterative development entailed first interviewing the expert and acquiring data, information, and knowledge beyond which we had learned thus far. Collecting examples of previous work, collecting forms and tools utilized by the expert and people the expert is trying to assist, and collecting personal notes utilized by the expert was performed for each iteration.

After acquiring this material, we went back and utilized it to see where the past iterations needed to be modified and changed, and then we made additions that were appropriate in accordance with the new information provided.

After the development and prior to testing, we documented our iterative development as far as we could. Any work we had done to delete previous logic was documented. Logic that we had modified because of information received since the last iteration, we documented as such.

The rest of the documentation had to do with additions. Additions documentation was usually of less volume than changes. The additions were related to only one interview and, until the point that we found that there was an error or problem with it, there was no other information about that particular instance.

For modifications and changes, we referred to previous notes, to the present information provided by the expert, to what the system was doing, to what the system was doing wrong, and to what the system should be doing. Therefore, the changes and modifications documentation tended to be more than the additions documentation.

At any time that we provided an iterative version of the advisor to users, we also needed to develop (or modify) a User Operating Manual in order to allow the users to readily install and utilize the prototype.

At the time that changes and additions were made, we did desk checking to see that the code was professionally developed, that it was easy to understand, and that there were no obvious errors in the knowledge base.

Following the desk check test scenario, we then ran the advisor at that version. We saw how it reacted to a set of responses that were provided by the expert, and determined if the response provided by the system was accurate from the view of the expert who provided the information (as to what the outcome should be in response to specific inputs provided).

Once the testing was completed, the next step that we did was to package the software, and documentation that might be appropriate to provide to the users at this point. We then mailed the software to them on a diskette, ready to install and run.

The diskette included the runtime version of the particular iteration as well as the source code. We also provided a cover letter. This cover letter or transmittal letter, amongst other things, always included two important pieces of information.

The first important piece of information in the transmittal letter was what this version did that the last version did not do, and what changes have been made as per information we had been provided from the expert. This information was the basis of the version description information that we maintain on systems that we have developed.

The second important piece of information was how to get the version loaded and started. Additional information on operating was provided either through the operating manual that we provide or during one-on-one instruction.

By providing the next version to the expert and possibly some of the potential users, they had a chance to privately and quietly look at the system and form their first impressions of it. After the expert and the users had time to look at this version, we then planned our next interview. We always attempted to have a full understanding and set of goals on what the next interview would be. We communicated this to the people who are going to be in the interview prior to the conducting the interview.

We found that interview planning is important because it assists the interviewees with knowing what subject area is going to be covered. It provides them with the ability to schedule or plan events that might have to occur in order to succeed in meeting the goals of the interview session.

With the interview planned, the interviewees have enough time to gather material they might need and have time to look at the present version of the system. We scheduled our interviews to occur as soon as it was possible for all parties to get together.

At the time the interview occurred, we were back to acquiring knowledge and going through our next development iteration.

This development methodology was iteratively done until such time as the expert and users declared that the system was complete with respect to knowledge from the expert's standpoint and was understandable and usable from the users' standpoints. At such time, we finalized our documentation, finalized the software, and made a complete delivery of the software, both source code and object code, as well as the documentation.

The documentation included two items. One was the operating instructions. The other was a copy of the developer's notebook, which included all of the historical information of the development of the system and any documentation that the programmer deemed necessary in order to develop and later maintain the system.

From the standpoint of how we generally develop expert systems, this development would come under what I term a classic expert system development methodology.

A point to be made is that for some of the systems we are developing today, we are using variations of this development methodology that include intermediate knowledge representations that are not directly coded into the expert system development tool but rather are documented in manual forms that include decision trees, decision tables, and flow charts, and then are coded.

This system did not have an intermediate knowledge representation performed on it. Therefore, that kind of documentation does not exist in the programmer's development notebook.

Whether having this intermediate documentation is a benefit is not something that I can definitely ascertain at this time. However, from experience, it does seem to help. Intuitively, it seems that this intermediate knowledge representation work is useful. We will probably be doing it more in the future. We find it seems to help not only in development but also in continued development (i.e., maintenance).

How Many Iterations?

When developing an expert system application using the iterative development technique, one question that naturally comes to mind is "how many iterations is it going to take to develop the expert system application?"

It should be intuitively obvious that this question has certain dependent attributes including how large the application is, how much of the application is going to be covered by the expert system, and how much work is going to be accomplished in each iteration.

We had estimated for this particular job, given the size that we thought the job was and the amount of work we would do per iteration, we would have six major iterations of development and possibly one or two minor refinement iterations prior to final delivery.

What came as a surprise to us is that we exceeded our six iterations by a large number. The number of iterations that we performed was 20. Looking back on the project, there were approximately six iterations done to accomplish the original project goals. Then there were four more iterations to develop the system with modified goals as per the domain expert's desire. Then there were ten more iterations that evolved through the changes of direction that this system would take from the expert, from potential users, and from the expert's management hierarchy. What finally was developed was a system that had capabilities that most likely would not be used and capabilities that had never been considered when the project started but most likely would get use.

What happened during the development of the system was that the client organization learned a lot more about what it was that they wanted to capture, how they wanted to do the product lot selection, and exactly whose heuristics (i.e., rules of thumb) were important.

Because of the change in direction during the project and, specifically, the increased scope in the system's capability, the original estimate was way off what the actual system costs became. Because the system is strategically important to the client and because the additions were client driven, they agreed to extending the project duration and cost.

Where the project was estimated to be a six month project, it actually had an elapsed time of one and a half years.

If additional systems would be developed for this client in the future, they would most likely take a lot less time to develop for two reasons. The first is that there are now some features in the advisor that will never be used given features added later. Therefore, these features would not have to be included in the next system. Additionally, the client now has a better idea of exactly what they want and where they are going. They also have one system that can be used as a template from which other systems can follow. Future systems would most likely take no longer than the original estimate of six months, yet would perform all the functions that would be necessary to keep the client user community happy.

A lesson that is learned from this is that it is generally difficult to provide a firm fixed price offer to an expert system development project. Either the firm fixed price would have to be offered in light of a specific set of functions whereby, during the development, other functions that would be identified and wanted would not be supported unless the client agreed to increase the firm fixed price offered. Alternatively, the firm fixed price would be null and void early on in the project as new functions became known and which the client wanted added to the system.

Most likely an important activity for a developer to perform would be to document what is being promised in the original version and then to document everything that is developed in order to assess if the system has gone beyond the original promised capability and to be able to provide the client with an early indication of the possibility of the project exceeding the original promised cost.

Tools Used During Development

The hardware tools used to develop this expert system included IBM personal computer compatible machines running PC-DOS 3.3. The machines included 640KB of memory, a hard disk, a floppy disk, a VGA color graphics controller and monitor, a PC keyboard, and a printer port to an IBM personal computer compatible printer.

The software tools utilized to develop this system included the LEVEL5 Expert System Development Tool, the PC PaintBrush Plus graphics development tool, and production rule development and database toolbox kits that assisted with making the system look more professional and provided extended database support.

The target machine was an IBM PS/2 model 70 with a hard disk, a floppy disk, a printer port, a PC compatible printer, a VGA graphics controller and display, and a card that directly connected the personal computer to an IBM mainframe of the 360/370 architecture type and, later, a card that connected the same personal computer to an AS400.

When connecting to a host machine , there was also the addition of software that allowed us to download mainframe or AS400 database data into the personal computer workstation.

Vital Statistics

The Product Lot Selection Advisor includes 33 knowledge bases. Each of these knowledge bases are programmed in the Production Rule Language (PRL) of the LEVEL5 Expert System development Tool. Additionally, there are 8 modules that are written in Turbo Pascal to provide extra functionality for the user system interface and database support.

This application also includes 8 databases that are of the dBase III+ database file format.

The Advisor is made up of 1,507 rules which comprises 30,749 lines of Production Rule Language source code.

The Turbo Pascal routines comprise approximately 5,000 lines of source code.

All of the development was done by one knowledge engineer who has additional capabilities in the area of database management systems and Turbo Pascal programming.

The project took one and a half calendar years of time to develop, of which time the first four months were full time with an average of 20 eight hour days spent per month. The rest of the development time had varying amounts of effort from month to month. There were a few months where there was little to no effort accomplished while waiting for feedback from the domain expert and potential users.

The total number of resources from the standpoint of manpower utilized included the one knowledge engineer, another knowledge engineer who performed auxiliary testing of the developing system, the domain expert, the domain expert's management who expended some time to see how the system was developing and providing guidance for how it best would fit the organizational requirements, and two potential users who gave some time to test the system and provide their inputs as to their belief of its usability and usefulness in supporting them in doing their work.

Basic Application Operation

This application basically ran off of multiple databases that were stored on corporate computing assets (i.e., the mainframe or the AS400). One of the databases provided inventory information for products available in the warehouses. This database included the product attribute information including the parameters necessary to select products for particular clients.

Another database available and used was the sales database that provided information on orders that are in need of being filled. This information was keyed by order number and provided such information as customer name and customer order information.

Another database used was on the personal computer workstation and not loaded on the corporate computing assets included information as to the requirements of each client. This information included product quality requirements as well as shipping instructions.

The expert system application is fired off from the personal computer by the user. The application asks the user for a particular order number that the user wants to fill. After replying with the order number, the expert system would look at the sales database that provided the order information and see if that order was a valid order that was yet to be filled. If there was something wrong with the number, the system would provide information back to the user. If the order number was correct, the expert system would go ahead and retrieve the information on the particular order including the customer. From the customer information, the system would then retrieve customer order information from the personal computer based database on customer requirements.

The system would then provide the user with the ability to select product for the client in various ways. One way was an automatic technique where the expert system would utilize its heuristics on selection and the databases in order to automatically fill the order and provide all the invoice, shipping, and inventory information.

Alternatively, the user had available the ability to manually select which lots will go to the client from more than one manual technique, each one providing the user with more and more capabilities to manually select the lots. Obviously the more manual the user got with the system, the less the user was utilizing the expertise that the expert system had available.

In any case, the system provides the information for an order and an updated list of what would then be available in the warehouses given that the order was filled. The system would then provide updated inventory information back to the corporate database.

Conclusion

The system has been completed, delivered to the client organization, and has been installed for use. It is now in what would be considered initial production acceptance testing prior to going on line as a standard application at the client location.

The client is hopeful that the application will support the goals of the project. At this time, all indications seem to support the contention that the system is able to provide the support for which the client organization was hoping.

If there is any potential problem with the system at this time, it might have to do with user acceptance given that it is introducing a different way of doing an important business activity. However, given the capability and understanding of the users, it is our belief that the system will be accepted and will go into productive use.

Author Biographical Data

Steven Oxman is President of OXKO Corporation which was founded in 1985. OXKO is a leading developer of expert systems in various industries including chemical, manufacturing, insurance and government. OXKO is a leader in expert system training and a supplier of third party production rule based knowledge engineering tools.

35

Packaging Advisor™: An Expert System for Rigid-Plastic Food Package Design

Alvin S. Topolski and Douglas K. Reece
E.I. duPont de Nemours & Company (Inc.)
Wilmington, Delaware

Introduction

In 1987, the DuPont Company entered the market for barrier resins, which are used in the fabrication of plastic food containers. The company was experiencing difficulty establishing a position against incumbent competitors. A new and technically superior product was about to be introduced, and a way was needed to induce customers to invest in qualifying the new material. The solution was Packaging Advisor™, an expert system for rigid food container design. Deployed in February of 1988, Packaging Advisor automates the design process, providing our customer, the package designer, with information on alternative materials, the quantities of these materials needed to meet performance specifications, and estimates of material costs.

Reprinted with permission by the American Association for Artificial Intelligence. Published in "Innovative Applications of Artificial Intelligence" edited by Herbert Schorr and Alan Rappaport., pp. 348-357. 1989. Second Annual Conference on Innovative Applications of Artificial Intelligence.©

Packaging Advisor was used in place of traditional marketing communications techniques to inform our customers and field sales staff about our new and existing products. Customer response, expressed in both words and orders, has been enthusiastic.

Packaging Advisor is an expert system which designs rigid plastic food containers. It was developed at the DuPont company to help both our customers and our own staff better understand the use and benefits of our barrier resin products in food containers. The barrier resins business was a new business for DuPont, and we had new and unique products to introduce to the marketplace. Packaging Advisor was the keystone of the marketing communications strategy for these new products, and has been recognized by management for a substantial contribution to the success of the new business venture. In this paper we will describe the business and technical environment in which the system was constructed, review the system and its development process, and describe how the system was used successfully to achieve our business objectives.

The Business Environment

Rigid plastic food containers offer many advantages to the consumer. They are light in weight, dentproof, rustproof, shatter resistant, and can be molded with such convenience features as handles and pouring spouts. These advantages have lead to extremely rapid growth of plastic containers as replacements for metal and glass on supermarket shelves.

The size and growth rate of the plastic food packaging market was therefore an attractive market for the DuPont company. Among our offerings to this market are several barrier resin products, introduced in 1987 and 1988. These plastic resins reduce the infiltration of oxygen, which can cause degradation or spoilage of package contents.

The products introduced in 1987 met with limited success for a number of reasons. We were competing with several established suppliers with similar products. the process which a package fabricator uses to qualify a new supplier is arduous: substantial design work must be followed by trial runs with the new material. Furthermore, as discussed below, the properties of barrier resins are complex, and many details of the application must be analyzed to predict performance.

DuPont's business plan called for the 1988 introduction of new and technically superior materials which should be very competitive. However, means were needed to establish DuPont as a credible and technically sophisticated supplier of barrier products and to communicate the value of the new products.

To address these needs the business team developed a concept for a computer system which would automate the process of designing a food package. The system would offer alternative designs using Du-Pont and competitive resins and showing corresponding costs. The system would show clearly the value of the new products, and its ability to automate tedious design calculations should make it appealing to customers.

The DuPont company has an aggressive program to implement artificial intelligence applications. Staff from the AI group agreed to provide the needed systems resources, and work was begun in May, 1987. The completed system was fielded nine months later in February, 1988.

The Food Package Design Problem

Food packages requiring an oxygen barrier are typically manufactured as multilayered structures. The bulk of the package will be composed of a structural material, selected for durability and low cost. One or more barrier layers will be used to achieve the required limits on oxygen infiltration, and layers of adhesive material will be used as necessary to prevent the package from delaminating. The design problem addressed by Packaging Advisor™ is to select appropriate barrier and structural materials for a given application, to determine how much of the (usually more costly) barrier material is required in order to achieve a specified limit on total oxygen permeation during the shelf life of the package, and to calculate the materials cost for the package. Selection of adhesive layers was left to a separate expert system.

The oxygen barrier properties of a material are measured by the rate at which oxygen infiltrates across a unit area and thickness of the material. This oxygen permeation parameter is a function of temperature and, for some materials, humidity. The humidity to which the material is exposed depends upon the humidity inside and outside the package and upon the water vapor transmission properties of other layers within the package. Some structural materials provide a significant amount of resistance to oxygen permeation. Thus, the requirements for the oxygen barrier material depend strongly on environmental factors and upon the other materials with which it is used.

Several other factors must also be considered. Packaging materials must be compatible with the intended fabrication process and must have whatever degree of optical clarity is required for the application. Since the layers of the package will be extruded together, they must all have similar processing temperature ranges. Federal regulations restrict or prohibit the use of some materials in food packages.

Another complication arises when packages are subjected to sterilization processes using steam. the steam will saturate the materials, altering the performance of humidity sensitive materials during the time the package is drying.

The package designer must choose among about 20 different structural resins and a similar number of barrier resins. The number of possible combinations is therefore in the hundreds—too many to analyze manually.

The complexity of the package design problem makes it difficult to adequately describe a new barrier material through printed specifications. However, we found that a personal computer does have enough power to perform the necessary analysis.

The Development Process

Packaging Advisor was placed in full commercial operation nine months after it was begun. A rapid prototyping strategy was used to develop the system; we did not attempt to develop a full functional specification up front. The first prototype was demonstrated to potential users only three months after the start of the project, refinements were made over the next five months, and a final month was required for duplication of diskettes and related materials.

The prototyping approach to system development proved to have a number of advantages. At first, the complexity of the problem seemed a bit overwhelming. In addition to all of the considerations described earlier, we had discussed a number of other factors which could influence package design: type and location of handles, the need to stack some kinds of packages during transit and storage, etc. The decision to start with the simplest plausible prototype helped us to identify and focus on the truly important factors.

The prototypes helped the experts to identify areas in which the system's knowledge was inaccurate or incomplete. The first prototype, for example, recommended combinations of materials which seemed implausible to one of the experts. After reflecting, he realized that the materials had very different processing temperatures: one of the materials would vaporize before the other melted. The fact that we needed to include knowledge regarding processing temperatures became apparent in this way.

The prototyping strategy also made it possible for the process of fielding the system to begin before the development process was complete. Prototype systems were shown to both customers and field sales representatives early in the development process. Several customers offered strong expressions of interest based on the prototypes and enabled the system to earn a positive reputation with the field sales force before it was formally introduced. By the time the system was placed in production, there were customers waiting to license it and sales representatives ready to communicate success stories to their peers. The early positive feedback from customers made it easier to maintain management's support for the project and helped build the developers' morale.

We estimate that a system such as Packaging Advisor would, at current prices, cost approximately $50,000 to develop and field. This estimate includes costs for system development, documentation, and duplication services but does not include a charge for services of domain experts.

Annual maintenance costs should also be considered. Since substantial enhancements to Packaging Advisor are planned, annual maintenance costs may approach development costs. In the case of Packaging Advisor, these costs are more than offset by license fees charged to customers and by expenses for other marketing communications activities which have been displaced by the Packaging Advisor system.

The System

Packaging Advisor is a stand-alone system which runs on the IBM PC and compatibles. It consists of two components. The expert system front end, written in the Level 5 shell from Information Builders, Inc., is responsible for interacting with the user to obtain a specification of the package to be designed. This subsystem will query the user for the dimensions of the package, fabrication process to be used, desired shelf life, maximum allowable oxygen infiltration, desired optical properties, and a number of similar parameters. A typical question is shown in Figure 35-1.

The front end is designed to minimize the number of questions routinely asked. Inferencing is done in order to determine typical values for a number of secondary parameters. The parameters specified by the user and inferred by the system are then presented in the display shown in Figure 35-2. In the example shown the system has inferred that scrap will be recycled and that the scrap rate will be about 50%. The user may make any needed changes at this point in the interaction.

After the user accepts the package specification, the other system component, written in the dBase III programming language and compiled with CLIPPER, is activated. This subsystem is responsible for performing the necessary design calculations, retrievals from resin property databases, and package cost calculations.

The output of the system is shown in Figure 35-3. Alternative designs are ranked with the least costly first, as long as the user's shelf life requirement is satisfied. If no available materials have adequate oxygen barrier properties to meet the user's shelf life requirement, the package with the longest shelf life is shown first.

Packaging Advisor Maximum Use Temperature

We need to know the maximum temperature the package will experience for a sustained period (over 2 minutes) during normal use. This maximum temperature will normally be attained either during package filling/sterilization or during heating of packages in microwave or conventional ovens.

{We will rule out materials which cannot tolerate the indicated maximum temperature. You may enter your temperature requirements directly if you wish}.

Room temperature or lower
Pasteurization (71oC)
Hot fill (85oC)
Retort sterilization and/or microwave oven heating (121oC)
Heating in conventional ovens (230oC)
Enter maximum usage temperature directly

Figure 35-1   Package Parameter Specification

The display in Figure 35-3 has been abbreviated; the actual system will present about 50 designs for the case shown.

Most of the barrier materials in the example are DuPont products, but the system does not treat DuPont materials preferentially. There are cases where competitive materials are more cost effective, and the system will present those materials first. The decisions to include competitive materials and to avoid preferential treatment of our own products reflect both faith in our product line and the desire to maximize the usefulness of the system to the customer. To further enhance the benefit of the system, it was designed to allow the customer to modify the databases to reflect his own resin costs and process economics.

The notes at the bottom of the screen in Figure 35-3 provide additional information which could not be handled easily elsewhere in the system. We found, for example, that FDA regulations were very complex. Some materials cannot be used with certain foods; other may be used with any food but only at certain temperatures, etc. We added this type of information to footnotes rather than ask all of the questions and code all of the rules that would be necessary to cover these cases.

After the user has analyzed a case, he may elect to return to the Package Requirements Summary screen shown in Figure 35-2 and make whatever changes he desires for a "what if" analysis. In this

Packaging Advisor Package Requirements Summary
Max. usage temp: 121 deg. C.   Humidity inside package: 100%
Storage temp: 23 deg. C.   Humidity outside package: 60%
Package area: 54.0 sq.   Retort sterilization: Yes
Package thickness: 30.0 mils
Shelf life: 365 days   Scrap recycled: Yes
Oxygen infiltration: 2.0 cc   Scrap rate: 50.0%
Processing method: Thermoforming
Optical properties required: Opaque, translucent, or clear materials
Location of barrier layer: Centered 7.5 mils from Outside
Must be covered by FDA food contact regulations:
 Structural resin: Yes
 Barrier resin: Yes
Maximum thickness of barrier layer: 8.0 mils
Minimum thickness of barrier layer: 0.4 mils
Do you wish to make any changes in these parameters?

⟶   Accept these parameter values
       Change a parameter value

Figure 35-2   The Package Requirements Summary

way he can gain a better understanding of the interrelationship between design criteria and package materials cost. For example, he
may find that he cost difference between clear and translucent containers is greater than threefold.

How The System Was Used

The system was deployed on laptop computers and used by field sales
personnel, with assistance from headquarters staff, to introduce the
new products to potential customers. The system was also made available for license to customers for their own use. A videotaped
demonstration of the system was prepared so that field sales personnel who did not yet have laptops or were not comfortable using them
could still demonstrate the system.

To our knowledge, Packaging Advisor™ is the first artificial intelligence system designed to be the keystone of the marketing
communications strategy for a new product line. Packaging Advisor
is also a product in its own right — one which offers substantial
benefits to customers. Since designers typically limit their analyses
to a few favored materials, Packaging Advisor will often suggest
lower cost alternatives than they would otherwise have considered.
Often, the designer is lead also to examine process changes involving
variables such as scrap rate, package wall thickness, etc. In fact, a

Packages for Consideration

Structural Resin (Thickness, mils)	Barrier Resin (Thickness, mils)	Need Bynel tie	Mat'l Cost $U.S./M	Shelf Life (Days)	Notes
PP (29.0)	SELAR OH (30%) (1.0)	Y	34.129	365	B6
PP (27.3)	SELAR OH (44%) (2.7)	Y	50.210	365	B7
CPET (29.3)7	SELAR OH (30%) (1.0)	Y	60.536	365	G3 S2 B6
PP talc filled (29.0)	SELAR OH (30%) (1.0)	Y	65.536	365	B6
PP (26.9)	PVDC (3.1)	Y	67.804	365	B5
CPET (28.0)	SELAR OH (44%) (2.7)	Y	70.175	365	G3 S2 B7
PP talc filled (27.3)	SELAR OH (44%) (2.7)	Y	77.804	365	B7
SELAR PT (29.1)	SELAR OH (30%)	Y	79.068	365	G3 S20 B6

NOTES AND CAUTIONS

G2   Structure does not meet your shelf life requirement.
G3   Resin processing temperatures may not be compatible.

Structural Resin Notes

S2   CPET: not suitable for 50% alcoholic beverages; other limitations apply.
S18   Polysulfone: FDA regs specify only frozen/refrigerated storage.
S20   SELAR PT: cannot withstand significant internal pressure at retort temps.

Barrier Resin Notes

B4   NYLON MXD6: FDA restrictions apply; Check regulations
B5   PVDC: Requires special fabrication eqpt; Barrier degrades at high temps.
B6   SELAR OH (30%): FDA regs specify max. 7 mil thickness & 100 Deg. C. Storage.
B7   SELAR OH (44%): FDA regs specify max. 7 mil thickness & 100 Deg. C. Storage.
B10   SELAR PA: FDA limited; not retortable in all cases.
B11   SELAR PT: cannot withstand significant internal pressure at retort temps.

Figure 35-3   System Output

number of customers have used the system to justify investments in process improvements. Clearly, the system is much more broadly useful than printed product literature.

The Results

Goals for the system were threefold: to establish DuPont as a technology leader in the eyes of barrier resin purchasers, to provide a means of demonstrating the value of DuPont products, and to increase resin sales. We judge the system a success by all three criteria. The system received favorable reviews in the trade press, and was well received by customers. The simultaneous introduction of technically advanced materials and Packaging Advisor established our position as leading-edge supplier. Moreover, we are now selling enough resin to justify an expansion in production capacity.

It is, of course, hard to estimate what sales of the new materials would have been without Packaging Advisor. However, management believes that about 30% of resin sales are attributable to accounts with whom we made contact via Packaging Advisor. Without the system, we might never have been able to open the door at these accounts.

The system enhanced the confidence of our sales representatives, enabled them to make more contacts, and improved the quality of their interactions with the customer.

Package designers often become deeply engrossed in their interactions with the system. On a number of occasions we have been them skip lunch or ask that a demonstration be extended so that they could complete their analysis. Few other marketing communications vehicles have been as successful at holding the attention of their target audience.

Conclusions

The Packaging Advisor™ case illustrates how expert systems technology may be used to codify technical knowledge and deliver it to the field to obtain a competitive advantage. The expert system provided the vehicle for transforming our knowledge from a possession to a high-yielding asset.

Author Biographical Data

Alvin S. Topolski and Douglas K. Reece are both members of the prestigious AI team at DuPont. This innovative department is responsible for overseeing the deployment of the over 600 expert systems at DuPont.

36

Neural Nets
for Custom Formulation

William H. VerDuin
General Manager,
AI WARE Incorporated
Cleveland, Ohio

Introduction

Neural nets and expert systems technologies combine effectively to solve custom formulation problems. The two technologies combine to provide rapid and robust solutions to problems that are difficult or expensive to solve by other approaches.

The task of formulation is to select or modify recipes for rubber, plastics, chemicals, or alloys to meet new performance or cost requirements. The challenge is in predicting how proportions of each ingredient will effect properties of the compound.

There is often no model or detailed understanding of these relationships. The product designer must call upon experience and much trial-and-error to find a formula that is sufficient, if not optimal. Product properties may be mutually exclusive: improving one degrades others. The constituents of the compound often interact: the

Reprinted with permission from the Proceedings of the Fourth Annual Export Systems Conference and Exposition, Engineering Society of Detroit, April 3, 1990.

effect on properties of varying the proportion of one component depends upon the specific proportions of other components.

Neural nets combined with expert systems have successfully solved this product design problem in several applications. In each case, the system automatically learned the correlation between product formula and product properties by being shown data on products for which both properties and formulas were known. The Custom Formulation System then assisted design of new products by estimating changes in formula required to accomplish desired changes in properties, and vice-versa.

In this paper, we will describe the implementation approach and user benefits of several applications of the Custom Formulation System.

CUSTOM FORMULATION is a challenging task. It is technically challenging due to the physical and chemical interactions between ingredients. In addition, the ability to successfully and rapidly develop new formulations can significantly impact both producers and users of formulated products. Neural net technology has been successfully applied to this challenging task, meeting a real-world application need and demonstrating the capabilities of this new technology.

The Formulation Task

PLASTICS, METALS, CERAMICS, PHARMACEUTICALS, RUBBER, CHEMICALS, FOODS, AND MANY OTHER PRODUCTS consist of many ingredients mixed in specific proportions with specific procedures. These products are called formulations because they are made according to specific formulas. The task of formulation, then, is to establish or modify a formula to meet product performance and cost requirements and preferences. Cost issues may include both raw material and processing costs.

Manufacturers of formulated products may be required to develop new formulations for a variety of reasons. Customers may request new or improved properties, or reduced cost. Environmental issues may discourage the use of certain ingredients. Processing capabilities or facilities constraints may exist. Operational issues such as a need to reduce raw material inventories or process set-ups may cause manufacturers to rationalize their product line by making a smaller number of different product varieties cover the same range of performance.

The Formulation Challenge

The performance characteristics of formulated products are a result of the physical and chemical interactions between the many ingredients. These interactions are often complex, and the relationships non-linear. Properties may be mutually exclusive, in that improving one degrades the other. Ingredients may counteract one another. The effect on properties of varying one ingredient may depend on the proportions of, for example, nineteen other ingredients. In this hypothetical case, if one knew the relationships and wished to plot these relationships a twenty dimensional space would be required. Of course, this is not feasible. But there are issues beyond the difficulty of representation. Few people can take the time to explore such complex relationships sufficiently to model them.

Without a model, estimation is difficult. In fact, traditional computer based approaches are of little benefit without a model. For this reason, computers have until now not provided assistance in formulation beyond the traditional arithmetic and data manipulation.

Current Formulation Approaches

Formulation is now typically done by some combination of the following methods:

- Rules of thumb

- Closest fit to previous formulations

- Experimental validation of trial and error

Each method has its strengths and failings. Rules of thumb are often useful to estimate required changes in formulation. Expert formulators are often able to correctly determine the direction formulations must change to move properties in a particular direction. However, there may be problems with this approach:

- Accurate estimation is difficult, since a multi-dimensional model is typically not available either explicitly or as an understanding of the nature of the multi-dimensional interactions.

- At least one expert is required. None may be available, due to retirements or transfers.

■   Multiple experts may not agree. In this case, the challenge is to ensure that the best expertise is used to provide consistently good results.

New formulations may be based on previous formulations. One might find that a previous product is a reasonably close fit to a new requirement. In this case, valuable information may already exist in company files and employee memories. However, this approach also has potential problems:

■   Memories or data may be fuzzy or non-existent. Complete, well organized, and error-free historical data is usually a goal but rarely a reality.

■   This approach imposes old performance capabilities on new performance requirements.

Experimental validation of trial and error is guaranteed to work, eventually. In fact, any new design should be validated experimentally before major commitments are made. The issue is rather one of degree: how many trials and errors are required to find a new formula. One must balance the need for products exactly meeting specifications against the need for timely answers and reasonable development expense.

The question arises: can computers assist the formulation task by providing design insight, rather than merely data retrieval and manipulation?

A Better Way: Neural Nets

Complex formulation problems in the rubber, plastics, and coatings industries have been solved with neural net technology. This technology is equally useful in other formulation applications such as metal alloys, ceramics, pharmaceuticals, food, and chemicals.

The fundamental capabilities of neural nets fit many of the critical requirements of computer-assisted formulation. These requirements include the need to develop a model that provides a basis to estimate new formulations capable of meeting new performance requirements. The neural net approach to this need is to automatically determine the formulation cause-and-effect relationship. "Cause" in this case is formulation and processing; "effect" is product properties. Determination of this relationship provides the model necessary to create new products. The power of neural net technology is its ability to uncover this relationship automatically.

Neural nets learn this relationship by example. The net is shown examples of this relationship by being shown examples of known formulations and known product properties. This information on existing or previous products typically includes records of formulations, laboratory and other test results, and product specifications. The relationship between formulation and performance is shown by example in this data. The neural net analyses this body of data, and finds the relationships 'common' throughout the data. This neural net activity is called supervised learning. The relationships found provide the formulation model.

The user is thus provided the functionality of a product design model without the need to develop one. The user need only collect data from which a model may be extracted. To provide a robust model, the data must cover all aspects of a problem. All relevant independent parameters should be present, with data illustrating the full range of values of each.

The data need not be exhaustive, since the neural net intelligently interpolates between known points as part of its ability to model even non-linear relationships. The data does need to cover as much of the full range of values as possible. Extrapolation beyond known points is possible with the neural net, and sometimes the only alternative, but the user must assume that relationships known to be valid in a certain range are valid outside this range. This may or may not be true.

The data provided to the neural net must contain an unambiguous cause-and-effect relationship. If a relevant parameter is missing, the net will not find an unambiguous relationship, and will tell the user so. The user must then identify the proper parameter and supply appropriate data.

It is possible to overspecify the problem by providing data on dependent variables. This is undesirable because it slows the neural network and creates "noise" in the results. The net tells the user that this may be the case by assigning that parameter a very low weighting factor. The user may then run the net without that input and determine whether its absence helps or hinders.

Once the net has "learned" the design model, it is ready to solve problems. In a process called consulting, the net is given new inputs (now desired properties) for which it provides new outputs (now new formulation and processing estimates). At this point, the net's operation is essentially instantaneous. The ease and speed of use provide a useful "what-if" capability. The user can try a variety of design approaches, or explore tradeoffs such as the impact of reducing raw material cost on product performance.

Estimation of new formulations saves time staff time, experimentation time, and time to market. Staff and experimentation time cost real money, while time to market involves an opportunity cost in deferred or lost business.

The neural net-based design tool provides further benefits. Users receive consistently good advice, reflecting the best practice. Valuable company expertise is collected, organized, and made available for future use. This lessens the impact of retirement or transfer of key staff members.

The Nature of Neural Nets

Neural nets typically consist of software simulations of large numbers of simple processors connected in an architecture similar to that of neurons interconnected in the brain. Like the brain, neural networks exhibit learning, pattern recognition, and associative memory recall. Neural net technology is based on research in pattern recognition, psychology, and neural science, seeking to understand the nature of learning and how it is accomplished by the brain.

The goal of Artificial Intelligence can be defined as enabling computers to provide higher-level support of complex, open-ended tasks such as diagnosis and decision-making. This requires computers to move beyond mere number-crunching and manipulation of data. The computer must acquire, in some sense, knowledge of a problem and its solution.

Neural nets are one approach to Artificial Intelligence (AI). Neural nets' ability to learn by experience makes them particularly suited to situations with fuzzy data or unknown, complex, or time-varying relationships. This ability may be contrasted with the capability of another approach to AI known as Expert Systems. Expert Systems are rule-based. A task to be performed by the system is broken down into rules. These rules are acquired from experts who, presumably, fully understand the parameters and interactions in a particular system or situation. The challenges in the Expert System approach include the need to fully understand complex relationships, and describe them concisely in the form of rules. This is often a very difficult task. The opportunity presented by neural nets is the opportunity to eliminate this issue by automatically discovering relationships.

One current area of neural net research is in learning methods, seeking faster and more robust methods to learn these unknown relationships. Another research area is the unification of the several related but different tasks typically involved in neural net-based solutions.

Current neural net development work includes identification of appropriate applications of the technology, and resolution of the resulting application-specific performance and integration issues. Here, technical needs may be met with a combination of well-executed Neural Nets and complementary technologies such as Expert Systems.

Formulation has been such a successful neural net application for three reasons. Chemists and chemical engineers involved in formulation see the need for better tools. The fundamental capabilities of neural nets lend themselves to this application, and the real-world requirements for a formulation tool are solvable in a neural net context.

Application Issues

Two broad application issues arise when using neural nets for formulation. One major class of issues concerns data quality. "Learn by example" is an easy way to learn complex relationships, but it does impose certain requirements. The examples must fully represent the range of the problem. The data used must therefore be complete and accurate.

In many applications, this condition is not initially met. Data is incomplete, and contains gaps, errors, duplications, and so on. For this reason, many system implementations must include as explicit tasks:

- data quality assessment

- data sorting for errors and duplications

- "filling in the gaps"

These tasks may be accomplished by the neural net through an approach called unsupervised learning. In essence, one asks the net to find natural clusterings of input data. From this, one might find that there are several clusters of data, perhaps reflecting several different but equally acceptable relationships. One might also find that some data lie outside the prevalent clusters. These deviating points should be analyzed to determine whether they represent bad data, or a different but equally acceptable relationship.

Identification of clusters can also be used to eliminate gaps and duplications. If data points are tightly clustered, one may simply use the centroid of that cluster as a single data point. This approach may also be used to reduce unmanageably large data sets.

Another class of issues relate to differences between fundamental neural net capabilities and real-world formulation requirements. A trained neural net provides a set of output numbers in response to a set of input numbers. In formulation, input numbers might be desired properties, and output numbers would be proportions of each ingredient. This approach is sufficient only if each of the desired properties is equally important, and there are no constraints on proportions or absolute values of ingredients. This is rarely the case.

Real-world problems involve production and cost constraints. Within these constraints, producers and buyers of formulated products find that some properties are more important than others. The importance to the customer of some properties may be equal for property values anywhere within a certain range. In this case, the solution to the formulation problem would be unnecessarily constrained if the net was only allowed a single value for the property. In other cases, a property value may be acceptable anywhere within a range, but more desirable to the customer with some values in that range than others.

In short, real-world formulation tasks requires the following features in a computer-based tool:

- Each property can be assigned a relative weight.

- A desirability function can be defined for each property, expressing relative preference for a property through a specified range of property values.

- Constraints in costs and proportions of ingredients can be accommodated.

These needs have been addressed in the CAD/Chem™ Custom Formulation System by AI WARE Incorporated as shown in Figure 36-1. The first two features are provided with a combination of sophisticated graphical user interfaces driving an automatically reconfigurable hierarchy of neural networks. The user assigns weights and draws desirability functions for each of the product properties. The Custom Formulation System automatically configures itself, then performs an iterative solution. The user is given a suggested formula meeting stated preferences as closely as possible. The user may then perform "what-if" studies to explore design trade-offs and sensitivities between the many factors, as shown in the following flowchart.

Hard constraints may also be expressed. These may include factors such as mutual incompatibilities between ingredients, or ingredients to be avoided for environmental, cost, or production reasons. These constraints are provided by expert system technology in the Custom Formulation System. In this case, the rule-based expert system technology is used because it fits the application requirement exactly. This is in line with AI WARE's experience in other areas: applications often benefit from synergistic combinations of diverse technologies. The key is to call upon the strengths of each technology, mitigating the weaknesses wherever possible. The implementation details are hidden behind user interfaces designed to meet the needs and expectations of users expert in the particular domain, but not expert in computer science.

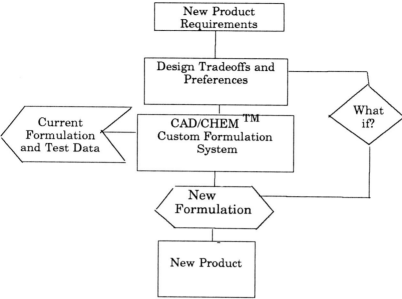

Figure 36-1   Custom Formulation System

Conclusion

Formulation is a complex task, combining technical challenges with significant marketing, production, and revenue implications for both producers and buyers. Traditional approaches to formulation rely heavily on testing and in-house expertise. Neural nets have been demonstrated as an effective computer-based method to assist this task. Properly-designed neural net-based formulation tools combine the "learn by example" capabilities of neural nets with other technologies as required to address the needs of formulating chemists. This application illustrates the capability of neural net technology to solve significant real-world problems. It also illustrates capabilities in other similarly complex manufacturing applications.

Author Biographical Data

Mr. VerDuin received a B.S. and M.S. degree in mechanical engineering from Carnegie-Mellon University, and has studied at the University of Michigan Graduate School of Business Administration, as well. His industrial experience includes working as project engineer at Ford Motor Co. and senior engineer at General Electric, and as section leader for Polytechniques Inc. Mr. VerDuin was awarded five patent disclosures in the course of his 14 years of work in product design, process design, and machine design and implementation, including robotics and other automation systems. He is also a Registered Professional Engineer in Ohio and Pennsylvania. In 1985, he joined the Center for Automation and Intelligent Systems Research at Case Western Reserve University in Cleveland, Ohio, as Associate Director. There, he worked with area industry representatives to establish research contracts and industry sponsors for the Center. Mr. VerDuin was directly involved in establishing AI WARE Incorporated, a successful spinoff business of the Center, and is currently General Manager there.

Expert System
for Specifying and Pricing
Process and Industrial Gauges

Joan B. Stoddard, Ph.D.
President, Stoddard Productivity Systems, Inc.
Canoga Park, California

Introduction

Expert systems are a means of gathering the highest level of corporate knowledge and putting it at the disposal of the personnel who are on the frontline in the daily battle for market share and profits.

By that definition it follows that one of the best applications for expert systems is in the area of specifying complex instruments, such as flowmeters and gauges. All too often detailed technical knowledge about such instruments has been accumulated by various members of the "high command" who are unable to package it effectively and make it available on a national (or international) scale.

The Instrument Division of Dresser Industries Inc. is one of the forward-looking organizations who have recognized the advantages in developing an expert system for a selection procedure. John W. Caldwell, Vice President and General Manager of the Instrument Division expects the Ashcroft Gauges Expert System to provide distributors, salespeople, applications engineers and sophisticated users the ability to select the best gauge from a spectrum of millions of unique configurations of gauges to satisfy an endless variety of applications. In reaching this goal the system must consider the use of the gauges with products ranging from air to zinc sulfate and having a wide range of concentration, temperature and pressure parameters.

The use of expert systems is in keeping with the theme of Dresser's Instrument Division in their effort to provide their customers with the Best Total Value in terms of service, reliability and quality of product.

Ashcroft Gauges—History of Leadership

In 1852 the leading edge of industrial technology was the use of steam power. Ocean-going vessels, river transport, railroad locomotives, and the machinery in factories all depended on steam as their energy source. Because of the importance of maintaining a balance between the amount of steam pressure necessary to do a job, and the amount of pressure that would exceed the strength of an engine's boiler, it became apparent that strong and accurate pressure gauges would be needed.

One of the pioneers in the field of pressure gauges was Edward Ashcroft who obtained the American rights to a French invention, the Bourdon tube. The Bourdon tube represented a way to translate pressure into controllable movement. With this as the basis of his design, Ashcroft's pressure gauges quickly set the standard for industrial gauges. Today, the Ashcroft gauge maintains a position of dominance in the field of pressure gauges throughout the world.

Background on the Application

In 1964, Manning, Maxwell and Moore, who had acquired the Ashcroft Gauge company, were in turn acquired by Dresser Industries, Inc. of Dallas, Texas. Since that time Ashcroft gauges have been produced at the headquarters plant in Stratford, Connecticut, and at its sister plants throughout the world, by the Instrument Division of Dresser Industries. Ashcroft pressure gauges are used in such diverse settings as food and chemical processing plants, petroleum refineries, commercial heating and air conditioning, and research and testing laboratories.

Every year Dresser's Stratford facility processes upwards of 50,000 orders from customers and distributors. While the order processing routine had been computerized and coordinated with other internal systems, such as purchasing and inventory, it was felt that Dresser's customers would be better served if a fast, accurate, computer-based method were available to assist in selecting the best Ashcroft gauge for a specific application. Having read, in my expert systems tutorial series (Measurements and Control, 1989-1990) an account of an expert system which I had designed to specify and price positive displacement flowmeters, the management of Dresser's Instrument Division requested that I develop a similar type of system for their Ashcroft line of pressure gauges.

The Original Plan

The original plan called for an expert system that could be distributed to users on two 5 1/4" floppy disks. One disk was to contain the compiled expert system. The second disk was to contain the runtime files and the database and text files which could be updated by Dresser when, for example, prices were changed.

The present expert system has grown considerably beyond this modest design. At present, the combined size of the databases, text files and compiled knowledge bases is 2.5 million bytes, not including the development system runtime files. It remains, however, a two-party maintenance arrangement. All files that contain information which Dresser personnel might want to modify, such as price lists, are accessible using dBase3 or a word processor. The expert system knowledge bases are compiled and, when a modification is desired, it must be performed by Stoddard Productivity Systems, Inc.

KnowledgePro

The Gauge Selection Expert System was designed using the development system, KnowledgePro™. KnowledgePro was the tool of choice because of its unique combination of the qualities of both a rule-based shell with an inference engine, and a language. Being able to mix rules with statements provided a highly flexible tool for dealing with an intricate series of relationships. KnowledgePro's ability to incorporate HyperText in the system was another important advantage since it offered a way in which directions for users, or definitions of terms, could be integrated with the gauge selection process.

Most importantly for a system designed to select one configuration of instrument from millions of possible combinations, KnowledgePro allows the developer to manipulate data in the form of lists and strings. This feature was invaluable in the many situations in which it was necessary to separate one item or a specific group of items from a mass of data.

The heart of KnowledgePro is a concept called a "TOPIC" which can take many forms. It is sometimes the equivalent of what other tools refer to as a variable. On another level, a TOPIC is an independent portion of a knowledge base that can be as short as one line, or up to several pages in length. Generic TOPICS can be created and used in much the same way as subroutines are in standard procedural type programs. A third use of TOPICS is similar to the use of FRAMES. Values can be inherited from parent TOPICS to child TOPICS, and parameters can be passed from one TOPIC to another.

Knowledge Acquisition

Creating an expert system to specify gauges was done, for the most part, from information contained in brochures, specifications and price lists, but inclusion of the knowledge and experience of company personnel was essential to the final success of the project.

At the outset of this project I made two visits to the Dresser Instrument Division's Stratford facility to do conventional recorded interviews with members of their top level of management and engineering team. After this period of orientation on the accepted procedures for specifying and ordering pressure gauges, I made a detailed study of each Dresser gauge publication. All additional information required for completion of the system beyond the above sources was the result of communications with Dresser by phone, fax and express mail.

This "information age" method of knowledge acquisition was facilitated by the fact that Nicholas E. Ortyl, Chief Engineer of Electro-Mechanical Instrumentation, was given direct oversight of the Gauge Expert System project. On a weekly basis, he supplied answers to technical questions, and kept me informed of Dresser personnel's reactions to the various facets of the developing system. The systematic advances in the system were conveyed to Dresser by means of a weekly shipment of floppy disks. Being computer knowledgeable, Mr. Ortyl was able to demonstrate new developments in the system to his colleagues at Dresser. This arrangement was beneficial to both Dresser and my company as I was saved the necessity of making numerous cross-country trips and Dresser, in turn, was spared the necessity of paying for them.

Storyboarding: A New Planning Tool for the Early Stage of an Expert System

One of the problems that is peculiar to a selection or specifying system is the need to establish an order of precedence for questions to be asked by the system. There are often good cases to be made for starting with any one of several variables (TOPICS in KnowledgePro) to identify the instrument that best fits the application. In addition, there are many paths from those initial starting points until a product—in this case, a gauge—has been chosen. This is especially true when, in practice, no single method has been followed systematically. A lack of customary procedure makes it difficult to get a consensus on the order in which information should be elicited from the user. It should always be the goal of the developer to present questions in a logical order, and at the same time to make the questions seem familiar and comfortable for the user. I solved the problem by using Post-it™ notes to make up a "storyboard" of questions to be included in the system.

On each of the Post-it notes I wrote one question and arranged them in a vertical sequence on a large pad of paper. In one of the meetings at Dresser, we discussed the questions and moved the notes around as we talked. At the end of the meeting, a pattern of questions had emerged that represented a consensus by the Dresser personnel on their preference for the order in which questions should be asked. As the system developed, and as people using the system became more familiar with the question and answer format of the system, some modifications were made. Modifications were also made for special groups of gauges that didn't fit the standard gauge ordering procedures. However, the story-board technique provided the underlying pattern for the system as a whole, and for the knowledge bases which make up the system.

Development of the Gauge Selection Expert System

Based on my experience with the meter selection expert system, I decided to create a modular system with each knowledge base dedicated to a family of gauges or to the performance of a specific task. I had found that having too many kinds of instruments in a knowledge base was not an efficient arrangement. The size of rules required in order to include or exclude all possible instrument selections slowed the system down. The use of a gateway module for macro routing, with dedicated modules for different groups of instruments produced a much more satisfactory system.

Initially, I selected several gauges as a starting point and laid out a simple design that took the user from the opening menu to the

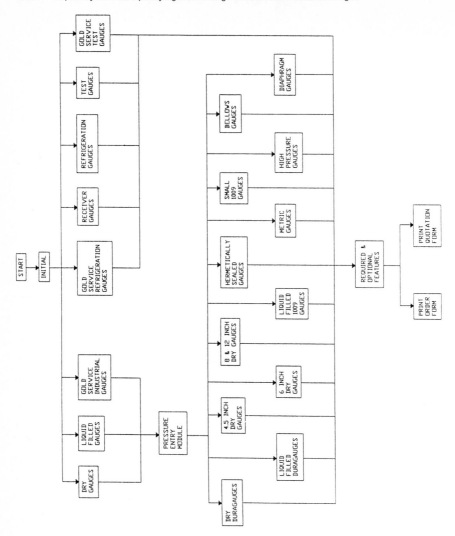

Figure 37-1   The knowledge base flow chart of the Ashcroft Gauge
Selection Expert System. Knowledge base modules on the
upper level consist of two types—complete unit modules
which select from a specialized group of gauges and then
exit to the Accessory module, and gateway modules which
lead first to the Gauge Pressure Entry module, and from
there to a module dedicated to a specific family of gauges,
after which the user is also taken to the Accessory module.
All users have the option of completing their transaction by
printing either an order form or a quote form.

selection of a gauge. The knowledge bases use a combination of stored knowledge about the gauges drawn from databases and text files. Based on the users responses to questions posed by the system, the choice of gauges is narrowed down to the gauge that will best satisfy the customer.

This initial effort was not in the nature of building a prototype; rather, it produced a small, practical system that was methodically enlarged to encompass the entire line of Ashcroft gauges. Figure 37-1 is a diagram of the modularized Ashcroft Gauge Expert System in its final form.

The Development of the Gauge Selection Expert System

The first step in the development of the system was to find a series of dichotomies along which lines the gauges would be divided into logically identifiable groups that could be systematically accessed by the users.

I began with the discovery that there were two basic groups of gauge orders which needed to be distinguished from each other. They presented a potential source of confusion since one group (Gold Service gauges) was a subset of other groups which, for the sake of identification, I called the Standard gauges. Actually the term Gold Service gauges refers to certain gauges among the Ashcroft gauge line which are available on an expedited delivery schedule at a premium price. This first dichotomy, then, was based on the willingness of customers to accept a limited choice of gauges at a slightly higher price in return for a rapid delivery schedule.

The next decision point was in the form of a menu offering a choice of two types of gauges, dry gauges or gauges that are liquid-filled.

Both of these questions dealt with mutually exclusive qualities. The Ashcroft line of gauges was at this point divided into four sub-categories.

As mentioned above, other factors might have preceded this decision. The accuracy of a gauge is important to the user, and the solid front Ashcroft gauges are often preferred in some applications, since their design affords greater safety to persons in the work area.

However, both these factors were deferred. For the first type of customer, speed of delivery was the critical consideration. If the person did not elect to choose a rapid delivery, then whether he chose a dry or liquid gauge, he would have the same range of choices of solid front gauges or accuracy ranges.

This was a necessary part of planning a specification system that I had learned from previous experience. Questions that made significant divisions among products to be selected by the system, questions which did not preclude other important choices, and questions which fit into the story-board order of precedence, were placed ahead of less significant or more limiting questions in the system.

These questions were the nucleus of the INITIAL module and routed the user from that point to the various knowledge bases.

All the knowledge bases were originally conceived as self-contained units devoted to "families" of gauges, but this concept required some deviations during the development process.

The process of specifying and pricing a gauge can be broken into five major steps.

1. Selecting the Gauge

From interviews and the story board sessions, it was clear that the interaction of the internal parts of the gauge with the product was a major concern. The Bourdon tube, and other types of pressure sensors may be subject to stress or damage from certain types of corrosive chemicals. In most installations pressure is measured by entry of a fluid or gas into the gauge's Bourdon tube. Given that the same range of accuracies are available for gauges with different metal construction, compatibility of the gauge and the fluid or gas product must be established first.

This is done by presenting the user with an alphabetized list of approximately 150 chemical products. After the user has indicated the product in the line or tank where pressure is to be measured, the system accesses a text file where it obtains a list of recommended metals that are compatible with the customer's product. Other environmental factors are evaluated and then the user is asked to indicate the level of accuracy necessary for his application. When he has made his selection, he is offered a list of all the gauge sizes meeting the selected accuracy. On the theory that a particular size gauge might be needed that didn't match his choice of accuracy, the sizes of all gauges of higher levels of accuracy are also included.

This is a case where, although accuracy may require a more important decision than the size of the gauge, all users are always given the greatest measure of latitude in selecting gauge features throughout the system.

In another case where the answer to a question might unduly limit later choices, I chose an alternate plan. I worded the question so that the person answering would be warned that he could be setting limits on his future options. For example, the mounting of a gauge may be an important factor in choosing a gauge, but logical order dictates that the user should choose the gauge before the mounting. Here is how the choice of gauges was displayed when one of them had only one mounting option:

1279 - Phenolic case, Stem/Flush/Surface mounting

1377 - Aluminum case, Flush mount only.

If a user were to choose the 1377 gauge, he would do so with full knowledge that it must be flush mounted, as in a control panel application.

After the questions about accuracy and size, the user is asked if he wants a pressure, vacuum or compound gauge. Almost all the Ashcroft gauges can be ordered in one of these three styles. For the sake of an example, we will suppose that the user selects a pressure gauge. He is first asked whether or not he knows what pressure range he wants (if, perhaps, he is buying a replacement for another gauge). If so, he is shown a list of measurement units (psi, kPa, etc.) and asked to choose the pressure range of the gauge. He might enter one of the standard ranges, such as 100 psi. If he entered a range that was not a standard range, such as 250 psi, the system would select the next higher standard range and display it for his approval.

The system also translates entries such as 100 bar into 1450 psi since, within the system, psi (pounds per square inch) is the universal measurement unit for pressure. However, the user would be given a choice of whether he wanted the gauge dial to read in psi or one of the other standard measurement units such as kPa. The final selection screen of the program and the printed form would reflect his choice of pressure units.

For users who are not sure of the required range of their gauge, the system will ask for their normal and maximum operating pressures and apply safety factor multipliers. The pressure range recommended will reflect whichever of the two figures is higher, a prudent decision that Dresser has always endorsed. A list of available metal Bourdon tubes is stored in the system for each type of gauge. The list varies depending on the pressure range for which the gauge is intended. For example, Bronze tubes are not available in gauges when the pressure exceeds 1000 psi, and steel tubes are not available when pressure exceeds 5000 psi. The tube metal list is now intersected with the list of metals compatible with the customer's product. All of the metals, common to both lists, are carried forward to the end of the knowledge base.

The system now allows the user to enter a price discount factor, if any applies to the transaction. Then a database search is instituted. Based on the pressure range, gauge size and style, and the tube metal, prices are determined and the system displays, for each available metal type, the list price and discounted price. The user can then select one of the choices offered.

At that point the module writes the pertinent TOPICS and their values to a SAVEFACTS file and the user is transferred to the ACCESSORIES module.

2. Adding Required Parts and Services

The first function of the ACCESSORIES module is to add on to the customer's order any parts or services that may be required for the selected gauge. A check list is consulted to see what additional parts, services or accessories should be included. Certain gauges require special rings for flush mounting in control panels. The system checks to see if the selected gauge requires such rings, and if so, it stores a two-letter code and the price of the rings. In some cases, if a product is an oxidizing agent, a special cleaning procedure may be required. The system will check to see if the product is one of those on the list with this requirement. If it is, the system will check to see how many gauges are being ordered since the number of gauges involved is a factor is determining the price of the cleaning service. The two letter code for cleaning and the price per gauge for the cleaning is also stored for later use.

3. Choosing Optional Features

The next task of the ACCESSORIES module is to assemble a list of optional features. This list varies depending on the kind of gauge selected and by previous choices made by the system. For example, the cleaning procedure mentioned above will appear on most of the option lists, unless, of course, the system has already determined that cleaning is a required feature. The system is designed only to offer options to users which fit previous choices.

Once the list is assembled, the user is allowed to select any desired options from the menu. In some cases, certain combinations of choices are not allowed. If a user were to select both a minimum and a maximum pointer for his gauge, he would be informed that he must choose between them. The system would then re-evaluate his original set of choices, eliminating one of the two pointers.

4. Calculating the Final Price

A final price is now calculated, taking into consideration the basic gauge, any required features or services and any options selected. This figure is then multiplied by the discount factor, if one applicable, and a display is presented to the user of both the total list price and the total discounted net price.

Also displayed is a summary of all the information that has been developed by the system, relating to the gauge. The size and type of gauge are shown, along with the type of connection, the pressure range and all the two letter codes signifying variations from the basic gauge.

The name of the product whose pressure is being monitored is displayed as well, along with any special notes that the system has determined are pertinent to the selected gauge and accessories, (e.g. "A diaphragm seal is required").

5. Generating the Printout

The user now has a the choice of using the data accumulated by the system to generate either an Order form, or a Quotation form. Depending on their choice they are routed to the appropriate PRINT module.

Both PRINT modules have many features in common and may therefore be discussed together. The first item of interest to the user is a message regarding the validity of the prices in the system. A file is read that contains the expiration date for the gauge and accessories pricing contained in the system databases. If the date obtained from the computer's calendar is later than the date on file, the user is warned that the system's pricing may be out of date. If the calendar date is within the period during which the prices are valid, this message will not appear. Any gauge selections performed by the system with expired pricing databases will have a warning note to this effect included on the final printout form.

Customer information can be entered manually or it can be accessed from a database if it is a repeat customer. The customer database can be generated and updated by the users at each site where the gauge selection expert system is installed. This customer database can include not only billing and shipping addresses, but preferred shipping methods, package markings and other customer specific details. An editing window command allows the user to enter multiple lines of special instructions for an order or a quote.

After entry of customer information is complete, the gauge data developed during the consultation is accessed by the PRINT module and the form is printed.

A copy of the form is also automatically saved to a file and given a unique number by which it can be identified. Later this form can be faxed, called up and modified, or simply consulted as a reference when a new order is received from the same customer.

One of the latest additions to the gauge expert system is the feature, "Order by Part Number". Added in response to a user's suggestion, this module will allow a user with sufficient information about a gauge to enter the data needed to order or quote the gauge without going through the normal selection system process. It provides an "Express Line" for regular customers who know what they want to order and who don't want to answer a lot of questions to get it. In this way the benefits of using standard order and quotation forms generation can be available for all types of customer inquiries.

The Ashcroft Gauge Expert System was designed to operate on IBM (or compatible) computers running DOS 3.3 or above. Due to the size of the knowledge bases and the complexity of some of the rule bases, a 386-based PC operating with at least 20 MHz speed is recommended for satisfactory response times. An EGA or VGA color monitor is strongly recommended because color is used extensively for producing the system's windows and menu-driven user-friendly displays.

Conclusion

The Dresser personnel associated with the Ashcroft Gauge Expert System are very pleased with the system and feel that it has exceeded their original expectations. They are confident that it will produce an accurate selection of gauges to provide the best service, safety and customer satisfaction, and consider it to be an important part of their Best Total Value theme.

Author Biographical Data

Joan B. Stoddard, Ph.D., is President of Stoddard Productivity Systems, Inc., a company specializing in the development of expert systems for the process control industry. Dr. Stoddard earned her B.A. in Psychology at the University of California at Los Angeles, and her M.A. and Ph. D in Psychology at the University of Southern California. She is an author of articles that have appeared in national and international AI Journals, and has authored a two year tutorial series on writing expert systems in *Measurements and Control* and *Medical Electronics*. Dr. Stoddard is a member of the American Association for Artificial Intelligence, the American Psychological Association and Mensa. She is also a member of the University of Southern California Productivity Network.

Appendixes

edited by
Jessica Keyes

APPENDIX A. AI VENDORS

AICorp. Inc.
138 Technology Drive
Waltham, MA 02254
(617) 891-6500 *[(617) 891-3500]*

AION Corp.
101 University Avenue
Palo Alto, California 94301
(415) 328-9595

AI Ware Inc.
11000 Cedar Avenue
Cleveland, Ohio 44106
(216) 421-2380

Artificial Intelligence Technologies
40 Saw Mill River Road
Hawthorne, New York 10532
(914) 347-3182 *[(914) 347-6860]*

Arity Corp.
29 Domino Drive
Concord, Mass 01742
(508) 371-1243

California Intelligence
912 Powell Street #8
San Francisco, Calif 94108
(415) 391-4846

California Scientific Software
10141 Evening Star Drive #6
Grass Valley, Calif 95945
(916) 477-8656

CAM Software Inc.
750 North 200 West, Suite 208
Provo, UT 84601
(801) 373-4080

Carnegie Group
5 PPG Place
Pittsburgh PA 15222
(412) 642-6900

CIM Solutions
P.O. Box 7041
Provo, Utah 84604
(801) 374-5626

Emerald Intelligence Inc.
3915-A1 Research Park Drive
Ann Arbor, MI 48108
(313) 663-8757

Expert Systems International
1700 Walnut Street
Philadelphia, Pa. 19103
(215) 735-8510

ExperTelligence Inc.
5638 Hollister Avenue, Suite 302
Goleta, Calif 93117
(805) 967-1797

EXSYS Inc.
P.O. Box 75158/Station 14
Albuquerque, N.M. 87194
(505) 256-8356

Hecht Nielson Corp.(HNC)
5501 Oberlin Dr.
San Diego, Calif. 92121
(619) 546-8877

IBM
P.O. Box 10
Princeton, N.J.
(201) 329-7000

Inference Corp.
550 N. Continental Blvd.
El Seguendo, Calif 90245
(213) 322-0200

Information Builders Inc.
1250 Broadway
NY NY 10001
(212) 736-4433

IntelliCorp
1975 El Camino Real W.
Mountain View, Calif. 94040
(415) 965-5500

IntelligenceWare Inc.
9800 S. Sepulveda Blvd., Suite 730
Los Angeles, Calif 90045
(213) 417-8896

Intelligent Environments Inc.
2 Highwood Drive
Tewksbury, MA 01876
(508) 640-1080

Jeffrey Perrone and Associates
3685 17th Street
San Francisco, Calif 94114
(415) 431-9562

Knowledge Garden Inc.
473A Malden Bridge Road
Nassau, N.Y. 12123
(518) 766-3000

Logicware International
2065 Dundas St. E, Suite 204
Mississauga, Ontario
Canada L4V 1T1
(416) 672-0300

Lucid Inc.
707 Laurel Street
Menlo Park, Calif 94025
(415) 329-8400

mdbs Inc.
P.O. Box 248
2 Executive Drive
Lafayette, Ind. 47902
(317) 463-4561

MegaKnowledge Inc.
One Kendall Square, Bldg. 600
Cambridge, MA 02139
(617) 494-9234

Micro Analysis & Design Inc.
3300 Mitchell Lane, Suite 175
Boulder, CO 80301
(303) 442-6947

Micro Devices
5695B Beggs Road
Orlando, FL 32810
(407) 299-0211

NeuralWare Inc.
103 Buckskin Court
Sewickley, Pa. 15143
(412) 741-5959

Nestor Inc.
1 Richmond Square
Providence, R.I. 02906
(401) 331-9640

Neuron Data
444 High Street
Palo Alto, California 94301
(415) 321-4488

Paperback Software International
2830 Ninth St.
Berkeley, Calif 94710
(415) 644-2116

ParcPlace Systems
1550 Plymouth Street
Mountain View, Calif 94043
(415) 691-6700

Pritsker Corporation
8910 Purdue Road, Suite 500
Indianapolis, IN 46268
317) 879-1011

Rawson Technologies Inc.
727 Charles Street, P.O. Box 352
Wellsburg, WV 26070
(304) 737-0090

Softsync Inc.
162 Madison Ave
NY NY 10016
(212) 695-2080

Software Architecture and Engineering
1600 Wilson Blvd.
Suite 500
Arlington, Va. 22209
(703) 276-7910

Texas Instruments
P.O. Box 809063
Dallas, Texas 75380
(800) 527-3500

Gensym Corp. (Cambridge, Mass.) Bruce Crane
(617) 577-9606
(617) 547-2500

Expert Edge Corp.
Palo Alto, Calif.
(415) 969-2800

AbTech Corp.
Charlottsville, Va.
(804) 977-0686

Symbologic ~~Adep~~ Corp.
Redmond, Wash.
(206) 851-3938

Craig Chelius
marketing

APPENDIX B.   TRADE GROUPS, OTHER NON-PROFIT
GROUPS, and MARKET RESEARCH FIRMS INVOLVED IN AI

Alberta Research Council
6815 8th Street NE, 34d floor
Calgary, Alberta
Canada, T2E 7H7
(403) 297-2600

American Association for Artificial Intelligence
445 Burgess Drive
Menlo Park, CA 94025
(415) 328-3123

Association for Computing Machinery
11 West 42nd Street,
New York, NY 10036
(212) 869-7440

BIRL (Basic Industry Research Laboratory)
Northwestern University
1901 Maple Avenue
Evanston, IL 60201
(708) 491-4619

CAM-I
1250 E. Copeland Road, Suite 500
Arlington, TX 76011
(817) 860-1654

Cooperative R&D firm for CIM
Environmental Research Institute of Michigan
P.O. Box 8618
Ann Arbor, MI 48107
(313) 994-1200

Frost & Sullivan
106 Fulton Street
New York, NY 10038
(212) 233-1080

IEEE Computer Society
10662 Los Vaqueros Circle
P.O. Box 3014
Los Alamitos, Calif 90720
(714) 821-8380

New Art Inc.
2170 Broadway, Suite 2290
New York, NY 10024
(212) 362-0559

New Science Associates Inc.
167 Old Post Road
Southport, CT 06490
(203) 259-1661

Robotic Industries Association
900 Victors Way, P.O. Box 3724
Ann Arbor, MI 48106
(313) 994-6088

Society of Manufacturing Engineers
One SME Drive
P.O. Box 930
Dearborn, MI 48121
(313) 271-1500

Handwritten notes:

Nina Buck
Gen
Rockwell Science Center
Allen Bradley
Sujeet Chand
(805) 373-4545

Cleveland, Ohio

☆ Lubrisol / regulations
ES in Chemical
Giorgio   Sorani
Robert   Lauer
☆

Allen Bradley / Rockwell Science Center
Peter Schmidt   Calif
Thousand Oaks,

P.

APPENDIX C. COMPANIES INVOLVED IN MANUFACTURING AI CONSULTING.

Amerinex Artificial Intelligence Inc.
115 Route 46, Building F
Mt. Lakes, NJ 07046
(201) 402-4090

Andersen Consulting
69 West Washington Street
Chicago, Il 60602
(312) 580-0069

Applied Intelligent Systems Inc.
110 Parkland Plaza
An Arbor, MI 48103
Autoflex Inc.

445 Enterprise Court
Bloomfield, MA 48018
(313) 253-9500

Automated Technology Systems Corporation
25 Davids Drive
Hauppauge, NY 11788
(516) 231-7777

Automated Inspection Devices Inc.
P.O. Box 6295
Toledo, OH 43614
(419) 536-1983

Automatix Inc.
755 Middlesex Turnpike
Billerica, MA 01821
(508) 667-7900

Battelle
505 King Avenue
Columbus, OH 43201
(614) 424-6424

E-KE Ltd.
301 Monterey Drive
Dublin, OH 43017
(614) 424-3624

Cap Gemini America
1034 S. Brentwood Blvd., Suite 1780
St. Louis, MO 63117
(314) 221-0123

Cimflex Teknowledge Inc.
P.O. Box 10119
1815 Embarcadero Road
Palo Alto, CA 94303
(415) 424-0500

Dialog Systems Division, A.T. Kearney Inc.
2842 E. Grand River Avenue
East Lansing, MI 48823
(517) 351-1147

Dynalytics Corp.
260 North Broadway
Hicksville, NY 11801
(516) 822-1760

Expert Implementations Corp.
34115 West Twelve Mile Road, Suite 107
Farmington Hills, MI 48331
(313) 553-3333

Gensym Corp.
125 Cambridge Park Drive
Cambridge, MA 02140
(617) 547-9606

Intelligent Applications Ltd.
Kirkton Business Center
Kirk Lane
Livingston Village
West Lothian, Scotland
EH54 7AY
(011 44) (506) 410242

Liu & Associates
11991 Nugent Drive
Granada Hills, CA 91344
(818) 366-1535

Stone and Webster Engineering
P.O. Box 2325
Boston, MA 02107
(617) 589-1567

Intellisys
4641 Crossroads Park Drive
Liverpool, NY 13088
(315) 454-2300

KnowledgeBase Group
2713 Farnswood Circle
Austin, TX 78704
(512) 440-8025

LTV Missiles & Electronics Group
P.O. Box 650003 M.S WT-50
Dallas, TX 75265
(214) 266-0239

OXKO Corporation
P.O. Box 6674
Annapolis, MD 21401
(301) 266-1671

Stoddard Productivity Systems Inc.
23825 Kittridge Street
Canoga Park, Calif 91307
(818) 883-5592

Texas Instruments Information Technology Group
12501 Research Blvd. MS 2222
Austin, TX 78759
(512) 250-6679

Universal Robot Systems
P. O. Box 3236
Laguna Hills, Calif 92654
(714) 951-7024

APPENDIX D. AI GLOSSARY.

Algorithm. A procedure that is systematized which results in a correct outcome. In developing a conventional program the programmer must specify the algorithms that the program will follow.

Artificial Intelligence. A subfield of computer science aimed at pursuing the possibility that a computer can be made to behave in ways that humans recognize as "intelligent" human behavior.

Attribute. A property of an object. For instance, cold and creamy is an attribute of ice cream.

Automatic Programming. The field of AI dealing with creation of programs that in turn write other programs.

Backward chaining. A control strategy that regulates order in which inferences are drawn. In Backward chaining the system attempts to determine if the goal rule is correct. It backs up to the IFs and tries to determine if they are correct.

Certainty Factor. A numerical weight given to a fact or relationship to indicate confidence in that fact or relationship.

Class. A set of information similar to a file.

Class member. The elements of information within a class, similar to a record within a file.

Common LISP. The standardized version of the most prevalent AI language.

Consultation Paradigm. These paradigms describe generic types of problem solving situations.

Control. Within the context of a knowledge based system, Control refers to the regulation of the ordering in which reasoning occurs. Examples are backward or forward chaining.

Demon. A procedure that is automatically triggered when a value is changed within an object.

Domain. An area of knowledge.

Embedded expert system. An expert system run under the control of a conventional data processing system.

Encapsulation. Refers to the fact that an object can be considered a mini-program. It is independent from other objects, with its own attributes, values and procedures.

End-user. Ultimate user of expert system.

Expert System. A computer system that can perform at or near the level of an expert.

Forward chaining. One of several control strategies that regulate the order in which inferences are drawn. It begins by asserting all of the rules whose if clauses are true. It continues this process, checking on what additional rules are true until the program reaches a goal or runs out of possibilities.

Frame. A knowledge representation scheme that associates an object with a collection of facts about that object.

Fuzzy Logic. Knowledge representation techniques that deals with uncertainty.

Garbage collection. How a programming language manages the storage of unused variables.

Heuristic. A rule of thumb.

Heuristic Rules. Rules written to capture the "rules of thumb" of an expert. Usage of these rules does not always lead to correct solutions,

High-level Languages. FORTRAN, Cobol, C and others written for more conventional data processing.

Induction. Some expert system shells have the capability of using examples to gather the knowledge that goes into the knowledge base. This eliminates the need to go through the formal knowledge acquisition process.

IF..THEN Rule. A statement of relationship in the form of IF A THEN B.

Inference. The process by which new facts are derived from known facts.

Inference Engine. The working program of the knowledge system that contains inference and control strategies. The term has also become linked with the attributes of user interface, external file interface, explanation features as well as other attributes.

Inheritance. Attributes of the parent object can be inherited by the child. For example, in the object CAR a parent attribute is that it has four wheels. The child object, MERCEDES, inherits this attribute from its parent CAR.

Instantiations. The rule and the list of class members that satisfy the left-hand side of the rule.

Interface. The linkage between the computer program and the outside world.

Knowledge Acquisition. The laborious process of collecting, documenting, verifying and refining knowledge.

Knowledge Base. That portion of an Expert System that consists of the facts and heuristics about an area called a domain. Can be composed of rules, objects and other methodologies for storing knowledge.

Knowledge Engineer. That individual whose role is to assess problems, acquire knowledge, build expert systems.

Knowledge Representation. Methods used to encode and store facts and relationships. Examples are Rules, Frames, Objects-Values-Attributes.

LISP. A programming language, favored by American AI researchers.

Meta-rule. A rule about a rule rather than about the knowledge application. This type of rule is used to control how the system operates.

Mid-run Explanation. Expert systems have the ability to stop, during a consultation run, and explain what it is doing.

MYCIN. One of the first backwards-chaining expert systems. Developed at Stanford in the middle of the 1970s, it was developed as a research tool in the domain of diagnosis and treatment of meningitis and bacteremia infectious diseases. When researchers emptied all of the medical information out of MYCIN, the first shell was born. This was called EMYCIN.

MIPS. Acronym for millions of instructions per second. This is a measurement of computer speed.

Neural Net. Another branch of AI, where software tries to mimic the interconnected neurons of the human brain. Here the net is trained by use of examples.

Object. Building block of the newer AI-based systems. Similar to frame in concept.

OPS5. One of the original AI languages. Used to build DEC's XCON. Most terminology used within Knowledgetool is derived from OPS5.

PROLOG. A symbolic programming language based on predicate calculus. Most popular AI language outside of US.

Prototype. Initial version of an expert, or for that matter any, system.

Pseudocode. Writing rules in English language.

RETE. Algorithm which optimizes the searching process of rule based systems. Systems that have this feature are much more efficient than systems that don't.

Robotics. Branch of AI dealing with Robots. Robots can see and manipulate their environment.

Rule. A conditional statement of two parts.
         IF... is the condition
         THEN..is the premise

Rule based system. A program that represents knowledge by means of rules.

Shell. A pre-written expert system tool. Includes interfaces and inference engine.

Speech Recognition. Ability of computer to understand human speech.

Uncertainty. Conventional systems cannot deal with uncertainty. Expert Systems, on the other hand, can deal with the vagueness and fuzziness that comes with the processing of human judgement.

APPENDIX E. SELECTED BIBLIOGRAPHY:

MAGAZINES

AI MAGAZINE. Official journal of AAAI.
    AAAI
    445 Burgess Drive
    Menlo Park, CA 94025

EXPERT SYSTEMS. Auerbach publication.
    Auerbach Publications
    Division of Warren, Gorham & Lamont
    210 South Street
    Boston, Mass 02111

IEEE EXPERT. Publication of IEEE.
    IEEE COMPUTER SOCIETY
    10662 LOS VAQUEROS CIRCLE
    LOS ALAMITOS, CA 90720

AI EXPERT. Commerical publication, available at newstands.
    Miller Freeman Publications
    500 Howard Street
    San Francisco, CA 94105

PC AI. Commercial AI magazine.
    PC AI
    3310 West Bell Rd.
    Suite 119
    Phoenix, AZ 85023

NEWSLETTERS THAT DISCUSS AI.

AI TRENDS.
    The Relayer Group - AI Trends
    8232 E. Buckskin Road
    Scottsdale, AZ 85255
AI WEEK.
    AIWEEK INC.
    2555 Cumberland Parkway, Suite 299
    Atlanta, GA 30339
ARTIFICIAL INTELLIGENCE REPORT.
    Booz-Allen Hamilton
    4330 East-West Highway
    Bethesda, MD 20814

CIM STRATEGIES NEWSLETTER.
Cutter Information Corp.
1100 Massachusetts Avenue
Arlington, VA 02124.
EXPERT SYSTEM STRATEGIES.
Cutter Information Inc.
1100 Mass Avenue
Arlington, Mass 02174

SPANG-ROBINSON REPORT.
830 Menlo Avenue, Suite 100
Menlo Park, CA 94025
TECHINSIDER.
2170 Broadway
Suite 2290
N.Y., N.Y. 10024

BOOKS AND ARTICLES.

Barr, Avron, Edware A. Feigenbaum and Paul Cohen. 1981. The Handbook of Artificial Intgelligence, Vols. I, II and III. Los Altos, CA: William Kaufmann.

Buchanan, Bruce G. and Edward H. Shortliffe. 1984. Rule-Based Expert Systems. Reading, MA: Addison-Wesley.

Cox, C.J. 1986. Object Oriented Programming: An Evolutionary Approach. Reading, MA: Addison-Wesley.

Ernst, Christian J. 1988. Management Expert Systems. Wokingham, England: Addison-Wesley.

Feigenbaum, Edward A. and Pamela McCorduck. 1984. The Fifth Generation. New York: Signet.

Frenzel, Louis E. 1987. Understanding Expert Systems. Indianapolis, Indiana: Howard W. Sams.

Graf, R.F. 1984. Modern Dictionary of Electronics. Indianapolis. IN: Howard W. Sams & Co., Inc.

Harmon, Paul Rex Maus and William Morrissey. 1988. Expert Systems. Tools and Applications. New York, NY: Wiley.

Harmon, Paul and David King. 1984. Expert Systems. New York, NY: Wiley.

Hayes-Roth Frederick, Donald A. Waterman and Douglas B. Lenat. 1983. Building Expert Systems. Reading, MA: Addison-Wesley.

Keyes, Jessica. 1990. The New Intelligence. New York, NY: Harper Business.

Mishkoff, Henry C. 1985. Understanding Artificial Intelligence. Dallas, Texas: Texas Instruments.

Nagy, Tom Dick Gault, and Monica Nagy. 1983 Building Your First Expert System. New York, NY: Halstead Press.

Peat, F. David. 1988. Artificial Intelligence: How Machines Think. New York: Bean Publishing.

Popolizio, John J. and William S. Cappelli. 1989. "New Shells For Old Iron." Datamation, (April): 41-48.

Prietula, Michael J. and Herbert A. Simon. 1989. "The Experts in Your Midst." Harvard Business Review, (January-February): 120-124.

Pritsker, A.A.B. 1984. Introduction to Simulation and SLAM II. 2nd ed. West Lafayette, ID: Systems Publishing Corp.

Rich, Elaine. 1983. Artificial Intelligence. New York, NY: McGraw-Hill.

Schaffer, G.H. 1986. "Artificial Intelligence: A Tool for Smart Manufacturing". American Machinist & Automated Manufacturing, (August): 91-94.

Schoen Sy and Wendell Sykes. 1987. Putting Artificial Intelligence To Work. New York, NY: Wiley.

Stock, Michael. 1988. AI Theory and Applications in the Vax Environment. New York, NY: Multiscience Press.

Tanimoto, Steven L. 1987. The Elements of Artificial Intelligence. Rockville, Maryland: Computer Science Press.

Tello, Ernest R. 1988. Mastering AI Tools and Techniques. Indianapolis, Indiana: Howard W. Sams & Company.

Waterman, Donald A. 1985. A Guide To Expert Systems. Reading, MA: Addison-Wesley.

Winston, P.H. 1984. Artificial Intelligence. 2nd ed. Reading MA: Addison-Wesley Publishing Company.

Index

A

B

C